£4.89

FORM AND FUNCTION

FORM AND FUNCTION

A CONTRIBUTION TO THE HISTORY OF ANIMAL MORPHOLOGY

By E. S. RUSSELL,
M.A., B.Sc., F.Z.S.

ILLUSTRATED

LONDON
JOHN MURRAY, ALBEMARLE STREET, W.
1916

ISBN 0 576 29177 3

Republished in 1972 by Gregg International Publishers Limited
Westmead, Farnborough, Hants., England

Printed in offset by Franz Wolf, Heppenheim/Bergstrasse
Western Germany

PREFACE

THIS book is not intended to be a full or detailed history of animal morphology: a complete account is given neither of morphological discoveries nor of morphological theories. My aim has been rather to call attention to the existence of diverse typical attitudes to the problems of form, and to trace the interplay of the theories that have arisen out of them.

The main currents of morphological thought are to my mind three—the functional or synthetic, the formal or transcendental, and the materialistic or disintegrative.

The first is associated with the great names of Aristotle, Cuvier, and von Baer, and leads easily to the more open vitalism of Lamarck and Samuel Butler. The typical representative of the second attitude is E. Geoffroy St Hilaire, and this habit of thought has greatly influenced the development of evolutionary morphology.

The main battle-ground of these two opposing tendencies is the problem of the relation of function to form. Is function the mechanical result of form, or is form merely the manifestation of function or activity? What is the essence of life—organisation or activity?

The materialistic attitude is not distinctively biological, but is common to practically all fields of thought. It dates back to the Greek atomists, and the triumph of mechanical science in the 19th century has induced many to accept materialism as the only possible scientific method. In biology it is more akin to the formal than to the functional attitude.

In the course of this book I have not hidden my own sympathy with the functional attitude. It appears to me probable that more insight will be gained into the real nature of life and organisation by concentrating on the active response of the animal, as manifested both in behaviour and in morphogenesis, particularly in the post-embryonic stages, than by giving attention exclusively to the historical aspect of structure, as is the custom of "pure morphology." I believe we shall only make progress in this direction if we frankly adopt the simple everyday conception of living things —which many of us have had drilled out of us—that they are active, purposeful agents, not mere complicated aggregations of protein and other substances. Such an attitude is probably quite as sound philosophically as the opposing one, but I have not in this place attempted any justification of it. I have touched very lightly upon the controversy between vitalism and materialism which has been revived with the early years of the present century. It hardly lends itself as yet to historical treatment, and I could hardly hope to maintain with regard to it that objective attitude which should characterise the historian.

The main result I hope to have achieved with this book is the demonstration, tentative and incomplete as it is, of the essential continuity of animal morphology from the days of Aristotle down to our own time. It is unfortunately true that modern biology, perhaps in consequence of the great advances it has made in certain directions, has to a considerable extent lost its historical consciousness, and if this book helps in any degree to counteract this tendency so far as animal morphology is concerned, it will have served its purpose.

I owe a debt of gratitude to my friends Dr James F. Gemmill and Prof. J. Arthur Thomson for much kindly encouragement and helpful criticism. The credit for the illustrations is due to my wife, Mrs Jehanne A. Russell. One is from Nature; the others are drawn from the original figures.

E. S. R.

CHELSEA, 1916.

CONTENTS

ILLUSTRATIONS

FORM AND FUNCTION

CHAPTER I

THE BEGINNINGS OF COMPARATIVE ANATOMY

THE first name of which the history of anatomy keeps record is that of Alcmaeon, a contemporary of Pythagoras (6th century B.C.). His interests appear to have been rather physiological than anatomical. He traced the chief nerves of sense to the brain, which he considered to be the seat of the soul, and he made some good guesses at the mechanism of the organs of special sense. He showed that, contrary to the received opinion, the seminal fluid did not originate in the spinal cord. Two comparisons are recorded of his, one that puberty is the equivalent of the flowering time in plants, the other that milk is the equivalent of white of egg.[1] Both show his bias towards looking at the functional side of living things. The latter comparison reappears in Aristotle.

A century later Diogenes of Apollonia gave a description of the venous system. He too placed the seat of sensation in the brain. He assumed a vital air in all living things, being in this influenced by Anaximenes whose primitive matter was infinite air. In following out this thought he tried to prove that both fishes and oysters have the power of breathing.[2]

A more strictly morphological note is struck by a curious saying of Empedocles (4th century B.C.), that "hair and foliage and the thick plumage of birds are one."[3]

[1] E. Zeller, *Greek Philosophy*, Eng. trans., i., 522 f.n., London 1881. Other particulars as to Alcmaeon in T. Gomperz, *Greek Thinkers*, Eng. trans., i., London, 1901.

[2] Zeller, *loc. cit.*, i., p. 297.

[3] Gomperz, *loc. cit.*, i., p. 244.

In the collected writings of Hippocrates and his school, the *Corpus Hippocraticum*, of which no part is later than the end of the 5th century, there are recorded many anatomical facts. The author of the treatise "On the Muscles" knew, for instance, that the spinal marrow is different from ordinary marrow and has membranes continuous with those of the brain. Embryos of seven days (!) have all the parts of the body plainly visible. Work on comparative embryology is contained in the treatise "On the Development of the Child."[1]

The author of the treatise "On the Joints," which Littré calls "the great surgical monument of antiquity," is to be credited with the first systematic attempt at comparative anatomy, for he compared the human skeleton with that of other Vertebrates.

Aristotle (384-322 B.C.)[2] may fairly be said to be the founder of comparative anatomy, not because he was specially interested in problems of "pure morphology," but because he described the structure of many animals and classified them in a scientific way. We shall discuss here the morphological ideas which occur in his writings upon animals —in the *Historia Animalium*, the *De Partibus Animalium*, and the *De Generatione Animalium*.

The *Historia Animalium* is a most comprehensive work, in some ways the finest text-book of Zoology ever written. Certainly few modern text-books take such a broad and sane view of living creatures. Aristotle never forgets that form and structure are but one of the many properties of living things; he takes quite as much interest in their behaviour, their ecology, distribution, comparative physiology. He takes a special interest in the comparative physiology of reproduction. The *Historia Animalium* contains a description of the form and structure of man and of as many animals as Aristotle was acquainted with—and he was acquainted with an astonishingly large number. The later *De Partibus Animalium* is a treatise on the causes of the form and

[1] R. Burckhardt, *Biologie u. Humanismus*, p. 85, Jena, 1907.

[2] See the interesting account of Aristotle's biological work in Prof. D'Arcy W. Thompson's Herbert Spencer lecture (1913) and his translation of the *Historia Animalium* in the Oxford series.

structure of animals. Owing to the importance which Aristotle ascribed to the final cause this work became really a treatise on the functions of the parts, a discussion of the problems of the relation of form to function, and the adaptedness of structure.

Aristotle was quite well aware that each of the big groups of animals was built upon one plan of structure, which showed endless variations "in excess and defect" in the different members of the group. But he did not realise that this fact of community of plan constituted a problem in itself. His interest was turned towards the functional side of living things, form was for him a secondary result of function.

Yet he was not unaware of facts of form for which he could not quite find a place in his theory of organic form, facts of form which were not, at first sight at least, facts of function. Thus he was aware of certain facts of "correlation," which could not be explained off-hand as due to correlation of the functions of the parts. He knew, for instance, that all animals without front teeth in the upper jaw have cotyledons, while most that have front teeth on both jaws and no horns have no cotyledons (*De Gen.*, ii. 7).

Speaking generally, however, we find in Aristotle no purely morphological concepts. What then does morphology owe to Aristotle? It owes to him, *first*, a great mass of facts about the structure of animals; *second*, the first scientific classification of animals;[1] *third*, a clear enunciation of the fact of community of plan within each of the big groups; *fourth*, an attempt to explain certain instances of the correlation of parts; *fifth*, a pregnant distinction between homogeneous and heterogeneous parts; *sixth*, a generalisation on the succession of forms in development; and *seventh*, the first enunciation of the idea of the *Échelle des êtres*.

(1) What surprises the modern reader of the *Historia Animalium* perhaps more than anything else is the extent and variety of Aristotle's knowledge of animals. He

[1] On Aristotle's forerunners, see R. Burckhardt, "Das koïsche Tiersystem, eine Vorstufe des zoologischen Systematik des Aristoteles." *Verh. Naturf. Ges. Basel*, xx., 1904.

describes more than 500 kinds.[1] Not only does he know the ordinary beasts, birds, and fishes with which everyone is acquainted, but he knows a great deal about cuttlefish, snails and oysters, about crabs, crawfish (*Palinurus*), lobsters, shrimps, and hermit crabs, about sea-urchins and starfish, sea-anemones and sponges, about ascidians (which seem to have puzzled him not a little!). He has noticed even fish-lice and intestinal worms, both flat and round. Of the smaller land animals, he knows a great many insects and their larvæ. The extent of his anatomical knowledge is equally surprising, and much of it is clearly the result of personal observation. No one can read his account of the internal anatomy of the chameleon (*Hist. Anim.*, ii.), or his description of the structure of cuttlefish (*Hist. Anim.*, iv.), or that touch in the description of the hermit crab (*Hist. Anim.*, iv.)—"Two large eyes . . . not . . . turned on one side like those of crabs, but straight forward"—without being convinced that Aristotle is speaking of what he has seen. Naturally he could not make much of the anatomy of small insects and snails, and, to tell the truth, he does not seem to have cared greatly about the minutiæ of structure. He was too much of a Greek and an aristocrat to care about laborious detail.

Not only did he lay a foundation for comparative anatomy, but he made a real start with comparative embryology. Medical men before him had known many facts about human development; Aristotle seems to have been the first to study in any detail the development of the chick. He describes this as it appears to the naked eye, the position of the embryo on the yolk, the palpitating spot at the third day, the formation of the body and of the large sightless eyes, the veins on the yolk, the embryonic membranes, of which he distinguished two.

(2) Aristotle had various systems of classifying animals. They could be classified, he thought, according to their structure, their manner of reproduction, their manner of life, their mode of locomotion, their food, and so on. Thus you

[1] T. E. Lones, *Aristotle's Researches in Natural Science*, pp. 82-3, London, 1912.

might, in addition to structural classifications, divide animals into gregarious, solitary and social, or land animals into troglodytes, surface-dwellers, and burrowers (*Hist. Anim.*, i.).

He knew that dichotomous classifications were of little use for animals (*De Partibus*, i. 3) and he explicitly and in so many words accepted the principle of all "natural" classification, that affinities must be judged by comparing not one but the sum total of characters. As everyone knows, he was the first to distinguish the big groups of animals, many of which were already distinguished roughly by the common usages of speech. Among his Sanguinea he did little more than define with greater exactitude the limits of the groups established by the popular classification. Among the "exsanguineous" animals, however, corresponding to our Invertebrates, he established a much more definite classification than the popular, which is apt to call them indiscriminately "shellfish," "insects," or "creeping things." He went beyond the superficialities of popular classification, too, in clearly separating Cetacea from fishes. He had some notion of species and genera in our sense. He distinguished many species of cuttlefish—*Octopus* (*Polypus*) of which there were many kinds, *Eledone* (*Moschites*) which he knew to have only one row of suckers while *Octopus* has two, *Argonauta*, *Nautilus*, *Sepia*, and apparently *Loligo media* (= his Teuthis) and *L. vulgaris* (or *forbesii*) which seems to be his Teuthos. He had a grasp of the principles which should be followed in judging of the natural affinities of species. For example, he knew that the cuckoo resembles a hawk. "But," he says, "the hawk has crooked talons, which the cuckoo has not, nor does it resemble the hawk in the form of its head, but in these respects is more like the pigeon than the hawk, which it resembles in nothing but its colour; the markings, however, upon the hawk are like lines, while the cuckoo is spotted" (*Hist. Anim.*, Cresswell's trans., p. 147, London, 1862).

The groups he distinguished were—man, viviparous quadrupeds, oviparous quadrupeds, birds, fishes, Cetacea, Cephalopoda, Malacostraca (=higher Crustacea), Insecta (= annulose animals), Testacea (= molluscs, echinoderms, ascidians). A class of Acalephæ, including sea-anemones and sponges, was grouped with the Testacea. The first five groups were classed together as sanguineous, the others as exsanguineous, from the presence or absence of red blood.

Besides these classes "there are," he says, "many other creatures in the sea which it is not possible to arrange in any class from their scarcity" (Creswell, *loc. cit.*, p. 90).

(3) Aristotle's greatest service to morphology is his clear recognition of the unity of plan holding throughout each of the great groups.

He recognises this most clearly in the case of man and the viviparous quadrupeds, with whose structure he was best acquainted. In the *Historia Animalium* he takes man as a standard, and describes his external and internal parts in detail, then considers viviparous quadrupeds and compares them with man. "Whatever parts a man has before, a quadruped has beneath; those that are behind in man form the quadruped's back" (Cresswell, *loc. cit.*, p. 26). Apes, monkeys, and Cynocephali combine the characteristics of man and quadrupeds. He notices that all viviparous quadrupeds have hair. Oviparous quadrupeds resemble the viviparous, but they lack some organs, such as ears with an external pinna, mammæ, hair. Oviparous bipeds, or birds, also "have many parts like the animals described above." He does not, however, seem to realise that a bird's wings are the equivalent of a mammal's arms or fore-legs. Fishes are much more divergent; they possess no neck, nor limbs, nor testicles (meaning a solid ovoid body such as the testis in mammals), nor mammæ. Instead of hair they have scales.

Speaking generally, the Sanguinea differ from man and from one another in their parts, which may be present or absent, or exhibit differences in "excess and defect," or in form. Unity of plan extends to all the principal systems of organs. "All sanguineous animals have either a bony or a spinous column. The remainder of the bones exist in some animals; but not in others, for if they have the limbs they have the bones belonging to them" (Cresswell, *loc. cit.*, p. 60). "Viviparous animals with blood and feet do not differ much in their bones, but rather by analogy, in hardness, softness, and size" (Cresswell, *loc. cit.*, p. 59).

The venous system, too, is built upon the same general plan throughout the Sanguinea. "In all sanguineous animals, the nature and origin of the principal veins are the same, but the multitude of smaller veins is not alike in all, for neither are the parts of the same nature, nor do all possess the same parts" (Cresswell, *loc. cit.*, p. 56). It will be noticed in the first and last of these three quotations that Aristotle recognises the fact of correlation between systems of organs—between limbs and bones, and between blood-vessels and the parts to which they go.

Sanguineous animals all possess certain organs—heart, liver, spleen, kidneys, and so on. Other organs occur in most of the classes—the œsophagus and the lungs. "The position which these parts occupy is the same in all animals [sc. Sanguinea]" (Cresswell, *loc. cit.*, p. 39).

Unity of plan is observable not only in the Sanguinea, but also within each of the other large groups. Aristotle recognises that all his cuttlefish are alike in structure. Among his Malacostraca he compares point by point the external parts of the carabus (*Palinurus*), and the astacus (*Homarus*), and he compares also the general internal anatomy of the various "genera" he distinguishes. As regards Testacea, he writes, "The nature of their internal structure is similar in all, especially in the turbinated animals, for they differ in size and in the relations of excess; the univalves and bivalves do not exhibit many differences" (Cresswell, *loc. cit.*, p. 83). There is an interesting remark about "the creature called carcinium" (hermit-crab), that it "resembles both the Malacostraca and the Testacea, for this in its nature is similar to the animals that are like carabi, and it is born naked" (Cresswell, *loc. cit.*, p. 85). In the last phrase we may perhaps read the first recognition of the embryological criterion.

With the recognition of unity of plan within each group necessarily goes the recognition of what later morphology calls the homology of parts. The parts of a horse can be compared one by one with the parts of another viviparous quadruped; in all the animals belonging to the same class the parts are the same, only they differ in excess or defect—these remarks are placed in the forefront of the

Historia Animalium. Generally speaking, parts which bear the same name are for Aristotle homologous throughout the class. But he goes further and notes the essential resemblance underlying the differences of certain parts. He classes together nails and claws, the spines of the hedgehog, and hair, as being homologous structures. He says that teeth are allied to bones, whereas horns are more nearly allied to skin (*Hist. Anim.*, iii.). This is an astonishingly happy guess, considering that all he had to go upon was the observation that in black animals the horns are black but the teeth white. One cannot but admire the way in which Aristotle fixes upon apparently trivial and commonplace facts, and draws from them far-reaching consequences. He often goes wrong, it is true, but he always errs in the grand manner.

While Aristotle certainly recognised the existence of homologies, and even had a feeling for them, he did not clearly distinguish homology from analogy. He comes pretty near the distinction in the following passage. After explaining that in animals belonging to the same class the parts are the same, differing only in excess or defect, he says, "But some animals agree with each other in their parts neither in form nor in excess and defect, but have only an analogous likeness, such as a bone bears to a spine, a nail to a hoof, a hand to a crab's claw, the scale of a fish to the feather of a bird, for that which is a feather in the bird is a scale in the fish" (Cresswell, *loc. cit.*, p. 2). One of these comparisons is, however, a homology not an analogy, and the last phrase throws a little doubt upon the whole question, for it is not made clear whether it is position or function that determines what are equivalent organs.

In the *De Partibus Animalium* there occurs the following passage:—"Groups that only differ in degree, and in the more or less of an identical element that they possess, are aggregated under a single class; groups whose attributes are not identical but analogous are separated. For instance, bird differs from bird by gradation, or by excess and defect; some birds have long feathers, others short ones, but all are feathered. Bird and Fish are more remote and only agree in having analogous organs; for what in the bird is feather, in the fish is scale. Such analogies can scarcely, however, serve universally as indications for the formation of groups, for almost all animals present analogies in their corresponding parts."[1] It is thus similarity in form and structure which determines the formation of the main groups. Within each group the parts differ only in degree, in largeness or smallness, softness and hardness, smoothness or roughness, and the like (*loc. cit.*, i., 4, 644[b]). These passages show that Aristotle had some conception of homology as distinct from analogy. He did not, however, develop the idea. What Aristotle sought in the variety of animal structure, and what he found, were not homologies, but rather communities of function, parts with the same attributes. His interest was all in *organs*, in functioning parts, not in the mere spatial relationship of parts.

This comes out clearly in his treatise *On the Parts of Animals*, which is subsequent to, and the complement of, his *History of Animals*. The latter is a description of the variety of animal form, the former is a treatise on the functions of the parts. He describes the plan of the *De Partibus Animalium* as follows:—"We have, then, first to describe the common functions, common, that is, to the whole animal kingdom, or to certain large groups, or to members of a species. In other words, we have to describe the attributes common to all animals, or to assemblages, like the class of Birds, of closely allied groups differentiated by gradation, or to groups like Man not differentiated into subordinate groups. In the first case the common attributes may be called analogous, in the second generic, in the third specific" (i., 5, 645[b], trans. Ogle). The alimentary canal is a good example of a part which is "analogous" throughout the animal kingdom, for "all animals possess in common those parts by which they take in food, and into which they receive it" (Cresswell, *loc. cit.*, p. 6).

The *De Partibus Animalium* becomes in form a comparative organography, but the emphasis is always on function and community of function. Thus he treats of bone, "fish-spine," and cartilage together (*De Partibus*, ii., 9, 655[a]), because they have the same function, though he says

[1] *De Partibus Animalium*, i., 4, 644[a], trans. W. Ogle, Oxford, 1911.

elsewhere that they are only analogous structures (ii., 8, 653[b]). In the same connection he describes also the supporting tissues of Invertebrates—the hard exoskeleton of Crustacea and Insects, the shell of Testacea, the "bone" of *Sepia* (ii., 8, 654[a]). Aristotle took much more interest in analogies, in organs of similar function, than in homologies. He did recognise the existence of homologies, but rather *malgré lui*, because the facts forced it upon him.

His only excursion into the realm of "transcendental anatomy" is his comparison of a Cephalopod to a doubled-up Vertebrate whose legs have become adherent to its head, whose alimentary canal has doubled upon itself in such a way as to bring the anus near the mouth (*De Partibus*, iv., 9, 684[b]). It is clear, however, that Aristotle did not seek to establish by this comparison any true homologies of parts, but merely analogies, thus avoiding the error into which Meyranx and Laurencet fell more than two thousand years later in their paper communicated to the Académie des Sciences, which formed the starting-point of the famous controversy between Cuvier and E. Geoffroy St Hilaire (see Chap. V., below).

Moreover, Aristotle did not so much compare a Cephalopod with a doubled-up Vertebrate as contrast Cephalopods (and also Testacea) with all other animals. Other animals have their organs in a straight line; Cephalopods and Testacea alone show this peculiar doubling up of the body.

(4) Aristotle was much struck with certain facts of correlation, of the interdependence of two organs which are not apparently in functional dependence on one another. Such correlation may be positive or negative; the presence of one organ may either entail the presence of the other, or it may entail its absence. Aristotle has various ways of explaining facts of correlation. He observed that no animal has both tusks and horns, but this fact could easily be explained on the principle that Nature never makes anything superfluous or in vain. If an animal is protected by the possession of tusks it does not require horns, and *vice versa*. The correlation of a multiple stomach with deficient

development of the teeth (as in Ruminants) is accounted for by saying that the animal needs its complex stomach to make up for the shortcomings of its teeth! (*De Partibus*, iii., 14, 674[b].) Other examples of correlation were not susceptible of this explanation in terms of final causes. He lays stress on the fact, in the main true, of the inverse development of horns and front teeth in the upper jaw, exemplified in Ruminants. He explains the fact in this way. Teeth and horns are formed from earthy matter in the body and there is not enough to form both teeth and horns, so "Nature by subtracting from the teeth adds to the horns; the nutriment which in most animals goes to the former being here spent on the augmentation of the latter" (*De Partibus*, iii., 2, 664[a], trans. Ogle). A similar kind of explanation is offered of the fact that Selachia have cartilage instead of bone, "in these Selachia Nature has used all the earthy matter on the skin [*i.e.*, on the placoid scales]; and she is unable to allot to many different parts one and the same superfluity of material" (*De Partibus*, ii., 9, 655[a], trans. Ogle). Speaking generally, "Nature invariably gives to one part what she subtracts from another" (*loc. cit.*, ii., 14, 658[a]).

This thought reappears again in the 19th century in E. Geoffroy St Hilaire's *loi de balancement* and also in Goethe's writings on morphology. For Aristotle it meant that Nature was limited by the nature of her means, that finality was limited by necessity. Thus in the larger animals there is an excess of earthy matter, as a necessary result of the material nature of the animal; this excess is turned by Nature to good account, but there is not enough to serve both for teeth and for horns (*loc. cit.*, iii., 2, 663[b]).

But there are other instances of correlation which seem to have taxed even Aristotle's ingenuity beyond its powers. Thus he knew that all animals (meaning viviparous quadrupeds) with no front teeth in the upper jaw have cotyledons on their fœtal membranes, and that most animals which have front teeth in both jaws and no horns have no cotyledons (*De Generatione*, ii., 7). He offers no explanation of this, but accepts it as a fact.

We may conveniently refer here to one or two other ideas of Aristotle regarding the causes of form. He makes the profound remark that the possible range of form of an organ is limited to some extent by its existing differentiation. Thus he explains the absence of external (projecting) ears in birds and reptiles by the fact that their skin is hard and does not easily take on the form of an external ear (*De Partibus*, ii., 12). The fact of the inverse correlation is certain; the explanation is, though very vague, probably correct.

In one passage of the *De Partibus* Aristotle clearly enunciates the principle of the division of labour, afterwards emphasised by H. Milne-Edwards. In some insects, he says, the proboscis combines the functions of a tongue and a sting, in others the tongue and the sting are quite separate. "Now it is better," he goes on, "that one and the same instrument shall not be made to serve several dissimilar ends; but that there shall be one organ to serve as a weapon, which can then be very sharp, and a distinct one to serve as a tongue, which can then be of spongy texture and fit to absorb nutriment. Whenever, therefore, Nature is able to provide two separate instruments for two separate uses, without the one hampering the other, she does so, instead of acting like a coppersmith who for cheapness makes a spit and lampholder in one" (iv., 6, 683[a]).

(5) The first sentence of the *Historia Animalium* formulates, with that simplicity and directness which is so characteristic of Aristotle, the distinction between homogeneous and heterogeneous parts, in the mass the distinction between tissues and organs. "Some parts of animals are simple, and these can be divided into like parts, as flesh into pieces of flesh; others are compound, and cannot be divided into like parts, as the hand cannot be divided into hands, nor the face into faces. All the compound parts also are made up of simple parts—the hand, for example, of flesh and sinew and bone" (Cresswell, *loc. cit.*, p. 1).

In the *De Partibus Animalium* he broadens the conception by adding another form of composition. "Now there are," he says, "three degrees of composition; and of these the first in order, as all will allow, is composition out of what some call the elements, such as earth, air, water, fire. . .

The second degree of composition is that by which the homogeneous parts of animals, such as bone, flesh, and the like, are constituted out of the primary substances. The third and last stage is the composition which forms the heterogeneous parts, such as face, hand, and the rest" (ii., 1, 646[a], trans. Ogle).

In the *Historia Animalium* the homogeneous parts are divided into (1) the soft and moist (or fluid), such as blood, serum, flesh, fat, suet, marrow, semen, gall, milk, phlegm, fæces and urine, and (2) the hard and dry (or solid), such as sinew, vein, hair, bone, cartilage, nail, and horn. It would appear from this enumeration that Aristotle's distinction of simple and complex parts does not altogether coincide with our distinction of tissues and organs. We should not call vein a tissue, nor do we include under this heading non-living secretions. But in the *De Partibus Animalium* Aristotle, while still holding to the distinction set forth above, is alive to the fact that his simple parts include several different sorts of substances. He distinguishes among the homogeneous parts three sets. The first of these comprises the tissues out of which the heterogeneous parts are constructed, *e.g.*, flesh and bone; the second set form the nutriment of the parts, and are invariably fluid; while the third set are the residue of the second and constitute the residual excretions of the body (ii., 2, 647[b]). He sees clearly the difficulty of calling vein or blood-vessel a simple part, for while a blood-vessel and a part of it are both blood-vessel, as we should say vascular tissue, yet a part of a blood-vessel is not a blood-vessel. There is form superadded to homogeneity of structure (ii., 2, 647[b]). Similarly for the heart and the other viscera. "The heart, like the other viscera, is one of the homogeneous parts; for, if cut up, its pieces are homogeneous in substance with each other. But it is at the same time heterogeneous in virtue of its definite configuration" (ii., 1, 647[a], trans. Ogle).

Aristotle, therefore, came very near our conception of tissue. He was of course not a histologist; he describes not the structure of tissues, which he could not know, but rather their distribution within the organism; his section on the homogeneous parts of Sanguinea (*Historia Animalium*, iii., second half) is largely a comparative topographical anatomy; in it, for instance, he describes the venous and skeletal systems.

This distinction which Aristotle drew plays an important part in all his writings on animals, particularly in his theory of development. It was a distinction of immense value, and is full of meaning even at the present day. No one has ever given a better definition of organ than is implied in Aristotle's description of the heterogeneous parts—"The capacity of action resides in the compound parts" (Cresswell, *loc. cit.*, p. 7). The heterogeneous parts were distinguished by the faculty of doing something, they were the active or executive parts. The homogeneous parts were distinguished mainly by physical characters (*De Generatione*, i., 18), but certain of them had other than purely physical properties, they were the organs of touch (*De Partibus*, ii., 1, 647[a]).

(6) In a passage in the *De Generatione* (ii, 3) Aristotle says that the embryo is an animal before it is a particular animal, that the general characters appear before the special This is a foreshadowing of the essential point in von Baer's law (see Chap. IX. below).

He considers also that tissues arise before organs. The homogeneous parts are anterior genetically to the heterogeneous parts and posterior to the elementary material (*De Partibus*, ii., 1, 646[b]).

(7) We meet in Aristotle an idea which later acquired considerable vogue, that of the *Échelle des êtres* (or "scale of beings"), that organisms, or even all objects organic or inorganic, can be arranged in a single ascending series. The idea is a common one; its first literary expression is found perhaps in primitive creation-myths, in which inorganic things are created before organic, and plants before animals. It may be recognised also in Anaximander's theory that land animals arose from aquatic animals, more clearly still in Anaxagoras' theory that life took its origin on this globe from vegetable germs which fell to earth with the rain. Anaxagoras considered animals higher in the scale than plants, for while the latter participated in pleasure (when they

grew) and pain (when they lost their leaves), animals had in addition "Nous." In Empedocles' theory of evolution, the vegetable world preceded the animal. Plato, in the *Timaeus*, describes the whole organic world as being formed by degradation from man, who is created first. Man sinks first into woman, then into brute form, traversing all the stages from the higher to the lower animals, and becoming finally a plant. This is a reversal of the more usual notion, but the idea of gradation is equally present.

Aristotle seems not to have believed in any transformation of species, but he saw that Nature passes gradually from inanimate to animate things without a clear dividing line. "The race of plants succeeds immediately that of inanimate objects" (Cresswell, *loc. cit.*, p. 94). Within the organic realm the passage from plants to animals is gradual. Some creatures, for example, the sea-anemones and sponges, might belong to either class.

Aristotle recognised also a natural series among the groups of animals, a series of increasing complexity of structure. He begins his study of structure with man, who is the most intricate, and then takes up in turn viviparous and oviparous quadrupeds, then birds, then fishes. After the Sanguinea he considers the Exsanguinea, and of the latter first the most highly organised, the Cephalopods, and last the simplest, the lower members of his class of the Testacea. In treating of generation (in *Hist. Animalium*, v.) he reverses this order. In the *De Generatione* (Book ii., 1) there is given another serial arrangement of animals, this time in relation to their manner of reproduction. There is a gradation, he says, of the following kind:—

1. Internally viviparous Sanguinea } producing a perfect
2. Externally viviparous Sanguinea } animal.
3. Oviparous Sanguinea—producing a perfect egg.
4. Animals producing an imperfect egg (one which increases in size after being laid).
5. Insects, producing a scolex (or grub).

In Aristotle's view the gradation of organic forms is the consequence, not the cause, of the gradation observable in their activities. Plants have no work to do beside nutrition, growth, and reproduction; they possess only the nutritive soul. Animals possess in addition sensation and the sensitive or perceptive soul—"their manner of life differs in their having pleasure in sexual intercourse, in their mode of parturition and rearing their young" (*Hist. Anim.*, viii., trans. Cresswell, p. 195). Man alone has the rational soul in addition to the two lower kinds.

As it is put in the *De Partibus* (ii., 10, 656[a], trans. Ogle), "Plants, again, inasmuch as they are without locomotion, present no great variety in their heterogeneous parts. For, where the functions are but few, few also are the organs required to effect them. . . . Animals, however, that not only live but feel, present a greater multiformity of parts, and this diversity is greater in some animals than in others, being most varied in those to whose share has fallen not mere life but life of high degree. Now such an animal is man."

With the great exception of Aristotle, the philosophers of Greece and Rome made little contribution to morphological theory. Passing mention may be made of the Atomists—Leucippus, Democritus, and their great disciple Lucretius, who in his magnificent poem "De Natura Rerum" gave impassioned expression to the materialistic conception of the universe. But the full effect of materialism upon morphology does not become apparent till the rise of physiology in the 17th and 18th centuries, and reaches its culmination in the 19th century. The evolutionary ideas of Lucretius exercised no immediate influence upon the development of morphology.

CHAPTER II

COMPARATIVE ANATOMY BEFORE CUVIER

FOR two thousand years after Aristotle little advance was made upon his comparative anatomy. Knowledge of the human body was increased not long after his death by Herophilus and Erasistratus, but not even Galen more than four centuries later made any essential additions to Aristotle's anatomy.

During the Middle Ages, particularly after the introduction to Europe in the 13th century of the Arab texts and commentaries, Aristotle dominated men's thoughts of Nature. The commentary of Albertus Magnus, based upon that of Avicenna, did much to impose Aristotle upon the learned world. Albertus seems to have contented himself with following closely in the footsteps of his master. There are noted, however, by Bonnier certain improvements made by Albertus on Aristotle's view of the seriation of living things. "He is the first," writes Bonnier, "to take the correct view that fungi are lower plants allied to the most lowly organised animals. From this point there start, for Albertus Magnus, two series of living creatures, and he regards the plant series as culminating in the trees which have well-developed flowers."[1]

Aristotle's influence is predominant also in the work of Edward Wotton (1492-1555), who in his book *De differentiis animalium* adopted a classification similar to that proposed by Aristotle. He too laid stress upon the gradation shown from the lower to the higher forms.

In the 16th century, two groups of men helped to lay foundations for a future science of comparative anatomy—

[1] *Le Monde végétal*, p. 41, Paris, 1907.

the great Italian anatomists Vesalius, Fallopius and Fabricius, and the first systematists (though their "systems" were little more than catalogues) Rondeletius, Aldrovandus and Gesner.

The anatomists, however, took little interest in problems of pure morphology; the anatomy of the human body was for them simply the necessary preliminary of the discovery of the functions of the parts — they were quite as much physiologists as anatomists.

One of them, Fabricius, made observations on the development of the chick (1615). Harvey, who was a pupil of Fabricius, likewise published an account of the embryology of the chick.[1] In his philosophy and habit of thought Harvey was a follower of Aristotle. It is worth noting that in his *Exercitationes anatomicae de motu cordis* (1628) there is a passage which dimly foreshadows the law of recapitulation in development which later had so much vogue.[2]

A stimulating contribution to comparative anatomy was made by Belon,[3] who published in 1555 a *Histoire de la nature des Oyseaux*, in which he showed opposite one another a skeleton of a bird and of a mammal, giving the same names to homologous bones. The anatomy of animals other than man was indeed not altogether neglected at this time. Coiter (1535-1600) studied the anatomy of Vertebrates, discovering among other things the fibrous structure of the brain. Carlo Ruini of Bologna wrote in 1598 a book on the anatomy of the horse.[4] Somewhat later Severino, professor at Naples, dissected many animals and came to the conclusion

[1] *Exercitationes de generatione animalium*, 1651. For an account of Harvey's work on generation and development, see Em. Rádl's masterly *Geschichte der biologischen Theorien*, i., pp. 31-8, Leipzig, 1905.

[2] The passage runs:—"Sic natura perfecta et divina nihil faciens frustra, nec quipiam animali cor addidit, ubi non erat opus, neque priusquam esset ejus usus, fecit; sed iisdem gradibus in formatione cujuscumque animalis, transiens per omnium animalium constitutiones (ut ita dicam) ovum, vermem, fœtum, perfectionem in singulis acquirit."

[3] See I. Geoffroy St Hilaire, *Essais de Zoologie générale*, p. 71, Paris, 1841.

[4] M. Foster, *Lectures on the History of Physiology*, Cambridge, p. 53, 1901.

that they were built upon the same plan as man.[1] Willis, of Oxford and London, in his *Cerebri Anatome* (1659) recognised the necessity for comparative study of the structure of the brain. He found out that the brain of man is very like that of other mammals, the brain of birds, on the contrary, like that of fishes![2] He described the anatomy of the oyster and the crayfish. He had, however, not much feeling for morphology.

The foundation of the Jardin des Plantes at Paris in 1626 and the subsequent addition to it of a Museum of Natural History and a menagerie gave a great impulse to the study of comparative anatomy by supplying a rich material for dissection. Advantage was taken of these facilities, particularly by Claude Perrault and Duverney.[3] In a volume entitled *De la Mécanique des Animaux*, Perrault recognises clearly the idea of unity of type, and even pushes it too far, seeking to prove that in plants there exists an arterial system and veins provided with valves.[4]

The beginning of the 17th century saw the invention of the microscope, which was to have such an enormous influence upon the development of biological studies. It did not come into scientific use until well on in the middle of the century. Just before it came into use Francis Glisson (1597-1677), an Englishman, gave in the introduction to his treatise on the liver an account of the notions then current on the structure of organic bodies. He classifies the parts as "similar" and "organic," the former determined by their material, the latter by the form which they assume. The similar parts are divided into the sanguineous or rich in blood and the spermatic. Both sets are further subdivided according to their physical characters,[5] the latter, for instance, into the hard, soft, and tensile tissues. The classification resembles greatly that propounded by Aristotle, though it is notably inferior in the details of its working out.

[1] *Zootomia democritea*, Nuremberg, 1645; *Antiperipatias, seu de respiratione piscium*, Amsterdam, 1661. [2] Rádl, *loc. cit.*, i., p. 50.

[3] Perrault et Duverney, *Mémoires pour servir à l'histoire des Animaux*, Paris, 1699.

[4] F. Houssay, *Nature et Sciences naturelles*, Paris, p. 76, n.d.

[5] Foster, *loc. cit.*, p. 85.

For Aristotle, as for all anatomists before the days of the microscope, the tissues were not much more than inorganic substances, differing from one another in texture, in hardness, and other physical properties. They possessed indeed properties, such as contractility, which were not inorganic, but as far as their visible structure was concerned there was little to raise them above the inorganic level. The application of the microscope changed all that, for it revealed in the tissues an organic structure as complex in its grade as the gross and visible structure of the whole organism. Of the four men who first made adequate use of the new aid, Malpighi, Hooke, Leeuenhoek, and Swammerdam, the first-named contributed the most to make current the new conceptions of organic structure. He studied in some detail the development of the chick. He described the minute structure of the lungs (1661), demonstrating for the first time, by his discovery of the capillaries, the connection of the arteries with the veins. In his work, *De viscerum structura* (1666), he describes the histology of the spleen, the kidney, the liver, and the cortex of the brain, establishing among other things the fact that the liver was really a conglomerate gland, and discovering the Malpighian bodies in the kidney. This work was done on a broad comparative basis. "Since in the higher, more perfect, red-blooded animals, the simplicity of their structure is wont to be involved by many obscurities, it is necessary that we should approach the subject by the observation of the lower, imperfect animals."[1] So he wrote in the *De viscerum structura*, and accordingly he studied the liver first in the snail, then in fishes, reptiles, mammals, and finally man. In the introduction to his *Anatome plantarum* (1675), in which he laid the foundations of plant histology, he vindicates the comparative method in the following words :—" In the enthusiasm of youth I applied myself to Anatomy, and although I was interested in particular problems, yet I dared to pry into them in the higher animals. But since these matters enveloped in peculiar mystery still lie in obscurity, they require the comparison of simpler conditions, and so the investigation of insects[2]

[1] Trans. by Foster, *loc. cit.*, p. 113.

[2] He made a careful study of the silkworm.

at once attracted me; finally, since this also has its own difficulties I applied my mind to the study of plants, intending after prolonged occupation with this domain, to retrace my steps by way of the vegetable kingdom, and get back to my former studies. But perhaps not even this will be sufficient; since the simpler world of minerals and the elements should have been taken first. In this case, however, the undertaking becomes enormous and far beyond my powers."[1] There is something fine in this life of broad outlines, devoted whole-heartedly to an idea, to a plan of research, which required a lifetime to carry out.

An important histological discovery dating from this time is that of the finer structure of muscle, made by Stensen (or Steno) in 1664. He described the structure of muscle-fibres, resolving them into their constituent fibrils.

To the microscope we owe not only histology but the comparative anatomy of the lower animals. Throughout the 17th and 18th centuries the discovery of structure in the lower animals went on continuously, as may be read in any history of Zoology.[2] We content ourselves here with mentioning only some representative names.

In the 17th century Leeuenhoek, applying the microscope almost at random, discovered fact after fact, his most famous discovery being that of the "spermatic animalcules."

Swammerdam studied the metamorphoses of insects and made wonderfully minute dissections of all sorts of animals, snails and insects particularly. He described also the development of the frog. It is curious to see what a grip his conception of metamorphosis had upon him when he

[1] "Etenim, ferventi aetatis calore, Anatomica aggressus, licet circa peculiaria fuerim solicitus, in *perfectioribus* tamen haec rimari sum ausus. Verum, cum haec propriis tenebris obscura jaceant, simplicium analogismo egent; inde *insectorum* indago illico arrisit; quae cum et ipsa suas habeat difficultates ad Plantarum perquisitionem animum *postremo* adjeci, ut diu hoc lustrato mundo gressu retroacto Vegetantis Naturae gradu, ad prima studia iter mihi aperirem. Sed nec forte hoc ipsum sufficiet cum simplicior *Mineralium Elementorumque* mundus praeire debeat. At in immensum excrescit opus, et meis viribus omnino impar," *Opera Omnia*, i., p. 1, London, 1686.

[2] See particularly E. Rádl, *loc. cit.*, 1 Teil. J. V. Carus, *Geschichte der Zoologie*, München, 1872.

homologises the stages of the frog's development with the Egg, the Worm, and the Nymph of insects (*Book of Nature*, p. 104, Eng. trans., 1785). He even speaks of the human embryo as being at a certain stage a Man-Vermicle.

In the 18th century, Réaumur and Bonnet continued the minute study of insects, laying more stress, however, on their habits and physiology than upon their anatomy. Lyonnet made a most laborious investigation of the anatomy of the willow-caterpillar (1762). John Hunter (1728-93) dissected all kinds of animals, from holothurians to whales. His interest was, however, that of the physiologist, and he was not specially interested in problems of form. It is interesting to note a formulation in somewhat confused language of the recapitulation theory. The passage occurs in his description of the drawings he made to illustrate the development of the chick. It is quoted in full by Owen (J. Hunter, *Observations on certain Parts of the Animal Œconomy*, with Notes by Richard Owen. London, 1837. Preface, p. xxvi). We give here the last and clearest sentence—" If we were to take a series of animals from the more imperfect to the perfect, we should probably find an imperfect animal corresponding with some stage of the most perfect."

The tendency of the time was not towards morphology, but rather to general natural history and to systematics, the latter under the powerful influence of Linnæus (1707-1778). The former tendency is well represented by Réaumur (1683-1757) with his observations on insects, the digestion of birds, the regeneration of the crayfish's legs, and a hundred other matters. To this tendency belong also Trembley's famous experiments on Hydra (1744), and Rösel von Rosenhof's *Insektenbelustigungen* (1746-1761).

Bonnet (1720-1793) deserves special mention here, since in his *Traité d'Insectologie* (1745), and more fully in his *Contemplation de la Nature* (1764), he gives the most complete expression to the idea of the *Echelle des êtres*.

This idea seems to have taken complete possession of his imagination. He extends it to the universe. Every world has its own scale of beings, and all the scales when joined together form but one, which then contains all the possible orders of perfection. At the end of the Preface to his *Traité d'Insectologie* (Œuvres, i., 1779) he gives a long table, headed "Idée d'une Échelle des êtres naturels," and rather resembling a ladder, on the rungs of which the following names appear:—

MAN.	Tube-worms.	STONES.
Orang-utan.	Clothes-moths.	Figured stones.
Ape.	INSECTS.	Crystals.
QUADRUPEDS.	Gall insects.	SALTS.
Flying squirrel.	Taenia.	Vitriols.
Bat.	Polyps.	METALS.
Ostrich.	Sea Nettles.	HALF-METALS.
BIRDS.	Sensitive plant.	SULPHURS.
Aquatic birds.	PLANTS.	
Amphibious birds.	Lichens.	Bitumens.
Flying Fish.		EARTHS.
FISH.	Moulds.	Pure earth.
Creeping fish.	Fungi, Agarics.	WATER.
Eels.	Truffles.	AIR
Water serpents.	Corals, and Coralloids.	
SERPENTS.	Lithophytes.	FIRE.
Slugs.	Asbestos.	
Snails.	Talcs, Gypsums.	More subtile matter.
SHELL FISH.	Selenites, Slates.	

The nature of the transitional forms which he inserts between his principal classes show very clearly his entire lack of morphological insight—the transitions are functional. The positions assigned to clothes-moths and corals are very curious! The whole scheme, so fantastic in its details, was largely influenced by Leibniz's continuity philosophy, and is in no way an improvement on the older and saner Aristotelian scheme.

Robinet, in the fifth volume of his book *De la nature* (1761-6), foreshadows the somewhat similar views of the German transcendentalists. "All beings," he writes, "have been conceived and formed on one single plan, of which they are the endlessly graduated variations: this prototype is the human form, the metamorphoses of which are to be considered as so many steps towards the most excellent form of being."[1]

[1] For a good historical account of the gradation theories see Thienemann's paper in the *Zoologische Annalen* (Würzburg) iii., pp. 185-274, 1910, from which the quotation from Robinet is taken.

The idea of a gradation of beings appears also in Buffon (1707-1788), but here it takes more definitely its true character as a functional gradation.[1] "Since everything in Nature shades into everything else," he says, "it is possible to establish a scale for judging of the degrees of the intrinsic qualities of every animal."[2]

He is quite well aware that the groups of Invertebrates are different in structural plan from the Vertebrates—"The animal kingdom includes various animated beings, whose organisation is very different from our own and from that of the animals whose body is similarly constructed to ours."[3]

He limits himself to a consideration of the Vertebrates, deeming that the economy of an oyster ought not to form part of his subject matter! He has a clear perception of the unity of plan which reigns throughout the vertebrate series.[4] What is new in Buffon is his interpretation of the unity of plan. For the first time we find clearly expressed the thought that unity of plan is to be explained by community of origin.

Buffon's utterances on this point are, as is well known, somewhat vacillating. The famous passage, however, which occurs in his account of the Ass shows pretty clearly that Buffon saw no theoretical objection to the descent of all the varied species of animals from one single form. Once admit, he argues, that within the bounds of a single family one species may originate from the type species by "degeneration," then one might reasonably suppose that from a single being Nature could in time produce all the other organised beings.[5] Elsewhere, *e.g.*, in the discourse *De la Dégéneration des Animaux*,[6] Buffon expresses himself with more caution. He finds that it is possible to reduce the two hundred species of quadrupeds which he has described to

[1] *Histoire naturelle*, i., p. 13; ii, p. 9; iv., p. 101; and xiv., pp. 28-9, 1749 and later.

[2] No translation can render the beauty of the original—"Comme tout se fait et que tout est par nuance dans la Nature . . ." (iv., p. 101).

[3] *Hist. nat.*, iv., p. 5.

[4] See particularly his comparison of the skeleton of the horse with that of man. *Hist. Nat.*, iv., p. 381, also p. 13.

[5] *Loc. cit.*, p. 382.

[6] Tome xiv., pp. 311-374.

quite a small number of families "from which it is not impossible that all the rest are derived."[1] Within each of the families the species branch off from a parent or type species. This we may note is a great advance on the linear arrangement implied in the idea of an *Échelle des êtres.*[2]

It is a mistake to suppose that Buffon was par excellence a maker of hypotheses. On the contrary he saw things very sanely and with a very open mind. He expressly mentions the great difficulties which one encounters in supposing that one species may arise from another by "degeneration." How does it happen that two individuals "degenerate" just in the right direction and to the right stage so as to be capable of breeding together? How is it that one does not find intermediate links between species? One is reminded of the objections, not altogether without validity, which were made to the Darwinian theory in its early days. I cannot agree with those who think that Buffon was an out-and-out evolutionist, who concealed his opinions for fear of the Church. No doubt he did trim his sails—the palpably insincere "Mais non, il est certain, par la révélation, que tous les animaux ont également participé à la grace de la création,"[3] following hard upon the too bold hypothesis of the origin of all species from a single one, is proof of it. But he was too sane and matter-of-fact a thinker to go much beyond his facts, and his evolution doctrine remained always tentative. One thing, however, he was sure of, that evolution would give a rational foundation to the classification which, almost in spite of himself, he recognised in Nature. If, and only if, the species of one family originated from a single type species, could families be founded rationally, *avec raison.*

Buffon was, curiously enough, rather unwilling to recognise any systematic unit higher than the species. Strictly speaking there are only individuals in Nature; but there

[1] Tome xiv., p. 358.

[2] See also "Oiseaux," Tome i., pp. 394, 395. Pallas in 1766 adopted for the whole animal kingdom this branching arrangement.

[3] "But this cannot be, for it is certain by revelation that all animals have equally participated in the grace of creation."

are also groups of individuals which resemble one another from generation to generation and are able to breed together. These are species—Buffon adheres to the genetic definition of species—and the species is a much more definite unit than the genus, the order, the class, which are not divisions imposed by us upon Nature. Species are definitely discontinuous,[1] and this is the only discontinuity which Nature shows us. Buffon put his views into practice in his *Histoire Naturelle*, where he describes species after species, never uniting them into larger groups. We have seen, however, how the facts forced upon him the conception of the "family."

Buffon was no morphologist. He left to Daubenton what one might call the "dirty work" of his book, the dissection and minute description of the animals treated.

But Buffon was a man of genius, and accordingly his ideas on morphology are fresh and illuminating. Few naturalists have been so free from the prejudices and traditions of their trade. He makes in the *Discours sur la Nature des Animaux*[2] a distinction, which Bichat and Cuvier later developed with much profit, between the "animal" and the "vegetative" part of animals.[3] The vegetative or organic functions go on continuously, even in sleep, and are performed by the internal organs, of which the heart is the central one. The active waking life of the animal, that part of its life which distinguishes it from the plant, involves the external parts—the sense-organs and the extremities. An animal is, as it were, made up of a complex of organs performing the vegetative functions, assimilation, growth, and reproduction, surrounded by an envelope formed by the limbs, the sense-organs, the nerves and the brain, which is the centre of this "envelope."[4] Animals may differ from one another enormously in the external parts, particularly in the appendicular skeleton, without showing any great difference in the plan and arrangement of their internal organs.

[1] iv., p. 385. [2] iv., pp. 3-110.

[3] It has been revived in our own days by Bergson, *Matière et Mémoire*, p. 57.

[4] iv., pp. 7-15.

Quadrupeds, Cetacea, birds, amphibians and fish are as unlike as possible in external form and in the shape of their limbs; but they all resemble one another in their internal organs. Let the internal organs change, however—the external parts will change infinitely more, and you will get another animal, an animal of a totally different nature. Thus an insect has a most singular internal economy, and, in consequence, you find it is in every point different from any vertebrate animal.

In this contrast, on the whole justified, between the importance of variations in the "vegetative" and variations in the "animal" parts, one may see without doing violence to Buffon's thought, an indication of the difference between homology and analogy. It is usually in the external parts, in the organs by which the animal adapts itself to its environment, that one meets with the greatest number of analogical resemblances. This contrast of vegetative and animal parts and their relative importance for the discovery of affinities was at any rate a considerable step towards an analysis of the concept of unity of plan.

To Xavier Bichat (1771-1802) belongs the credit of working out in detail the distinction drawn by Aristotle and Buffon between the animal and the vegetative functions. Bichat was not a comparative anatomist; his interest lay in human anatomy, normal and pathological. So his views are drawn chiefly from the consideration of human structure.

He classifies functions into those relating to the individual and those relating to the species. The functions pertaining to the individual may be divided into those of the animal and those of the organic life.[1] "I call *animal life* that order of functions which connects us with surrounding bodies; signifying thereby that this order belongs only to animals" (p. lxxviii.). Its organs are the afferent and efferent nerves, the brain, the sense-organs and the voluntary muscles; the brain is its central organ. "Digestion, circulation, respiration, exhalation, absorption, secretion, nutrition, calorification, or production of animal heat, compose organic life, whose principal and central organ is the heart" (p. lxxix.).

The contrast of the animal and the organic life runs

[1] *Anatomie Générale*, Paris, 1801, Eng. trans. 1824.

through all Bichat's work; it receives classical expression in his *Recherches Physiologiques sur la Vie et la Mort* (1800). The plant and the animal stand for two different modes of living. The plant lives within itself, and has with the external world only relations of nutrition; the animal adds to this organic life a life of active relation with surrounding things (3rd ed., 1805, p. 2). "One might almost say that the plant is the framework, the foundation, of the animal, and that to form the animal it sufficed to cover this foundation with a system of organs fitted to establish relations with the world outside. It follows that the functions of the animal form two quite distinct classes. One class consists in a continual succession of assimilation and excretion; through these functions the animal incessantly transforms into its own substance the molecules of surrounding bodies, later to reject these molecules when they have become heterogeneous to it. Through this first class of functions the animal exists only within itself; through the other class it exists outside; it is an inhabitant of the world, and not, like the plant, of the place which saw its birth. The animal feels and perceives its surroundings, reflects its sensations, moves of its own will under their influence, and, as a rule, can communicate by its voice its desires and its fears, its pleasures or its pains. I call organic life the sum of the functions of the former class, for all organised creatures, plants or animals, possess them to a more or less marked degree, and organised structure is the sole condition necessary to their exercise. The combined functions of the second class form the 'animal' life, so named because it is the exclusive attribute of the animal kingdom" (pp. 2-3).

In both lives there is a double movement, in the animal life from the periphery to the centre and from the centre to the periphery, in the organic life also from the exterior to the interior and back again, but here a movement of composition and decomposition. As the brain mediates between sensation and motion, so the vascular system is the go-between of the organs of assimilation and the organs of dissimilation.

The most essential structural difference between the organs of animal life and the organs of organic life is, in man and the higher animals at least, the symmetry of the one set and the irregularity of the other—compare the symmetry of the nerves and muscles of the animal life with the asymmetrical disposition of the visceral muscles and the sympathetic nerves, which belong to the organic life.

Noteworthy differences exist between the two lives with respect to the influence of habit. Everything in the animal life is under the dominion of habit. Habit dulls sensation, habit strengthens the judgment. In the organic life, on the contrary, habit exercises no influence. The difference comes out clearly in the development of the individual. The organs of the organic life attain their full perfection independently of use; the organs of the animal life require an education, and without education they do not reach perfection (*loc. cit.*, p. 127).

Bichat was the founder of what was known for a time as General Anatomy—the study of the constituent tissues of the body in health and disease. His classification of tissues was macroscopical and physiological; he relied upon texture and function in distinguishing them rather than upon microscopical structure. The tissues he distinguished are as follows:—[1]

1. The cellular membrane.
2. Nerves of animal life.
3. Nerves of organic life.
4. Arteries.
5. Veins.
6. Exhalants.
7. Absorbents and glands.
8. Bones.
9. Medulla.
10. Cartilage.
11. Fibrous tissue.
12. Fibro-cartilage.
13. Muscles of organic life.
14. Muscles of animal life.
15. Mucous membrane.
16. Serous membrane.
17. Synovial membrane.
18. The Glands.
19. The Dermis.
20. Epidermis.
21. Cutis.

The "cellular membrane" seems to mean undifferentiated connective tissue; "exhalants" are imperceptible tubes arising from the capillaries and secreting fat, serum, marrow, etc.; the "absorbents and glands" are the lymphatics and the lymphatic glands.

In Bichat's eyes this resolution of the organism into

[1] *Anatomie Générale*, Eng. trans., i., p. lii.

tissues had a deeper significance than any separation into organs, for to each tissue must be attributed a *vie propre*, an individual and peculiar life. "When we study a function we must consider the complicated organ which performs it in a general way; but if we would be instructed in the properties and life of that organ we must absolutely resolve it into its constituent parts."[1] The tissues have, too, a great importance for pathology, for diseases are often diseases of tissues rather than of organs.[2]

[1] *Anatomie Générale*, Eng. trans., i., p. lviii.
[2] *Loc cit.*, i., sect. vii.

CHAPTER III

CUVIER

CUVIER was perhaps the greatest of comparative anatomists; his work is, in the best sense of the word, classical.

Like all his predecessors, like Aristotle, like the Italian anatomists, Cuvier studied structure and function together, even gave function the primacy.

Some functions, he says,[1] are common to all organised bodies—origin by generation, growth by nutrition, end by death. There are also secondary functions. Of these the most important, in animals at least, are the faculties of feeling and moving. These two faculties are necessarily bound up together; if Nature has given animals sensation she must also have given them the power of movement, the power to flee from what is harmful and draw near to what is good. These two faculties determine all the others. A creature that feels and moves requires a stomach to carry food in. Food requires instruments to divide it, liquids to digest it. Plants, which do not feel and do not move, have no need of a stomach, but have roots instead. Thus the "Animal Functions" of feeling and moving determine the character of the organs of the second order, the organs of digestion. These in their turn are prior to the organs of circulation, which are a means to the end of distributing the nutrient fluid or blood to all parts of the body. These organs of the third order are not only dependent on those of the second order, but are also not even necessary, for many animals are without them. Only animals with a circulatory system can have definite breathing organs—lungs or gills.

[1] *Leçons d'Anatomie Comparée*, tome i., pp. 10 *et seq.*, 1800.

Plants, and animals without a circulation, breathe by their whole surface.

There is accordingly a rational order of functions, and therefore of the systems of organs which perform them. The most important are the Animal Functions, with their great organ-system, the neuro-muscular mechanism. Then come the digestive functions, and after them, and in a sense accessory to them, the functions and organs of circulation and respiration. The last three may be grouped as the Vital Functions.

The Animal Functions not only determine the character of the Vital Functions, but influence also the primary faculty of generation, for animals' power of movement has rendered their mode of fecundation more simple, has therefore had an effect on their organs of generation.

This division into "Animal" and "Vital" functions recalls Buffon's and Bichat's distinction of the "animal" and the "vegetative" lives. Cuvier apparently took this idea from Buffon, for he says that a plant is an animal that sleeps.[1] But the idea is as old as Aristotle, who discusses the "sleep" of embryos and of plants in the last book of the *De Generatione animalium.* The distinction between animal and vegetative life is, of course, based for Aristotle in the difference between the ψυχὴ αἰσθητική and the ψυχὴ θρεπτική. Cuvier, like Aristotle, Buffon, and Bichat, makes the heart the centre of the "vegetative" organs.

It is important to note that Cuvier puts function before structure, and infers from function what the organ will be. "Plants," he writes, "having few faculties, have a very simple organisation."[2] It is only after having discussed and classified functions that Cuvier goes on to examine organs

First his views on the composition of the animal body. Aristotle distinguished three degrees of composition—the "elements," the homogeneous parts, and the heterogeneous parts or organs. Cuvier does the same. Some small advance has been made in the two thousand years' interval, due in the first place to the progress of chemistry, and in the second to the invention of the microscope. To the first circumstance Cuvier owes his knowledge that the inorganic

[1] *Leçons d'Anatomie Comparée*, i., p. 18. [2] *Loc. cit.*, i., p. 13.

substances forming the first degree of composition are principally C, N, H, O, and P, combined to form albumen, fibrine, and the like, which are in their turn combined to form the solids and fluids of the body. To the latter circumstance Cuvier owes the statement that the finest fragments into which mechanical division can resolve the organism are little flakes and filaments, which, joined up loosely together, form a "cellulosity." The discovery of the true cellular nature of animal tissues did not come till much later, till some years after Cuvier's death in 1832. Knowledge of histological detail was, however, considerable by the beginning of the 19th century. Cuvier knew, for example, that each muscle fibre has its own nerve fibre. But he gives no elaborate account of the homogeneous parts, no detailed histology. On the other hand his treatment of the heterogeneous parts or organs is detailed and masterly.[1]

The main systems of organs are, in order of importance, the nervous and muscular, the digestive, the circulatory, and the respiratory. Each organ or system of organs may have many forms. If any form of any organ could exist in combination with any form of all the others there would be an enormous number of combinations theoretically possible. But these combinations do not all exist in Nature, for organs are not merely assembled (*rapprochés*), but act upon one another, and act all together for a common end. Accordingly only the combinations that fulfil these conditions exist in Nature. Cuvier thus dismisses the question of a science of possible organic forms and considers only the forms or combinations actually existing. This question of the possibility of a "theoretical" morphology of living things, after the fashion of the morphology of crystals with their sixteen possible types, was raised in later years by K. G. Carus, Bronn, and Haeckel.

Organisms, then, are harmonious combinations of organs, and the harmony is primarily a harmony of functions. Every function depends upon every other, and all are necessary. The harmony of organs and their mutual dependence are the results of the interdependence of function. This thought, the recognition of the functional unity of the

[1] *Leçons d'Anatomie Comparée*, tome i., Articles iii.-iv., 1800.

organism, is the fundamental one at the base of all Cuvier's work. Before him men had recognised more or less clearly the harmony of structure and function, and had based much of their work upon this unanalysed assumption. Cuvier was the first naturalist to raise this thought to the level of a principle peculiar to natural history. "It is on this mutual dependence of the functions and the assistance which they lend one to another that are founded the laws that determine the relations of their organs; these laws are as inevitable as the laws of metaphysics and mathematics, for it is evident that a proper harmony between organs that act one upon another is a necessary condition of the existence of the being to which they belong."[1]

This rational principle, peculiar to natural history, Cuvier calls the principle of the conditions of existence, for the following reason:—"Since nothing can exist that does not fulfil the conditions which render its existence possible, the different parts of each being must be co-ordinated in such a way as to render possible the existence of the being as a whole, not only in itself, but also in its relations with other beings, and the analysis of these conditions often leads to general laws which are as certain as those which are derived from calculation or from experiment."[2]

By "conditions of existence" he means something quite different from what is now commonly understood. The idea of the external conditions of existence, the environment, enters very little into his thought. He is intent on the adaptations of function and organ within the living creature—a point of view rather neglected nowadays, but essential for the understanding of living things. The very condition of existence of a living thing, and part of the essential definition of it, is that its parts work together for the good of the whole.

The principle of the adaptedness of parts may be used as an explanatory principle, enabling the naturalist to trace out in detail the interdependence of functions and their organs. When you have discovered how one organ is adapted to another and to the whole, you have gone a certain way towards understanding it. That is

[1] *Leçons d'Anatomie Comparée*, i., p. 47.

[2] *Le Règne Animal*, i., p. 6, 1817.

using teleology as a regulative principle, in Kant's sense of the word. Cuvier was indeed a teleologist after the fashion of Kant, and there can be no doubt that he was influenced, at least in the exposition of his ideas, by Kant's *Kritik der Urtheilskraft*, which appeared ten years before the publication of the *Leçons d'Anatomie Comparée*. Teleology in Kant's sense is and will always be a necessary postulate of biology. It does not supply an explanation of organic forms and activities, but without it one cannot even begin to understand living things. Adaptedness is the most general fact of life, and innumerable lesser facts can be grouped as particular cases of it, can be, so far, understood.

Cuvier's famous principle of correlation, the corner-stone of his work, is simply the practical application to the facts of structure of the principle of functional adaptedness. By the principle of correlation, from one part of an animal, given sufficient knowledge of the structure of its like, you can in a general way construct the whole. " This must necessarily be so: for all the organs of an animal form a single system, the parts of which hang together, and act and re-act upon one another; and no modifications can appear in one part without bringing about corresponding modifications in all the rest."[1] The logical basis of the principle is sound. The functions of the parts are all intimately bound up with one another, and one function cannot vary without bringing in its train corresponding modifications in the others. Structure and function are bound up together; every modification of a function entails therefore the modification of an organ. Hence from the shape of one organ you can infer the shape of the other organs—if you have sufficiently extensive empirical knowledge of functions, and of the relation of structure to function in each kind of organ. Given an alimentary canal capable of digesting only flesh, and possessing therefore a certain form, you know that the other functions must be adapted to this particular function of the alimentary canal. The animal must have keen sight, fine smell, speed, agility, and strength in paws and jaws. These particular functions must have correspondingly modified organs,

[1] *Histoire des Progrès des Sciences naturelles depuis* 1789, i., p. 310, 1826.

well-developed eyes and ears, claws and teeth. Further, you know from experience that such and such definitely modified organs are invariably found with the carnivorous habit, carnassial teeth, for example, and reduced clavicles. From a "carnivorous" alimentary canal, then, you can infer with certainty that the animal possessed carnassial teeth and the other structural peculiarities of carnivorous animals, *e.g.*, the peculiar coronoid process of the mandible. From the carnassial tooth you can infer the reduced clavicle, and so on. "In a word, the form of the tooth implies the form of the condyle; that of the shoulder blade that of the claws, just as the equation of a curve implies all its properties."[1]

Similarly the great respiratory power of birds is correlated with their great muscular strength, and renders necessary great digestive powers. Hence the correlated structure of lungs, muscles and their attachments, and alimentary canal, in birds.

Not only do systems of organs, by being adjusted to special modifications of function, influence one another, but so also do parts of the same organ. This is noticeably the case with the skeleton, where hardly a facet can vary without the others varying proportionately, so that from one bone you can up to a certain point deduce all the rest.

We deduce the necessity, the constancy, of these co-existences of organs from the observed reciprocal influence of their functions. That being established, we can argue from observed constancy of relation between two organs an action of one upon the other, and so be led to a discovery of their functions. But even if we do not discover the functional interdependencies of the parts, we can use the established fact of the constant co-existence of two parts as proof of a functional correlation between them.

Correlation is either a rational or an empirical principle, according as we know or do not know the interdependence of function of which it is the expression. Even when we apply the rational principle of correlation it would be useless in our hands if we had not extensive empirical knowledge; when we use an empirical rule of correlation we depend entirely upon observation. "There are a great many cases," writes

[1] *Recherches sur les Ossemens Fossiles*, i., p. 60, 1812.

Cuvier,[1] "where our theoretical knowledge of the relations of forms would not suffice, if it were not filled out by observation," that is to say, there are many cases of correlation not yet explicable in terms of function. From a hoof you can deduce the main characters of herbivores (with a certain amount of assistance from your empirical knowledge of herbivores), but could you from a cloven hoof deduce that the animal is a ruminant, unless you had observed the constancy of relation, not directly explicable in terms of function, between cloven hoofs and chewing the cud? Or could you deduce from the existence of frontal horns that the animal ruminates? "Nevertheless, since these relations are constant, they must necessarily have a sufficient cause; but as we are ignorant of this cause, observation must supplement theory; observation establishes empirical laws which become almost as certain as the rational laws, when they are based upon a sufficient number of observations. . . . But that there exist all the same hidden reasons for all these relations is partly revealed by observation itself, independently of general philosophy."[2] That is to say, even correlations for which no explanation in terms of function can be supplied are probably in reality functional correlations. This may, in some cases, be inferred from the graded correspondence of two sets of organs. For example, ungulates which do not ruminate, and have not a cloven hoof, have a more perfect dentition and more bones in the foot than the true cloven-hoofed ruminants. There is a correlation between the state of development of the teeth and of the foot. This correlation is a graded one, for camels, which have a more perfect dentition than other ruminants, have also a bone more in their tarsus. It seems probable, therefore, that there is some reason, that is, some explanation in terms of function, for this case of correlation.

Nevertheless, the fact remains that many correlations are not explicable in terms of function, and the substitution of correlation as an empirical principle for correlation as a rational principle marks for Cuvier a step away from his functional comparative anatomy towards a pure morphology. It is significant that in later times the term correlation

[1] *Ossemens fossiles*, i., p. 60. [2] *Loc. cit.*, i., p. 63.

has come to be applied more especially to the purely empirical constancies of relation, and has lost most of its functional significance. But the correlation of the parts of an organism is no mere mathematical concept, to be expressed by a coefficient, but something deeper and more vital.

Cuvier interpreted the functional dependence of the parts in terms of what we now call the general metabolism. He had a clear vision of the constant movement of molecules in the living tissue, combining and recombining, of the organism taking in and intercalating molecules from outside from the food and rejecting molecules in the excretions, a ceaseless *tourbillon vital.* "This general movement, universal in every part, is so unmistakably the very essence of life that parts separated from a living body straightway die."[1] The organisation of the body, the arrangement of its solids and liquids, is adapted to further the *tourbillon vital.* "Each part contributes to this general movement its own particular action and is affected by it in particular ways, with the result that, in every being, life is a unity which results from the mutual action and reaction of all its parts."[2]

Cuvier, however, did not resolve life into metabolism, nor reduce vital happenings to the chemical level. The form of organised bodies is more essential than the matter of which they are composed, for the matter changes ceaselessly while the form remains unchanged. It is in form that we must seek the differences between species, and not in the combinations of matter, which are almost the same in all.[3] The differences are to be sought at the level of the second and third degrees of composition.

The existence of differences of form introduces a new problem, the problem of diversity. There are only a few possible combinations of the principal organs, but as you get down to less important parts the possible scope of variation is greatly increased, and most of the possible variations do exist. Nature seems prodigal of form, of form which needs not to be useful in order to exist. "It needs only to be possible, *i.e.*, of

1 *Leçons d'Anatomie Comparée*, i., p. 6.

2 *Le Règne Animal*, i., p. 16.

3 *Hist. Prog. Sci. Nat.*, i., p. 187, 1826.

such a character that it does not destroy the harmony of the whole."[1] We seize here the relation of the principle of the adaptedness of parts to the problem of the variety of form. The former is in a sense a regulative and conservative principle which lays down limits beyond which variation may not stray. In itself it is not a fountain of change; there must be another cause of change. This thought is of great importance for theories of descent.

Cuvier has no theory to account for the variety of form: he contents himself with a classification. There are two main ways of classifying forms; you may classify according to single organs or according to the totality of organs. By the first method you can have as many classifications as you have organs, and the classifications will not necessarily coincide. Thus you can divide animals according to their organs of digestion into two classes, those in which the alimentary canal is a sac with one opening (zoophytes) and those in which the canal has two openings,[2] a curious forestalment, in the rough, of the modern division of Metazoa into Cœlentera and Cœlomata.

It is only by taking single organs that you can arrange animals into long series, and you will have as many series as you take organs. Only in this way can you form any *Échelle des êtres* or graded series; and you can get even this kind of gradation only within each of the big groups formed on a common plan of structure; you can never grade, for example, from Invertebrates to Vertebrates through intermediate forms[3] (which is perfectly true, in spite of Amphioxus and Balanoglossus!).

In the *Règne Animal* Cuvier restricts the application of the idea of the *Échelle* within even narrower limits, refusing to admit its validity within the bounds of the vertebrate phylum, or even within the vertebrate classes. This seems, however, to refer to a seriation of whole organisms and not of organs, so that the possibility of a seriation of organs within a class is not denied. Cuvier was, above all, a positive spirit, and he looked askance at all speculation which went beyond the facts. "The pretended scale of beings," he wrote, "is only

[1] *Leçons*, i., p. 58.

[2] *Loc. cit*, i., Article iii.

[3] *Loc. cit.*, i., p. 60.

D

an erroneous application to the totality of creation of partial observations, which have validity only when confined to the sphere within which they were made."[1] This remark, which is after all only just, perfectly expresses Cuvier's attitude to the transcendental theories, and was probably a protest against the sweeping generalisations of his colleague, Etienne Geoffroy St Hilaire.

A true classification should be based upon the comparison of all organs, but all organs are not of equal value for classification, nor are all the variations of each organ equally important. In estimating the value of variations more stress should be laid on function than on form, for only those variations are important which affect the mode of functioning. These are the principles on which Cuvier bases the classification of animals given in the *Leçons*, Article V., "Division des animaux d'après l'ensemble de leur organisation." The scheme of classification actually given in the *Leçons* recalls curiously that of Aristotle, for there is the same broad division into Vertebrates, with red blood, and Invertebrates, almost all with white blood. Nine classes altogether are distinguished — Mammals, Birds, Reptiles, Fishes, Molluscs, Crustacea, Insects, Worms, Zoophytes (including Echinoderms and Cœlenterates).

A maturer theory and practice of classification is given in the *Règne Animal* of seventeen years later. Here the principle of the subordination of characters (which seems to have been first explicitly stated by the younger de Jussieu in his *Genera Plantarum*, 1789,[2]) is more clearly recognised. The properties or peculiarities of structure which have the greatest number of relations of incompatibility and coexistence, and therefore influence the whole in the greatest degree, are the important or dominating characters, to which the others must be subordinated in classification. These dominant characters are also the most constant.[3] In deciding which characters are the most important Cuvier makes use of his fundamental classification of functions and organs into two main sets. "The heart and the organs of circulation are

[1] *Régne Animal*, i., p. xx.

[2] Cuvier, *Hist. Prog. Sci. Nat.*, i., p. 288, 1826.

[3] *Règne Animal*, i., p. 10.

a kind of centre for the vegetative functions, as the brain and the spinal cord are for the animal functions."[1] These two organ-systems vary in harmony, and their characters must form the basis for the delimitation of the great groups. Judged by this standard there are four principal types of form,[2] of which all the others are but modifications. These four types are Vertebrates, Molluscs, Articulates, and Radiates. The first three have bilateral, the last has radial symmetry. Vertebrates and Molluscs have blood-vessels, but Articulates show a functional transition from the blood-vessel to the tracheal system. Radiates approach the homogeneity of plants; they appear to lack a distinct nervous system and sense organs, and the lowest of them show only a homogeneous pulp which is mobile and sensitive. All four classes are principally distinguished from one another by the broad structural relations of their neuromuscular system, of the organs of the animal functions. Vertebrates have a spinal cord and brain, an internal skeleton built on a definite plan, with an axis and appendages; in Molluscs the muscles are attached to the skin and the shell, and the nervous system consists of separate masses; Articulates have a hard external skeleton and jointed limbs, and their nervous system consists of two long ventral cords; Radiates have ill-defined nervous and muscular systems, and in their lowest forms possess the animal functions without the animal organs.

This well-rounded classification of animal forms is in a sense the crown of Cuvier's work, for the principle of the subordination of characters, in the interpretation which he gives to it, is a direct application of his principle of functional correlation. Each of the great groups is built upon one plan. The idea of the unity of plan has become for Cuvier a commonplace of his thought, and it is tacitly recognised in all his anatomical work. But he never takes it as a hard-and-fast principle which must at all costs be imposed upon the facts.

Cuvier has become known as the greatest champion of the fixity of species, but it is not often recognised that his

[1] *Règne Animal*, p. 55.

[2] First propounded by Cuvier in 1812, *Ann. Mus. d'Hist. Nat.*, xix.

attitude to this problem is at least as scientific as that of the evolutionists of his own and later times. No doubt he became dogmatic in his rejection of evolution-theory, but he was on sure ground in maintaining that the evolutionists of his day went beyond their facts. He considered that certain forms (species) have reproduced themselves from the origin of things without exceeding the limits of variation. His definition of a species was, "the individuals descended from one another or from common parents, together with those that resemble them as much as they resemble one another."[1] "These forms are neither produced nor do they change of themselves; life presupposes their existence, for it cannot arise save in organisations ready prepared for it."[2]

He based his rejection of all theories of descent upon the absence of definite evidence for evolution. If species have gradually changed, he argued, one ought to find traces of these gradual modifications.[3] Palæontology does not furnish such traces. Again, the limits of variation, even under domestication, are narrow, and the most extreme variation does not fundamentally alter the specific type. Thus the dog has varied perhaps most of all, in size, in shape, in colour. "But throughout all these variations the relations of the bones remain the same, and the form of the teeth never changes to an appreciable extent; at most there are some individuals in which an additional false molar develops on one side or the other."[4] This second objection is the objection of the morphologist. It would be an interesting study to compare Cuvier's views on variation with those of Darwin, who was essentially a systematist.

Cuvier's first objection was of course determined to some extent by the imperfection of the palæontological knowledge of his time. But even at the present day the objection has a certain force, for although we have definite evidence of many serial transformations of one species into another along a single line, for example, Neumayr's *Paludina* series,

[1] *Règne Animal*, i., p. 19.

[2] *Loc. cit.*, p. 20.

[3] *Recherches sur les Ossemens Fossiles*, i., p. 74, 1812.

[4] *Loc. cit.*, p. 79.

yet at any one geological level the species, the lines of descent, are all distinct from one another.[1]

Cuvier recognised very clearly that there is a succession of forms in time, and that on the whole the most primitive forms are the earliest to appear. Mammals are later than reptiles, and fishes appear earlier than either. As Depéret puts it, "Cuvier not only demonstrated the presence in the sedimentary strata of a series of terrestrial faunas superimposed and distinct, but he was the first to express, and that very clearly, the idea of the gradual increase in complexity of these faunas from the oldest to the most recent" (p. 10).

He did not believe that the fauna of one epoch was transformed into the fauna of the next. He explained the disappearance of the one by the hypothesis of sudden catastrophes, and the appearance of the next by the hypothesis of immigration. He nowhere advanced the hypothesis of successive new creations. "For the rest, when I maintain that the stony layers contain the bones of several genera and the earthy layers those of several species which no longer exist, I do not mean that a new creation has been necessary to produce the existing species, I merely say that they did not exist in the same localities and must have come thither from elsewhere."[2] It was left to d'Orbigny to teach the doctrine of successive creations, of which he distinguished twenty-seven (*Cours élémentaire de palaeontologie stratigraphique*, 1849).

Cuvier, however, can hardly have believed that all species were present at the beginning, since he does admit a progression of forms. Probably he had no theory on the subject, for theories without facts had little interest for him. At any rate it is a mistake to think that Cuvier was a supporter of the theological doctrine of special creation. His philosophy of Nature was mechanistic, and he dedicated his *Recherches sur les Ossemens Fossiles* to his friend Laplace. He admitted the idea of evolution at least so far as to conceive of a development of man from a savage

[1] See C. Depéret, *Les transformations du Monde animal*, Paris, 1907, and G. Steinmann, *Die geologischen Grundlagen der Abstammungslehre*, Leipzig, 1908.

[2] *Recherches*, i., p. 81.

to a civilised state.[1] He refused to accept the extravagant evolutionary theory of Demaillet and the somewhat confused theory of Lamarck (whom he joins with Demaillet),[2] just as he rejected the transcendental theories of Geoffroy St Hilaire, because they seemed to him not based upon facts.

[1] *Règne Animal*, i., p. 91.

[2] *Ossemens Fossiles*, i., p. 26.

CHAPTER IV

GOETHE

SCIENCE, in so far as it rises above the mere accumulation of facts, is a product of the mind's creative activity. Scientific theories are not so much formulæ extracted from experience as intuitions imposed upon experience. So it was that Goethe, who was little more than a dilettante,[1] seized upon the essential principles of a morphology some years before that morphology was accepted by the workers.

Goethe is important in the history of morphological method because he was the first to bring to clear consciousness and to express in definite terms the idea on which comparative anatomy before him was based, the idea of the unity of plan. We have seen that this idea was familiar to Aristotle and that it was recognised implicitly by all who after him studied structure comparatively. In Goethe's time the idea had become ripe for expression. It was used as a guiding principle in Goethe's youth particularly by Vicq d'Azyr and by Camper. The former (1748-1794), who discovered[2] in the same year as Goethe (1784) the intermaxillary bone in man, pointed out the homology in structure between the fore limb and the hind limb, and interpreted certain rudimentary bones, the intermaxillaries and rudimentary clavicles, in the light of the theory that Vertebrates are built upon one single plan of structure.

"Nature seems to operate always according to an original and general plan, from which she departs with regret and

[1] *See* Kohlbrugge, "Hist. krit. Studien über Goethe als Naturforscher," *Zool. Annalen.* v., 1913, pp. 83-231.

[2] Or re-discovered, according to Kohlbrugge.

whose traces we come across everywhere" (Vicq d'Azyr, quoted by Flourens, *Mém. Acad. Sci.*, XXIII., p. xxxvi.).

Peter Camper (1722-1789), we are told by Goethe himself in his *Osteologie*, was convinced of the unity of plan holding throughout Vertebrates; he compared in particular the brain of fishes with the brain of man.

The idea of the unity of plan had not yet become limited and defined as a strictly scientific theory; it was an idea common to philosophy, to ordinary thought, and to anatomical science. We find it expressed by Herder (who perhaps got it from Kant) in his *Ideen zur Philosophie der Geschichte der Menschheit* (1784), and it is possible that Goethe became impressed with the importance of the idea through his conversations with Herder. Be that as it may, it is certain that Goethe sought for the intermaxillaries in man only because he was firmly convinced that the skeleton in all the higher animals was built upon one common plan and that accordingly bones such as the intermaxillaries, found well developed in some animals, must also be found in man. The idea was not drawn from the facts, but the facts were interpreted and even sought for in the light of the idea. "I eagerly worked upon a general osteological scheme, and had accordingly to assume that all the separate parts of the structure, in detail as in the whole, must be discoverable in all animals, because on this supposition is built the already long begun science of comparative anatomy."[1]

The principle comes to clear expression in his *Erster Entwurf einer allgemeinen Einleitung in die vergleichende Anatomie* (1795).[2] He writes:—"On this account an attempt is here made to arrive at an anatomical type, a general picture in which the forms of all animals are contained in potentia, and by means of which we can describe each animal in an invariable order."[3] His aim is to discover a general scheme of the constant in organic parts, a scheme into which all animals will fit equally well, and no animal better than the rest. When we remember that the type to which anatomists before him had, consciously or unconsciously, referred all

[1] Cotta ed., vol. ix., p. 448.

[2] "First Draft of a General Introduction to Comparative Anatomy."

[3] Cotta ed., ix., p. 463.

other structure was man himself, we see that in seeking after an abstract generalised type Goethe was reaching out to a new conception. The fact that only the structure of man and the higher animals was at all well-known in his time led Goethe to think that his general Typus would hold for the lower animals as well, though it was to be arrived at primarily from a study of the higher animals. All he could assert of the entire animal kingdom was that all animals agreed in having a head, a middle part, and an end part, with their characteristic organs, and that accordingly they might, in this respect at least, be reduced to one common Typus. Goethe's knowledge of the lower animals was not extensive.

Though Goethe did not work out a criterion of the homology of parts with any great clearness, he had an inkling of the principle later developed by E. Geoffroy St Hilaire, and called by him the "Principle of Connections." According to this principle, the homology of a part is determined by its position relative to other parts. Goethe expresses it thus:—"On the other hand the most constant factor is the position in which the bone is invariably found, and the function to which it is adapted in the organic edifice."[1] But from this sentence it is not clear that Goethe understood the principle as one of form independent of function, for he seems to consider that the homology of an organ is partly determined by the function which it performs for the whole. He wavers between the purely formal or morphological interpretation of the principle of connections and the functional. We find him in the additions to the *Entwurf* (1796), saying:—"We must take into consideration not merely the spatial relations of the parts, but also their living reciprocal influence, their dependence upon and action on one another."[2] But in seeking for the intermaxillary bone in man he was guided by its position relative to the maxillaries — it must be the bone between the anterior ends of the maxillaries, a bone whose limits are indicated in the adult only by surface grooves.

As a matter of fact Goethe's morphological views are neither very clearly expressed nor very consistent. This

[1] Cotta ed., p. 478. [2] *Loc. cit.*, p. 491.

comes out in his treatment of the relation between structure and function. Sometimes he takes the view that structure determines function. "The parts of the animal," he writes, "their reciprocal forms, their relations, their particular properties determine the life and habits of the creature."[1] We are not to explain, he says, the tusks of the *Babirussa* by their possible use, but we must ask how it comes to have tusks. In the same way we must not suppose that a bull has horns in order to gore, but we must investigate the process by which it comes to have horns to gore with. This is the rigorous morphological view. On the other hand he admits elsewhere that function may influence form. Apparently he did not work out his ideas on this point to logical clearness, and Rádl[2] is probably correct in saying that the following quotation with its double assertion represents most nearly Goethe's position:—

> "Also bestimmt die Gestalt die Lebensweise des Thieres,
> Und die Weise zu leben, sie wirkt auf alle Gestalten
> Mächtig zurück."[3]

His best piece of purely morphological work was his theory of the metamorphosis of plants. Stripped of its vaguer elements, and of the crude attempt to explain differences in the character of plant organs by differences in the degree of "refinement" of the sap supplied to them, the theory is that stem-leaves, sepals, petals, and stamens are all identical members or appendages. These appendages differ from one another only in shape and in degree of expansion, stem-leaves being expanded, sepals contracted, petals expanded, and so on alternately. It is equally correct to call a stamen a contracted petal, and a petal an expanded stamen, for no one of the organs is the type of the others, but all equally are varieties of a single abstract plant-appendage.

What Goethe considered he had proved for the appendages of plants he extended to all living things. Every living thing is a complex of living independent beings, which "der

[1] *Entwurf*, Cotta ed., ix., p. 465.

[2] *Geschichte der biologischen Theorien*, i., p. 266.

[3] "So the form determines the manner of life of the animal, and the manner of life in its turn reacts powerfully upon all forms."

Idee, der Anlage nach," are the same, but in appearance may be the same or similar, different or unlike.[1] Not only is there a primordial animal and a primordial plant, schematic forms to which all separate species are referable, but the parts of each are themselves units, which "der Idee nach," are identical *inter se.* This fantasy can hardly be taken seriously as a scientific theory; it seems, however, to have been what guided Goethe in his "discovery" of the vertebral nature of the skull. Just as the fore limb can be homologised with the hind limb, so, reasoning by analogy, the skull should be capable of being homologised with the vertebræ. To what ludicrous extremes this doctrine of the repetition of parts within the organism was pushed we shall see when we consider the theories of the German transcendentalists of the early nineteenth century.

Though Goethe's morphological views were lacking in definiteness he hit upon one or two ideas which proved useful. Thus he enunciated the "law of balance" long before Etienne Geoffroy St Hilaire, the law "that to no part can anything be added, without something being taken away from another part, and *vice versa.*"[2] He saw, too, what a help to the interpretation of adult structure the study of the embryo would be, for many bones which are fused in the adult are separate in the embryo.[3] This also was a point to which the later transcendentalists gave considerable attention.

So far we have spoken of Goethe as if he were merely the prophet of formal morphology; we have pointed out how he brought to clear expression the morphological principle implicit in the idea of unity of type, and how he seized upon some important guiding ideas, such as the principle of connections. But Goethe was not a formalist, and he was very far from the static conception of life which is at the base of pure morphology. His interest was not in *Gestalt* or fixed form, but in *Bildung* or form change. He saw that *Gestalt* was but a momentary phase cf *Bildung*, and could be considered apart and in itself only by an abstraction fatal to all understanding of the living thing. Mephistopheles

[1] *Bildung und Umbildung organischer Naturen*, 1807.

[2] Cotta ed., ix., p. 466.

[3] *Loc. cit.*, pp. 474-5.

scoffs at the scholars who would explain a living creature by anatomising it:

> "Dann hat er die Theile in seiner Hand,
> Fehlt leider! nur das geistige Band."[1]

Goethe kept clear of this mistake; he knew that the artist comes nearer to the truth than the analyst.

In the fragment entitled *Bildung und Umbildung organischer Naturen* (1807), introductory to a reprint of his paper on the "Metamorphosis of Plants," we get an exposition of his general views on living things. He points out there how we try to understand things by separating them into their parts. We can, it is true, resolve the organism into its structural elements, but we cannot recompose it or endow it with life by joining up the parts. Hence we require some other means of understanding it. "In all ages even among scientific men there can be discerned a yearning to apprehend the living form as such, to grasp the connection of their external visible parts, to interpret them as indications of the inner activity, and so, in a certain measure, to master the whole conceptually." This science which should discover the inner meaning of organic *Bildung* is called Morphology.[2] In Morphology we should not speak of *Gestalt* or fixed form, or if we do we should understand by it only a momentary phase of *Bildung*. Form is of interest not in itself but only as the manifestation of the inner activity of the living being. Over development, he says elsewhere, there presides a formative force, a *bildende Kraft* or *Bildungstrieb*, which works out the idea of the organism. Living things, in his view of them, strive to manifest an idea. They are Nature's works of art—and so, incidentally, they require an artist to interpret them.

This profound conception of the nature of life is applied not only to the growing changing individual but also to the whole changing world of organisms. They are all manifestations of a living shaping power which moulds them. This shaping power, immanent in all life, is conceived to work according to a general plan, and so we get an explanation of

[1] Then he has all the parts within his hand, excepting only, sad to say, the living bond.

[2] Goethe was the inventor of the word.

the fact that living things seem simply varieties of one common type.

"If we once recognise," says Goethe, "that the creative spirit brings into being and shapes the evolution of the more perfect organic creatures according to a general scheme, is it altogether impossible to represent this original plan if not to the senses at least to the mind. . . ?"[1]

Such an interpretation of the unity of plan reaches perhaps beyond the bounds of science.

[1] Cotta ed., ix., p. 490.

CHAPTER V

ETIENNE GEOFFROY SAINT-HILAIRE

E. GEOFFROY made an experiment, unsuccessful but instructive. He tried to found a science of pure morphology; he failed: his failure showed, once and for all, that a pure morphology of organic forms is impracticable.

Already, in 1796, in one of his earliest memoirs,[1] Geoffroy was guided by the idea that Nature has formed all living things upon one plan. Organs which seem anomalous are merely modifications of the normal; the trunk of an elephant is formed by the excessively prolonged nostrils, the horn of a rhinoceros is simply a mass of adhering hairs. In general, however varied their form, all organs are simply variations of a common scheme; Nature employs no new organs. Organs which are rudimentary, such as the clavicles in the ostrich and the nictitating membrane in man, bear witness to the unity of plan. In this Geoffroy goes no further than his predecessors. They too had recognised homologies of organs; they too had interpreted rudimentary organs as vestiges of an original plan.

In a series of papers published in 1807, Geoffroy took a further step, and sought to establish homologies which were not obvious—homologies, too, not so much of organs as of parts.

These memoirs (published in the *Annales du Muséum d'Histoire naturelle*, vols. ix. and x., 1807) dealt with the homology between the bones of the pectoral fin and girdle in fish and the bones of the arm and shoulder-girdle in higher

[1] "Mémoire sur les rapports naturels des makis," *Magasin Encyclopèdique*, vii.

Vertebrates, with the homologies of the bones of the sternum, and with the determination of the pieces of the skull, particularly in the crocodile. All Geoffroy's morphological doctrine is found in them, but for the full expression of his views we must take his chief work, the *Philosophie anatomique*, particularly the first volume (1818). This volume contains, beside the important "Discours préliminaire" and "Introduction" which we shall presently consider in detail, five memoirs, which deal with the various bones connected with the respiratory organs in fishes (the bones of the operculum, of the hyoid, of the branchial arches, of the pectoral girdle), and seek to discover their homologies with corresponding bones in air-breathing Vertebrates.

"Can the organisation of vertebrated animals be referred to one uniform type?" This is the question with which the *Philosophie anatomique* opens, the question to which the whole book is an answer. But is it not generally acknowledged by naturalists that Vertebrates are built upon one uniform plan, that, for instance, the fore limb may be modified for running, climbing, swimming, or flying, yet the arrangement of the bones remain the same? How else could there be a "natural method" of classification?[1]

But the homologies so drawn repose upon a vague and confused feeling for likenesses; they are not based upon an explicit principle. What general principle can be applied? "Now it is evident that the sole general principle one can apply is given by the position, the relations, and the dependencies of the parts, that is to say, by what I name and include under the term of *connections*." For instance, the part known as the hand in man and generally as the fore foot in other Vertebrates, is the fourth part in order in the anterior member, and its homologue can always be recognised by this fact of its connections (p. xxvi.). The principle of connections serves as a guide in tracing an organ through all its functional transformations, for "an organ can be deteriorated, atrophied, annihilated, but not transposed" (p. xxx.).

It is this principle which enables one to follow out in detail the further fundamental conception that in every

[1] Discours préliminaire, pp. xv.-xxiv.

Vertebrate there are found the same "organic materials," or units of construction. This conception, which Geoffroy calls the *Théorie des analogues* (p. xxxii.), is clearly one part of the old idea of the unity of type; it teaches the *unity of composition* of organic beings, while the *Principe des connexions* adds the *unity of plan.*

Both conceptions are logically implicit in the vague notion of unity of type; Geoffroy disengaged them, and pushed each to its logical extreme.

Most of the ordinary homologies of structure in air-breathing Vertebrates have already been seized, he continues, for they are more or less obvious, and many intermediate states exist (p. xxxiv.). But ordinary methods of comparison fail when the attempt is made to homologise the structure of fishes with that of air-breathing Vertebrates, for the homologies are anything but obvious and no intermediate organs are found.

Most air-breathing Vertebrates have a larynx, a trachea, and bronchi, which are absent in fish; and fish have many parts which seem to be absent in higher Vertebrates. But apply the "Theory of Analogues"; it teaches that there can be no organ peculiar to fish and not found in other Vertebrates; apply the "Principle of Connections," it will show which organs are homologous in the two types (p. xxxv.).

Comparative anatomists, with few exceptions, had hitherto taken man as the type, and referred all structure to his; Geoffroy's principles led him to give preference to no one animal in particular, but to seize upon each part in the species in which it reaches the maximum of its development (p. xxxvi.). He is thus led to refer all structures to a generalised abstract type. In this abstract type each organ exists at the maximum of its development, each organ shows all its potentialities realised. In a way, therefore, this type, this abstraction, gives the scheme of the possible transformations of each organ.

It is true Geoffroy does not refer to this "Archetype" in so many words, but it must always have been vaguely present in his mind. He has this idea in his head when he says in one of his later works, "There is, philosophically speaking, only a single animal."[1] The "single animal" is simply the generalised type.

Having laid down his two principles Geoffroy goes on to apply them to the difficult case of the comparison of the skeleton of fish with the skeleton of the higher Vertebrates. "My present task is to demonstrate that there is no part of the bony framework of fishes that cannot find its analogue in the other vertebrated animals."[2] It seems at first sight that many bones are peculiar to fish, formed expressly for performing the functions which fish do not share with higher animals. These are the bones connected with respiration—the operculum, the branchiostegal rays, the branchial arches, and others. That the peculiar bones should be connected with the respiratory functions is only natural, for the contrast between fish and higher Vertebrates is essentially a contrast between water-breathing and air-breathing animals. Considering first the general form of the skeleton in fish, we are met at once with a difficulty; there is no obvious homologue in fishes of the neck, the trunk, and the abdomen of higher animals. What apparently corresponds to the trunk is in fishes crowded close up under the head. But, after all, it is not of the essence of the vertebrate type to have the trunk and the abdomen attached at definite and invariable distances along the vertebral column—that is a notion surviving from the anatomy which made man its type. The "trunk" differs in position according to the class, in quadrupeds, birds, and fishes (p. 9). Now, says Geoffroy, allow me this one hypothesis, that the trunk with its organs can, as it were, move bodily along the vertebral column, so as to be found in one class near the front end of the vertebral column, in another about the middle, and in a third near the end, then I can show you in detail that the constituent parts of this trunk are found in all classes to be invariably in the same positions relatively to one another (p. 10). It is important to note this hypothesis of a "metastasis" which Geoffroy makes, for it is the key to the understanding of many of the far-fetched homologies which he tries to establish. It is, of course, clear that this hypothesis is in formal contradiction

[1] *Études progressives d'un Naturaliste*, p. 50, Paris, 1835.

[2] *Philosophie Anatomique.*, i., Introduction, p. 1.

with his principal hypothesis of the invariability of connections, and that he, so to speak, gets a hold on his fish to apply his principle of connections only by admitting at the very outset an exception to his primary principle. A further application of the hypothesis of metastasis will be noticed below in connection with the determination of the sternum of fishes. We note here an interpretation of the first metastasis in terms of functional adaptation. "The constant and violent action of the tail, if it does not go so far as actually to displace and move forward the internal organs, at least fits in well with an arrangement in which the organs are so disposed" (p. 99).

The first memoir deals with the homologies of the opercular bones. Geoffroy considers that the external opening of the ear corresponds to the external opening of the gill-chamber, which lies between the operculum and the pectoral girdle. The ear communicates with the buccal cavity by the Eustachian tube, so does the branchial chamber by means of the gill-slits. The auditory chamber of higher Vertebrates is, therefore, the homologue of the branchial chamber in fish; the opercular bones in fish and the ossicles of the ear in other Vertebrates stand in close relation to this chamber; therefore the opercular bones are the homologues of the ossicles of the ear, the interoperculum corresponding to the malleus, the suboperculum to the lenticular, the minute lower part of the suboperculum to the incus, the operculum to the stapes, and the pre-operculum to the tympanic ring. In making these particular determinations Geoffroy professes to be led by his principle of connections. The pre-operculum has, he says, the same connections with neighbouring bones as the tympanic bone in other Vertebrates, and the other pieces of the gill-cover are homologised with particular ear-ossicles according to the order in which they stand to one another. The second memoir in the book deals with the sternum, and affords a very good example of Geoffroy's method of dealing with the facts of structure. We shall omit here any detailed reference to the other three memoirs, which deal with the hyoid, with the branchial arches and the structures which correspond in air-breathing Vertebrates, and with the bones of the shoulder-girdle.

In the memoir on the sternum Geoffroy's first care is to arrive at a definition of what a sternum is. He defines it partly by its functions, partly by its connections, as the system of bones which covers and protects the thorax, and gives attachment to certain groups of muscles.

The most highly developed sternum (according to this definition) is the plastron of the tortoise, whose structure it dominates (p. 103). It is important, therefore, to determine of how many bones the plastron is composed, since the full number of elementary parts of which an organ is composed is best seen when the organ is at the maximum of its development. There are nine bones in the plastron of the tortoise. "The conclusion to be drawn from this is that every sternum, provided that it is not inhibited in its development by some obstacle, is composed of *nine elementary parts*" (p. 105). These nine bones are in Geoffroy's nomenclature, the episternals, the hyosternals, the hyposternals, the xiphisternals, which are all paired bones, and the entosternal, which is unpaired. The arrangement of them is in the tortoise:—

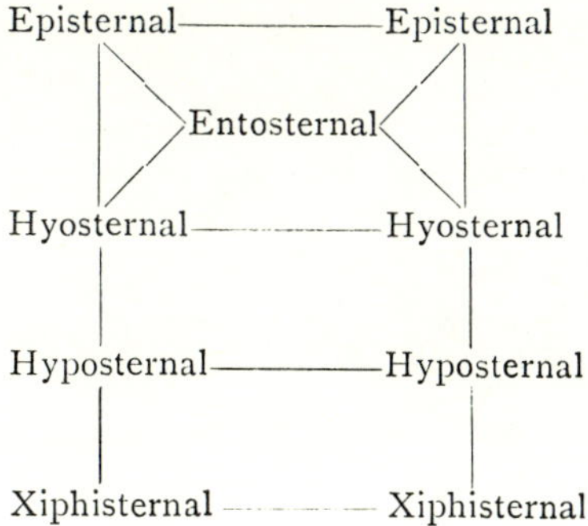

The articulations in the tortoise are indicated by the connecting lines. Geoffroy tries to show that the sternum in other animals is composed of these nine bones, or at least of a certain number of them, always in the same invariable relative positions. Thus in birds the sternum consists of five pieces, of a huge keeled entosternal, and of two "annexes" on either side, which are the hyo- and hyposternals. These are separate only in young birds. Occasionally, especially in young birds, rudiments of episternals and xiphisternals also occur. The minuteness of the episternals and the xiphi-

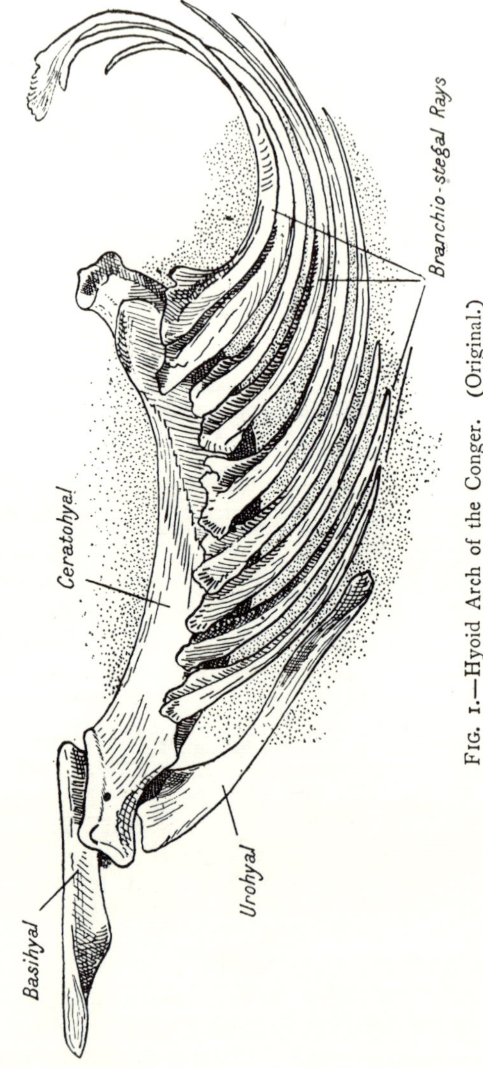

Fig. 1.—Hyoid Arch of the Conger. (Original.)

sternals may be attributed to the gigantic size of the entosternal, in accordance with the *Loi de balancement.* In the other air-breathing Vertebrates the nine sternal elements can according to Geoffroy be discovered without great difficulty. But when we come to the determination of the sternum in fishes, difficulties abound, which Geoffroy solves in the following way. He points out that between the clavicles (*cleithra*) and the hyoid bone (*basihyal*) in fishes there is a long median bone (*urohyal*) which is attached in front by two strong tendons to the horns of the hyoid and is free behind (see Fig. 1). Gouan (1720) had seen in this bone the homologue of the sternum. Geoffroy adopts this view, but considers that this bone alone cannot represent the whole sternum. He finds the representatives of other bones of the sternum in the large bones (*epihyal* and *ceratohyal*, or the two pieces of the *ceratohyal*) which are comprised in the hyoid arch. But he is immediately met by the difficulty that this complex of bones is situated in front of the pectoral girdle, whereas the sternum in higher Vertebrates lies behind the pectoral girdle. He reflects, however, that the gills of fish, situated in front of the clavicles, are merely the lungs under another name. The gills have become shifted forward by a metastasis similar to that which brought the whole thoracic organs far forward in fish. This being so, their supporting elements, the sternum and the ribs, must have moved with them, and are hence to be found in front of the pectoral girdle.

Geoffroy's next step is to point out that the only possible homologues of sternal ribs are the branchiostegal rays, which arise from the large bones of the hyoid arch. If these are sternal ribs, the bones to which they are attached must be the hyo- and hyposternals or "annexes," the bones from which in birds the ribs take their origin.

The unpaired sternal bone (*urohyal*) cannot be homologous with the entosternal, for it has no connections with the annexes. He decides that it must represent the episternals, for in some young birds there is a two-headed episternal to which two strong tendons are attached, just in the same way as the unpaired piece in fish is bound to the bones of the hyoid by two tendons. "Thus it is not the sternum as a whole that has shifted in front of the clavicles and covered with its side pieces the gills placed there; it is a piece exclusively piscine, in the sense that it is only in the class of fishes that it reaches the *maximum* of its development" (p. 83).

To sum up, the sternum in all four vertebrate classes is composed of the same elements, arranged always in the same way. "One is . . . led to the conception of an ideal type of sternum for all Vertebrates, which then, considered from a lower standpoint, resolves itself into several secondary forms according as the whole or the majority of the constituent materials are employed, or even as these elements come to change their respective dimensions or proportions" (p. 134). As to the elementary constituents, "they give proof of individuality, and sometimes even, in certain abnormalities, of independence, and rise to the level of primary organisatory materials" (p. 132). What holds good for the sternum holds good for other organs—and accordingly the unity of plan and composition can be demonstrated for all the organs of Vertebrates.

Soon after the publication of the *Philosophie anatomique* (1818) Geoffroy went further in his search for unity, and maintained that the structure of insects and Crustacea could be reduced to the vertebrate type.

He proposed to replace Cuvier's classification of the animal kingdom into the four large groups, Vertebrata, Mollusca, Articulata, and Radiata by the following classification :—[1]

Vertébrés { Hauts-Vertébrés (Vertebrata, Cuv.).
Dermo-Vertébrés (Articulata, Cuv.).

Invertébrés { Mollusques (Mollusca, Cuv.).
Rayonnés (Radiata, Cuv.).

The idea upon which is based the comparison of Articulates with Vertebrates is that each skeletal segment of Articulates is a vertebra. In the Hauts-vertébrés the vertebræ are internal; in the Dermo-vertébrés they are external. "*Every animal lives either outside or inside its vertebral column.*"[2] The essence of a vertebra is not its form, nor its function, but its composition from four

[1] "Sur une colonne vertébrale et ses côtes dans les insectes apiropodes," (*Acad. Sci.*, Feb. 12, 1820). Printed in *Isis*, pp. 527-52, 1820 (2).

[2] "Sur l'organisation des insectes," p. 458. *Isis*, pp. 452-62, 1820 (2).

elementary pieces which unite round a central space (*Isis, loc. cit.*, p. 532). Serres had shown that in the higher animals every vertebra is formed from four centres of ossification, that the body of the vertebra is at first tubular, and that afterwards it becomes filled up. In lobsters and crabs each segment is composed of four elementary pieces, as may be seen most easily in young ones. "Accordingly each segment corresponds to a true vertebra in composition: there is the same number of 'materials,' the same order in the course of ossification, the same kind of articulation, the same annular arrangement, the same empty space in the middle" (p. 534). The only difference is that in Articulates the central space is very great and contains all the organs of the body, whereas in the higher Vertebrates the body of the vertebra becomes completely filled up. In the thoracic region of Crustacea it is not the whole segment with part of the carapace which corresponds to a vertebra, but merely the part round the ventral nerve-cord (endophragmal skeleton).

FIG. 2.—"Vertebra" of a Pleuronectid. (After Geoffroy.)

If the skeleton of the segment in Articulates corresponds to the body of a vertebra and is here external, then the appendages of the Articulate must correspond to ribs (p. 538). The full development of this thought is found in a Memoir of 1822, "Sur la vertèbre."[1] He takes as the typical vertebra that of a Pleuronectid, probably the turbot. His original figure is reproduced (Fig. 2).

He includes as part of the vertebra not only the neural (e′, e″) and hæmal (o′, o″) arches, but also, above and below these, the radialia (a″, u′) and the fin-rays (a′, u″). (Neither the radialia nor the fin-rays are,

[1] *Mém. Mus. d'Hist. nat.*, ix., pp. 89-119, Pls. v-vii.

by the way, in the same transverse plane as the body of the vertebra). Every vertebra, he considers, contains these nine pieces—the cycleal (or body), the two perials (e′, e″) and the two epials (a′, a″) above, the two paraals (o′, o″) and the two cataals (u′, u″) below. The epials and the cataals are in reality paired bones which in fish mount one on top of the other to support the median fins. In the cranial region—the skull is formed of modified vertebræ—the epials and perials open out so as to form the walls and roof of the brain; in the thoracic region the paraals and cataals reach their maximum of development and perform the same service for the thoracic organs, the paraals becoming vertebral, and the cataals sternal, ribs.

We have seen that in Arthropods the body of the vertebra (cycleal) forms the open ring of the segment, which lies immediately under the skin, the vertebral tube coinciding with the epidermal tube. The homologues of the other eight pieces of the vertebra must accordingly be sought in the external appendages. At first sight there seems here a contradiction of the principle of connections, for the appendages in Arthropods are lateral, whereas the paired bones of the vertebra are dorsal and ventral. But there is in reality no contradiction, for "what our law of connections absolutely requires is that all organs, whether internal or external, should stand to one another in the same relations; but it is all one whether the box (*coffre*) that encloses them lies with this or that side on the ground. What similarities in the organisation of man and the digitate mammals, and yet what differences between their attitudes when standing! The same holds true as regards the normal attitudes of the pleuronectids and the other fishes" (p. 107).

The exact way in which Geoffroy homologised the parts of the appendages in Arthropods with the paired pieces of the typical vertebra is best shown by the reproduction of his figure of an abdominal segment of the lobster (Fig. 3), in which the parts homologous with those represented in the figure of the typical vertebra (Fig. 2) are indicated by the same letters. The ingenuity of the comparison is astonishing.

The comparison of the Arthropod with the Vertebrate is extended also to the internal organs. The internal organs of the Arthropod are shown to stand in the same order to one another as in the Vertebrate, only the organs are inverted. Thus the nervous system is dorsal in the Vertebrate, ventral in the Arthropod. Turn the Arthropod on its back and the relative positions of the systems of organs are the same as in the Vertebrate. The relation of the organs to the external tube is of course different in Arthropods and Vertebrates, but this is no contradiction of the principle of connections. "Such a tube, although it is the organs essential to life that it contains, can yet behave in different ways with regard to the mass of these organs: the principle of connections demands only that all the organs maintain with one another fixed and definite relations; but the principle would be in no way invalidated if the whole mass had rotated inside the tube" (p. 112).

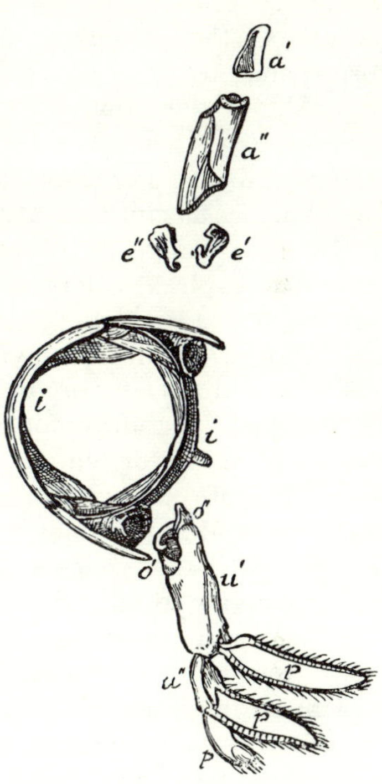

FIG. 3.—Abdominal Segment of the Lobster. (After Geoffroy.)

Geoffroy pushed the analogy between Arthropods and Vertebrates very far, for he asserted that every piece in the skeleton of an insect was homologous with some bone in Vertebrates, that it stood always in its proper place, and remained faithful to at least one of its connections.[1] It does not appear that he attempted to prove in detail this very big assumption, but the beginnings of a detailed comparison are found in the paper of 1820, *Sur l'organisation des insectes.* Six segments are distinguished in an insect—the head, the three divisions

[1] *Sur l'organisation des insectes*, p. 459.

of the thorax, the abdomen, and the terminal segment of the abdomen (p. 455).

The skeleton of the insect's head is said to correspond to the bones of the face, to the bones of the cerebrum and to the hyoid of higher Vertebrates, the skeleton of the prothorax to the bones of the cerebellum, of the palate, and the pieces of the larynx, the skeleton of the mesothorax to the parietals, interparietals, and opercular bones, and that of the metathorax to the skeleton of the thorax of Vertebrates. The pieces of the abdomen and of the terminal segment correspond to the bones of the abdomen and coccyx (p. 458). It does not need the subsequent likening of the hind wings of insects to the air bladder of fish, and of the stigmata to the pores of the lateral line, to convince one finally of the fancifulness of the whole comparison.

In 1830 two young naturalists, Meyranx and Laurencet, presented to the Académie des Sciences a memoir in which they likened a Cephalopod to a Vertebrate bent back at the level of the umbilicus, saying that the Vertebrate in this position had all its organs in the same order as in the Cephalopod. Geoffroy took up this idea with enthusiasm, seeing in it a further application of his master-idea of the unity of plan and composition. By means of this comparison Mollusca definitely took their place in the *Echelle des êtres*, after the Articulata, just as Geoffroy had maintained in 1820, saying that crabs formed a link between the other Crustacea and the molluscs.[1] The comparison brought him nearer to the end he had in view, the reference of all animal structure to one single type.

But in championing the memoir of Meyranx and Laurencet, Geoffroy found himself in direct antagonism with Cuvier, who held that his four "Embranchements" had each a separate and distinct plan of structure. In a paper read to the Academy in February 1830,[2] Cuvier easily demolished the crude comparison of the Cephalopod to the Vertebrate. He gave diagrams of the internal organs of a Cephalopod and of a Vertebrate bent back in the manner indicated by Meyranx and Laurencet, and he showed in

[1] *Isis*, p. 549.

[2] Published in *Ann. Sci. Nat.*, xix., pp. 241-59, 1830.

detail that the arrangement of the main organs was quite different, that the likeness would have been much greater if the Cephalopod had been likened to a Vertebrate doubled up the other way,[1] but that even then the arrangement of the organs would not be the same. The organs, too, of the Cephalopod are differently constructed. He sums up his criticism by saying:—"I give true and summary expression to all these facts when I say that Cephalopods have several organs in common with Vertebrates, which fulfil in either case similar functions, but that these organs are differently arranged with respect to one another, and often constructed in a different way; that they are in Cephalopods accompanied by several other organs which Vertebrates do not possess, whilst the latter on their side have many organs which Cephalopods lack" (p. 257). Geoffroy could not accept this commonsense view of the matter, but made a fight for his transcendental theories. This was the beginning of the famous controversy between Geoffroy and Cuvier which so excited the interest of Goethe. It was a struggle between "comparative anatomy" and "morphology," between the commonsense teleological view of structure and the abstract, transcendental. Geoffroy brought forward all his theories on the homology of the skeleton of fish with the skeleton of higher Vertebrates, and tried to prove by them his great principle of the unity of plan and composition; Cuvier took Geoffroy's homologies one by one, and showed how very slight was their foundation. Cuvier was on sure ground in insisting upon the observable diversities of structural type, and his vast knowledge enabled him to score a decisive victory.[2]

The controversy was not, as we are sometimes told, a controversy between a believer in evolution and an upholder of the fixity of species, although it raised a question upon which evolution theory was to throw some light.

[1] *Cf.* Aristotle (*supra*, p. 10).

[2] For an account of the controversy reference may be made to I. Geoffroy St Hilaire, *Vie Travaux et Doctrine scientifique d'Etienne Geoffroy St Hilaire*, Paris, 1847; also Semper, *Arb. zool. zoot. Instit. Würzburg*, iii., 1876-7, K. E. von Baer, *Lebensgeschichte Cuviers*, ed. L. Stieda, 1897, and J. Kohlbrugge, in *Zoolog. Annalen*, v., pp. 143-95, 1913.

In these Darwinian days Geoffroy has reaped a little posthumous glory as an early believer in evolution. That he did believe in evolution to a limited extent is certain; that his theory of evolution was, as it were, a by-product of his life-work, is also certain. Geoffroy was primarily a morphologist and a seeker after the unity hidden under the diversity of organic form. His theory of evolution had as good as no influence upon his morphology, for he did not to any extent interpret unity of plan as being due to community of descent. His morphological, non-evolutionary standpoint comes out quite clearly in several places in the *Philosophie anatomique.* He does not derive the structure of the higher Vertebrates from the simpler structure of the lower, but when he finds in fish a part at the maximum of its development, he speaks of the same part, rudimentary in the higher forms, as being, as it were, held in reserve for use in the fish. Thus, speaking of the episternal in fish which forms the central piece of its sternum, he says, "it is a bone that is rudimentary in birds (one might almost add a bone that is held in reserve in birds for this fate) which is destined to form in the centre the principal keel of this new machine" (p. 84). Again, with reference to the homology of the ossicles of the ear with the opercular bones in fish, "employing other resources equally hidden and rudimentary, Nature makes profitable use of the four tiny ossicles lodged in the auditory passage, and, raising them in fish to the greatest possible dimensions, forms from them these broad opercula. . . ." (p. 85). Or you may take it the other way about, and start from the organisation of fishes; opercular bones are of no use to air-breathing animals, so they dwindle away, and are pressed into the service of the ear, although they are of little use in hearing (p. 46).

There is here no thought of evolution; in later years, however, his researches upon fossil crocodilians led him to consider the possibility that the living species were descended from the antediluvian. For the factors of the transformation he refers to Lamarck's hypotheses.[1] In a memoir of 1828,[2]

[1] "Recherches sur l'organisation des Gavials," *Mém. Mus. d'Hist. nat.*, xii., 1825.

[2] *Mém. Mus. d'Hist. nat.*, xvii., pp. 209-29.

dealing with the possible genetic relation of living to fossil species, he still regards the question as more or less open. Although fossil species are mostly different from living species are we therefore to conclude, he asks, that they are not the ancestors of the present day forms? "The contrary idea arises more naturally in the mind; for otherwise the six-days' creation would have had to be repeated and new beings produced by a fresh creation. Now this proposition, contrary as it is to the most ancient historical traditions, is inadmissible" (p. 210). It is sufficiently clear from this quotation that Geoffroy was thinking only of a transformation of the antediluvian species created by God, and by no means of an evolution of all species from one primitive type. In matters of religion Geoffroy was orthodox. He goes on to point out how great a resemblance there is in essential structure between fossil and living species. All find their place in one scheme of classification; does it not seem that all are modifications "of one single being, of that abstract being or common type, which it is always possible to denote by the same name?" (p. 211). This type is abstract, not actual, and it is certainly not conceived as an original ancestor of all animals.

The fullest development of Geoffroy's views on evolution is found in his memoir "Le degré d'influence du monde ambiant pour modifier les formes animales."[1] Here the relation of his evolution-theory to his morphology is pointed out. The principle of unity of plan and composition cannot be the final goal of zoology; there must follow on it a philosophical study of the *differences* between organic forms. The causes of these differences are to be found in the environment (pp. 66-7). Geoffroy seems here to be moving from a pure to a causal morphology. It is probable, he continues, that living species have descended by uninterrupted generation from the antediluvian species (p. 74), and that they have in the process become modified through external influences.

Now of all functions respiration is the most important, and upon respiration everything is regulated. "If it be admitted that the slow progression of the centuries has

[1] *Mém. Acad. Sci.*, xii., pp. 63-92, 1833.

brought in its train successive changes in the proportion of the different elements of the atmosphere, it follows as a rigorously necessary consequence that the organisation has been proportionately influenced by them" (p. 76). The respiratory milieu changes, the species change with it, or are eliminated (p. 79). We may see, perhaps, in the stress which Geoffroy lays upon respiration and the respiratory milieu a result of his constant obsession with the comparison of fish with air-breathing Vertebrates.

In the first geological period, we read in another Memoir of the same year,[1] when ammonites and *Gryphæa* flourished, hot-blooded animals with lungs could not exist. "A lung constructed like that of mammals and birds would not have been adapted to the essence of the respiratory element such as in my conception of it the system of the environing air used to be"[2] (p. 58).

Geoffroy does not tell us exactly how the milieu is to act upon the organism; the whole theory is little more than a sketch and a pointing out of the way for future research—and in this prophetic enough. The action of external agents was apparently considered as physical, and no power of active adaptation was ascribed to the organism.

From a passage in the memoir "Sur la Vertèbre" we may perhaps infer that he believed increasing complexity of structure to be due to a realisation of potentialities, to the development of parts present in the lower animals only in potency — "the organisation . . . only awaits favourable conditions to rise, by addition of parts, from the simplicity of the first formations to the complication of the creatures at the head of the scale" (p. 112). Evolution takes place as the environment allows, and in a sense in opposition to the environment.

He believed in saltatory evolution, for he considered that the lower oviparous Vertebrates could not be transformed into birds by slow modification, but only by a sudden transformation of their lungs, which would bring about the other characteristics of birds (p. 80). He considered, too,

[1] *Mém. Acad. Sci.*, xii., pp. 43-61, 1833.

[2] Geoffroy's French style is at times incredibly bad, and more or less literal translations of his sentences are apt to read queerly!

that transformations could arise by means of monstrous development (p. 86). In this connection the experiments which he made on the hen's egg[1] in order to produce artificial monstrosities are significant, though his purpose was rather to obtain proof of the inadequacy of the preformation hypothesis.[2]

It seems probable enough that if Geoffroy had developed his views on evolution he would finally have been led to interpret unity of plan in terms of genetic relationship. But as it was he remained at his morphological standpoint. He did not interpret rudimentary organs as useless heritages of the past; he preferred to think that Nature had prepared double means for the same function, one or other being predominant according as the animal lived in the water or on the land. "To the animal that lives exclusively in the air Nature has granted an organisation suited to this mode of respiration, without however suppressing the other corresponding means, that is to say, without depriving it of a second system which is applicable only to the mode of respiration by the intermediary of water, and *vice versa*."[3]

He seems, in one instance at least, to have hit upon the root-idea of the biogenetic law, but he was far from appreciating its significance. He recognised that an amphibian in its development passed through a stage when it was in all essentials similar to a fish, and he saw in this visible transformation a picture of the evolutionary transformation. "An amphibian," he writes,[4] "is at first a fish under the name of tadpole, and then a reptile [*sic*] under that of frog. . . . In this observed fact is realised what we have above represented as an hypothesis, the transformation of one organic stage into the stage immediately superior." But it is not clear that he considered the development of the amphibian to be a *repetition* of its ancestral history.

He went, however, a certain length towards recognising the main principle of a law which was a commonplace of

[1] *Mém. Mus. d'Hist. nat.*, xiii., p. 289, 1826.

[2] *Mém. Mus. d'Hist. nat.*, xviii., p. 221, 1828. His teratological work is important, and is chiefly contained in the second volume of the *Philosophie anatomique*.

[3] *Phil. anat.*, i., p. 449.

[4] *Mém. Acad. Sci.*, xii., p. 82, 1833.

German transcendental thought, and was developed later by his disciple E. Serres, the law that the higher animals repeat during their development the main features of the adult organisation of animals lower in the scale. Thus he compared fish as regards certain parts of their structure with the fœtus of mammals. He compared also Articulates with embryonic Vertebrates in respect of their vertebræ, for in the higher Vertebrates the body of the vertebra is tubular at an early stage of development, and in Articulates the body of the vertebra remains tubular permanently (*supra*, p. 61). As regards their vertebræ, "insects occupy a place in the series of the ages and developments of the vertebrate animals, that is to say, they realise one of the states of their embryo, as fishes do one of the states of their fœtal condition."[1]

This idea was destined to exercise a great influence upon the development of morphology. A further development of the thought is that certain abnormalities in the higher animals, resulting from arrest of development, represent states of organisation which are permanent in the lower animals.[2]

So far we have considered Geoffroy's theories in their application to the facts. We go on to discuss the theories themselves, and the general conception of living things which underlies them.

The principle of unity of plan and composition is the keynote of Geoffroy's work. It states that the same materials of organisation are to be found in all animals, and that these materials stand always in the same general spatial relations to one another. The "materials of organisation" are not necessarily organs in the physiological sense, and indeed the principle of the unity of plan cannot be upheld if the unity has reference to organs only. This became clear to Geoffroy, especially in his later years. In 1835 he wrote, speaking of the principle of the unity of plan, "I have, moreover, regenerated this principle, and obtained for it universality of application, by showing that it is not always the organs as a whole, but merely the materials composing each

[1] *Mém. Mus. d'Hist. nat.*, ix., p. 101, 1822.

[2] *Cours de l'histoire naturelle des Mammifères*, i., Leçon 3, p. 13, 1829.

organ, that can be reduced to unity."[1] Even in the *Philosophie anatomique* he deals rather with parts than with organs; he deals, for instance, with the elementary parts of the sternum, not with the organ "sternum" in its totality. The functions of the sternum vary, and the primary protective function of the sternum may be assumed by quite other parts, *e.g.*, by the clavicles in fish, which protect the heart.[2]

True homologies can be established between materials of organisation but not always between organs, which may be composed of different "materials."

Almost as a corollary to this comes the further view that form is of little importance in determining homologies. An organ is essentially an instrument for doing a particular kind of work, and its form is determined by its function. Organs which perform the same function are usually similar in form though the elementary materials composing them may be different. This is seen in many cases of convergence. Organs, therefore, which perform the same function and are similar in external form are not necessary homologous. Conversely, the same complex of materials, say a fore limb, may take on the most varied shapes according as the function of the organ changes—but homology remains though form changes. Accordingly, form is one of the least important elements to be considered in determining a homology. "Nature," he wrote in one of his early papers, "tends to repeat the same organs in the same number and in the same relations, and varies to infinity only their form. In accordance with this principle I shall have to draw my conclusions, in the determining the bones of the fish's skull, not from a consideration of their form, but from a consideration of their connections."[3]

Again, after comparing a vertebra of the Aurochs with an abdominal segment of the crab, he says, "I have insisted upon an identity which has extended to the least important relation of all, that of form."[4]

[1] *Études progressives d'un Naturaliste*, p. 59, f.n., Paris, 1835.
[2] *Phil. Anat.*, i., p. 444.
[3] *Ann. Mus. d'Hist. nat.*, x., p. 344, 1807.
[4] *Isis*, p. 534, 1820 (2).

Geoffroy's morphological units or materials of organisation were in the case of the skeleton—with which his researches principally deal—the single bones. But the interesting point is that he sought his skeleton-units in the embryo, and considered each separate centre of ossification as a separate bone. Coalescence of bones originally separate is one of the most usual events in development, and it is an occurrence which, more than any other, tends to obscure homologies. Because of its coalescence with the maxillaries, the intermaxillary in man was not discovered until Vicq d'Azyr and Goethe found it separate in the embryo. Apparently quite independently of Goethe, Geoffroy hit upon this plan of seeking in the embryo the primary elements or materials of organisation. In an early paper on the skull of Vertebrates,[1] where he is concerned with showing that each bone of the fish's skull has its homologue in the skull of higher Vertebrates, he is faced with the difficulty that the skull of the fish has more bones than the skull of higher Vertebrates. "Having had the inspiration," he writes, "to reckon as many bones as there are distinct centres of ossification, and having made a consistent trial of this method, I have been able to appreciate the correctness of the idea: fish, in their earliest stages, are in the same conditions relatively to their development as the fœtuses of mammals, and hence bear out the theory" (p. 344). So, too, in dealing with the homologies of the sternal elements (*supra*, p. 57) he treats as separate bones the "annexes" of the sternum in birds, though these are separate only in the young.

If the same materials of organisation are present in all animals, and if they are arranged always in the same positions relatively to one another, how does it come about that animal forms are so varied, what explanation can be offered of the diversities of organic structure? Geoffroy's main answer to this question is his *Loi de balancement.* The law was enunciated by him already in 1807.[2] We take the following quotation, which represents his thought most nearly, from the *Cours de l'histoire naturelle des Mammifères* (1829). "According to our manner of regarding the

[1] *Ann. Mus. d'Hist. nat.*, x., pp. 342-65, 1807.

[2] *loc. cit.*, x., p. 343.

organisation of mammals, there is only a single animal modified by the inverse reciprocal variation of all or some of its parts. Now, from the fact that there is only one single general animal, it follows that for each section of its components or for each of its organs there is available only a given quantity of formative materials. Now suppose that the distribution of these materials has not been made in such a way as to ensure an exact equilibrium between all the parts concerned, one organ will get more than its share, another less. My law of the compensation of organs is founded on these principles" (i., *Leçon* 16, p. 12). "The atrophy of one organ turns to the profit of another; and the reason why this cannot be otherwise is simple, it is because there is not an unlimited supply of the substance required for each special purpose."[1] The nutritive material available is limited for each species; if one part gets more than its share the other parts must get less—that is all the law means. As an example, take the minuteness of the episternals and xiphisternals in birds, as contrasted with the huge size of the entosternal. "The minuteness of the episternals and xiphisternals might be imputed to this gigantic piece diverting to its own profit the nutritive fluid, since the bigger it is the smaller these are."[2]

One has constantly to remember in dealing with Geoffroy's theories that he was not an evolutionist, but purely a morphologist. It is therefore, perhaps, to ask too much to require of him an explanation of the causes of diversity. The morphologist describes, classifies, generalises; he does not seek for causes. But we must leave this question aside in order to discuss how far Geoffroy's theory of the unity of plan and composition fits the facts. As Geoffroy himself admitted on several occasions, his theory was an *à priori* one, a theory hit upon by hasty induction, then erected into a principle and imposed upon the facts. No more than Goethe did he extract his principle from a sufficient mass of data.

Now he found his theory to be in its pure form unworkable; he found, for example, that the skeleton of fishes

[1] *Phil. anat.*, i., 450, f.n. *Cf.* Aristotle (*supra*, p. 11).
[2] *Loc. cit.*, p. 136.

could not be compared directly, bone for bone, with the skeleton of higher Vertebrates; he had to admit differences of position of whole sets of organs in the two groups, he had to admit various *metastases*, before he could bring the skeleton of fish into line. And these metastases are due to functional requirements — for example, the forward position of sternum and thoracic organs in fish is an adaptation to swimming.

So he does not so much demonstrate the unity of plan of whole organisms as the unity of plan of particular corresponding parts of them. Thus he does not prove or attempt to prove that Articulates are in all points like Vertebrates, but simply that their skeleton is built upon the same plan as that of Vertebrates. The rest of the organs, while still comparable with the organs of Vertebrates, stand in different relations to the skeleton. An Articulate therefore, on his own showing, is not, *as a whole*, built upon the same general structural plan as a Vertebrate.

Further, he does not always remain true to his principles, for he does not establish homologies of parts entirely by their connections but sometimes by their functions as well. Thus the sternum, or rather the complex of sternal elements, is defined and discovered in particular cases not by its connections only but also by its functions. The framework of the gills is homologised part by part with the framework of the lungs, not because the relations of the framework to the rest of the skeleton are the same in fish and air-breathing Vertebrates, but simply because gills are considered the equivalents of lungs—a comparison which is purely physiological.

Even with these concessions to the functional view of living things, Geoffroy was unable to make good his contention that all animals are built upon the same plan. His arguments failed to carry conviction to his contemporaries, and Cuvier in particular subjected them to destructive, and indeed final, criticism.

The paper, already referred to, in which Cuvier disposed of the transcendentalists' comparison of Cephalopods and Vertebrates is of great significance, for it states in the clearest way the radical opposition between the functional and the formal attitudes to living things.

Cuvier points out that if by unity of composition is meant identity, then the statement that all animals show the same composition is simply not true—compare a polyp with a man!—on the other hand, if by unity is meant simply resemblance or homology, the statement is true within certain limits, but it has been employed as a principle since the days of Aristotle, and the theory of unity of composition is original only in so far as it is false. He admits, however, that Geoffroy has seized upon many hidden homologies, especially by his valuable discovery of the importance of fœtal structure. In all this Cuvier is undoubtedly right. Unity of plan and composition, as Geoffroy conceived it, simply does not exist. Cuvier goes on to say that this principle of Geoffroy's, in the greatly modified form in which it can be accepted, and has been accepted from the dawn of zoology, is not the sole and unique principle of the science. On the contrary, it is merely a subordinate principle, subordinate to a higher and more fruitful principle, that, namely, of the conditions of existence, of the adaptation (*convenance*) of the parts, of the co-ordination of the parts for the rôle which the animal is to play in Nature. "That is the true philosophical principle," he says, "whence may be deduced the possibility of certain resemblances, the impossibility of certain others; it is the rational principle from which follows the principle of the unity of plan and composition, and in which at the same time it finds those limits, which some would like to disregard" (p. 248).

Geoffroy's position is the direct contrary. He holds that the principle of the unity of plan and composition is the true base of natural history,[1] and that this unity limits the possible transformations of the organism. Thus, speaking of the influence of the respiratory medium, he says, "All the same this influence of the external world, if it has ever become a cause which disturbed organisation, must necessarily have been confined within fairly narrow limits; animals must have opposed to it certain conditions inherent to their nature, the existence of the same materials composing them, and a

[1] *Mammifères*, i., Discours prél., p. 18.

manifest tendency to resemble one another, and to reproduce invariably the same primordial type."[1] Unity of plan and composition is, on this view, prior to adaptation and limits adaptation. Cuvier's view, on the contrary, is that the necessity of functional and ecological adaptation accounts for the repetition of the same types of structure. There are, of all the possible combinations of organs, only a few viable types—those whose structure is adapted to their life. Therefore it is reasonable that these few types should be repeated in innumerable exemplars. One must remember, in order to appreciate Cuvier's view, that he was not obsessed, as we are, by the idea of evolution.

Cuvier thought in terms of organs, not in terms of "materials of organisation." He held that the resemblances between the organs of one class of animals and the organs of another were due to the similarity of their functions. "Let us conclude, then, that if there are resemblances between the organs of fish and those of other classes, it is only in the measure that there is a resemblance between their functions."[2] There are only a few kinds of organs, each adapted for a particular function, and these organs are necessarily repeated from class to class.—"As the animal kingdom has received only a limited number of organs, it is inevitable that some at least of these organs should be common to several classes."[3]

Geoffroy thought in terms of "materials," of parts of indefinite function, parts which might take on any function. He insists upon the necessity of disregarding function when tracing out the unity of composition. He considers, in direct opposition to Cuvier's interpretation of structural resemblance as due to similarity of function, that unity of composition is the primary fact, and similarity of function subsidiary. In his reply in the *Mammifères* (1829) to Cuvier's criticisms in the *Histoire naturelle des Poissons* (1828), he insists on the necessity of excluding function from consideration in any truly philosophical treatment of comparative anatomy (Discours prél., p. 25). Cuvier held that function determined structure, or at least that the necessity

[1] *Phil. anat.*, i., p. 208.

[2] Cuvier and Valenciennes, *Hist. nat. Poissons*, i., p. 550, 1828.

[3] Cuvier and Valenciennes, *loc. cit.*, p. 544.

of adaptation ruled the transformations of form. Geoffroy considered that structure determined function, that changes of structure, however they might arise, caused changes of function. "Animals," he writes, "have no habits but those that result from the structure of their organs; if the latter varies, there vary in the same manner all their springs of action, all their faculties and all their actions."[1]

Again, "a vegetarian régime is imposed upon the Quadrumana by their possession of a somewhat ample stomach, and intestines of moderate length."[2] The hand of the bat has become so modified as to constrain the bat to live in the air.[3]

The best example of Geoffroy's insistence upon the priority of structure to function, and so of his purely morphological attitude, is perhaps his interpretation, already alluded to, of the appendages of Articulates. The segments of the Articulate are, he says, the equivalents of the bodies of the vertebræ of higher forms. Now "from the circumstance that the vertebra is external, it results that the ribs must be so too; and, as it is impossible that organs of such a size can remain passive and absolutely functionless, these great arms, hanging there continually at the disposition of the animal, are pressed into the service of progression, and become its efficient instruments."[4] The ribs become locomotory appendages.

We may compare the similar thought that the ear ossicles are simply opercular bones reduced and turned to other uses.

Geoffroy could not but recognise the correlation of structure to function, for this is a fact which imposes itself upon every observer. He recognised also correlation between functions, as when he pointed out the connection between increased respiration and enhanced muscular activity in birds.[5] He interpreted structure at times in terms of function, the short, strong clavicle of the mole as an adaptation to digging, the keeled sternum of birds as an adaptation to flying, and so on. But we may say that his whole tendency was to disregard function, to look upon it as subsidiary. He protests against arguing from function and habits to structure, as an "abuse

[1] *Mammifères*, i., *Leçon* 4, p. 17.

[2] *Loc. cit.*, *Leçon* 5, p. 8.

[3] *Loc. cit.*, *Leçon* 13, p. 6.

[4] *Isis*, p. 539, 1820 (2).

[5] *Mammifères*, i., *Leçon* 4, p. 6.

of final causes."[1] He was not so convinced as Cuvier was of the all-importance of functional correlation; in this view he was probably confirmed by his work on teratology. It did not surprise him that Insects, in which lungs, heart and circulation have disappeared (!), should yet have a skeleton built upon the same plan as the skeleton of Vertebrates, which possess these organs; the correlation of organ-systems is not so close as to prevent this.[2] So too, although the other organs of the insect are all inside the body of the vertebræ, they are yet comparable with the organs of Vertebrates.[3] The existence of rudimentary organs also seemed to him an argument against too strict a correlation of parts.

The contrast between the teleological attitude, with its insistence upon the priority of function to structure, and the morphological attitude, with its conviction of the priority of structure to function, is one of the most fundamental in biology.

Cuvier and Geoffroy are the greatest representatives of these opposing views. Which of them is right? Is there nothing more in the unity and diversity of organic forms than the results of functional adaptation, or is Geoffroy right in insisting upon an element of unity which cannot be explained in terms of adaptation? If there be an irreducible element of unity, is there any truth in Geoffroy's suggestion that this unity results from a power which is exercised in the world of atoms where are elements of inalterable character?[4]

The problem as Geoffroy and Cuvier understood it was not an evolutionary one. But the problem exists unchanged for the evolutionist, and evolution-theory is essentially an attempt to solve it in the one direction or the other. Theories such as Darwin's, which assume a random variation which is not primarily a response to environmental changes, answer the problem in Geoffroy's sense. Theories such as Lamarck's, which postulate an active responsive self-adaptation of the organism, are essentially a continuation and completing of Cuvier's thought.

[1] *Mammifères*,, Discours prél., p. 7. [2] *Isis*, p. 460, 1820 (2).
[3] *Mém. Mus. d'Hist. nat.*, ix., p. 102, 1822.
[4] *Mém. Acad. Sci.*, xii., p. 76, 1833.

CHAPTER VI

THE FOLLOWERS OF ETIENNE GEOFFROY SAINT-HILAIRE

GEOFFROY'S theories were not generally accepted by his contemporaries, but his methods had considerable influence, especially in France, where many made essays in pure morphology.

His chief follower was Serres, who is mentioned indeed in the *Philosophie anatomique* as a fellow-worker. Serres was primarily a medical anatomist; his interest lay in human anatomy and embryology, normal and pathological.

His best early work was an *Anatomie comparée du cerveau* (1824-26), which met with a flattering reception from Cuvier.[1] He laid great stress upon the development of the brain and spinal cord in the different classes, and was quick to point out analogies not only between adult but also between embryonic structures. He paid much attention to cases of correlation, and noted a great many; he observed, for instance, a constant relation between the development of the spinal cord and of the corpora quadrigemina, and between the size of the corpora quadrigemina and the volume of the optic nerves and eyes. In this the influence of Cuvier is unmistakable.

Serres' early theoretical views are to be found in a series of papers in the *Annales des Sciences naturelles*,[2] under the general title *Recherches d'Anatomie transcendante, sur les Lois de l'Organogénie appliquées à l'anatomie pathologique*, also published separately. We follow these papers in our exposé of Serres' doctrine, reserving for a future chapter (Chap. XII.) the consideration of his matured views of thirty years later.

[1] *Mém. Acad. Sci.*, iv., pp. cclxxxiv.-ccci., 1824.

[2] *Ann. Sci. Nat.*, xi., xii., 1827; xvi., 1829; xxi., 1830.

In the first of them he points out how neither position nor function has proved altogether sufficient to establish homologies. In the early days anatomists were guided by form; when form failed them, they traced an organ in its changes throughout the series of animals by considering its function. This method was satisfactory enough as regards the organs of the nutritive life. But in the organs of the life of relation, in the nervous system, the functions of the parts were difficult to discover, and their form very changeful. Hence a new principle was required, and Serres found it in the thought which he probably owed to the German transcendentalists (see Chap. VII.), that the permanent structure of the lower animals could be compared with phases in the development of the higher, and particularly of man, or, as he put it, that comparative anatomy was often only a fixed and permanent anthropogeny, and anthropogeny a fugitive and transitory comparative anatomy (xi., p. 106).

"In rising towards the first formations," he writes, "transcendental anatomy recognised that one and the same organ, however complicated its definitive form might be, repeated in its transitory states the organic simplicities of the lower classes. Thus the primitive heart of birds was first of all a canal, then a pocket or single cavity, then finally the complex organ of the class. Comparative anatomy was thus seen to be repeated and reproduced by embryogeny" (xii., p. 85).

His explanation of the fact of repetition is that, "in animals belonging to the lower classes the *formative force*, whatever it may be, has a less energetic impulsion than in the higher animals, and hence the organs pass through only a part of the transformations which those of the higher forms undergo; and it is for this reason that they show permanently the organic dispositions which are only transitory in the embryo of man and the higher Vertebrates. Hence these double aortas, these double venæ cavæ which one observes more or less constantly among reptiles" (xxi., p. 48).

The number of stages in embryogeny is proportionate to the complexity of the adult; the younger the embryo the simpler its organs—such is the general formula of the relation between the embryo and the adult. But here in Serres' doctrine of parallelism a complication enters. He observed that embryonic organs did not always develop in a piece, by simple growth, but often were formed by the union of separately formed parts or layers. Thus the kidney in man is formed by the fusion of a number of "little kidneys," and the spinal cord reaches its full development by the laying down of successive layers within it. He was greatly impressed with this fact, which, as a convinced believer in epigenesis, he used with great effect against the preformistic theories. "This method of isolated formation," he wrote, "is noticed in early stages in the thyroid, the liver, the heart, the aorta, the intestinal canal, the womb, the prostate, the clitoris, and the penis" (xi., p. 69). So, too, in the development of the skeleton, ossification proceeds from separate centres, foramina are formed by the fusion of separate bones round them. In his memoir, *Lois d'Osteogénie* (1819), Serres established several laws of ossification based upon this principle of separate formation.[1]

How is the fact of multiple formation to be reconciled with the principle of repetition, according to which organs are simplest in the early embryo and in the lower animals? But observation shows that, as a rule, the further down the scale you go the more divided organs become—the more numerous the bones of the skull, for example. There is thus a parallel between multiple formation of organs in the embryos of the higher Vertebrates and their subdivided state in the lower. Take, for example, the kidney. In the genus *Felis*, and in birds, each kidney has two lobes, in the elephant four, in the otter ten, in the ox twelve to fourteen. The human kidney in its development starts with about a dozen lobes, and the number diminishes as the kidney grows. Thus the permanent state of the kidney in the animals mentioned is reproduced by the stages of its development in man (xii., p. 126).

So, too, at the second or third month the uterus of the human embryo is bicornuate, and afterwards passes through stages comparable to the adult and permanent uterus of rodents, ruminants, and carnivores. There is indeed a time in the development of the human embryo when it resembles

[1] See Rádl, *loc. cit.*, i., pp. 225-6.

in many of its organs the adult stage of various lower animals. It is about this time that it possesses a tail.

We note that Serres' theory of parallelism applies, strictly speaking, only to organs, not to organisms, although he, too, readily fell into the error of supposing that the organisation of an embryo could be compared as a whole with the adult organisation of an animal lower in the scale. Thus he wrote in one of his later papers[1]—"As our researches have made clear, an animal high in the organic scale only reaches this rank by passing through all the intermediate states which separate it from the animals placed below it. Man only becomes man after traversing transitional organisatory states which assimilate him first to fish, then to reptiles, then to birds and mammals." Serres was not altogether free from the besetting sin of the transcendentalists—hasty generalisation.

The law of parallelism applied not only to Vertebrates but also to Invertebrates. In a short paper[2] of 1824 Serres attempted an explanation of the nervous system of Invertebrates. Invertebrates, he considered, lacked the cerebro-spinal axis of Vertebrates, and their nervous system was the homologue of the sympathetic system of Vertebrates. The relation of the invertebrate to the vertebrate nervous system being thus fixed, can the nervous system of Invertebrates be reduced to one plan? It does not seem possible to establish a common plan for the adult nervous systems. But apply the principle of parallelism, which has proved so valuable within the limits of the vertebrate series. Taking insects as the highest class, we find that there are three stages in the development of their nervous system; in the first the nervous system is composed of two separate strands, in the second the strands unite round the œsophagus, in the third they unite also behind. Now in *Bulla aperta*, stage (1) is permanent; in *Clio*, *Doris*, *Aplysia*, *Tritonia*, *Sepia*, *Helix*, stage (2) is permanent, and in *Unio* stage (3). In fact, all the varieties of the nervous system of molluscs fall into one or other of these three classes. "It follows, then, that as regards their nervous system, the Mollusca are more or less advanced larvæ of insects" (p. 380). The law of parallelism

[1] *Ann. Sci. nat.* (2), ii., p. 248, 1834.

[2] *Ann. Sci. nat.*, iii., pp. 377-80, 1824.

is here applied to single organ-systems, but in later years Serres applied it to whole organisations also, saying that the lower Invertebrates were permanent embryos of the higher.

In the paper of 1834, already referred to, Serres pushed his speculations further and attempted to establish the unity of type of all animals, Vertebrates and Invertebrates alike—a favourite pastime of the transcendentalists. It is incontestable, he admits, that adult Invertebrates are quite different in structure from adult Vertebrates, "but if one regards them as what I take them to be, namely, *permanent embryos*, and if one compares their organisation with the embryogeny of Vertebrates, one sees the differences disappear, and from their analogies arise a crowd of unsuspected resemblances (*loc. cit.*, p. 247).

The last point of Serres' doctrine which calls for remark is his interpretation of abnormalities as being often comparable to grades of structure permanent in the lower animals. Thus the double aorta which may occur as an abnormality in man is the normal and permanent state in reptiles. This idea, of course, he got from Etienne Geoffroy St Hilaire. It is further developed in his "*Théorie des formations et des déformations organiques appliquée à l'anatomie comparée des monstruosités* (1832), and in his final large memoir of 1860 (see below, p. 205).

In 1816 appeared a fine piece of work by J. C. Savigny on the homologies of the appendages in Articulates. The standpoint was that of pure morphology. "I am convinced," he wrote, "that when a more complete examination has been made of the mouth of insects, properly so called, that is to say, having six legs and two antennæ, it will be found that whatever form it affects it is always essentially composed of the same elements. . . . The organ remains the same, only the function is modified or changed—such is Nature's constant plan."[1] In this the influence of Geoffroy can be traced ; but the work was very free from the exaggerations of the transcendentalists, and many of Savigny's homologies are accepted even to-day. The first memoir dealt with the mouth-parts of insects ; the

[1] *Mémoires sur les Animaux sans Vertèbres*, Part I., p. 10, Paris, 1816.

second with the anterior appendages of Articulates generally. Savigny shows that the mouth-parts of insects can be reduced to the type shown in Orthoptera, where there are clearly two mandibles, two maxillæ, and a lower lip formed by the fusion of two second maxillæ. All other insects have these same mouth-parts, disposed in the same order, however much their form may have been modified in response to new functions. He goes on to compare the anterior set of appendages in a long series of Articulates, in *Julus*, *Scolopendra*, *Cancer*, *Gammarus*, *Cyamus*, *Nymphon*, *Phalangium*, *Apus*, *Caligus*, *Limulus*, and a few others. For Crustacea he established the homologies now accepted, of the mandibles with the mandibles of insects, of the first and second pairs of maxillæ with the parts so named in insects, and so on. He is quite clear that the maxillipedes of Crustacea are the homologues of the feet of Hexapoda. "Their disposition must lead one to think that the six anterior feet of *Julus*, that is to say, all the feet of the Hexapoda, are here transformed into jaws" (*loc. cit.*, p. 48). In *Scolopendra* also there is a similar transformation of two pairs of legs into auxiliary jaws. In *Gammarus*, where there is only the first pair of maxillipedes, the other two pairs have become "retransformed" into feet. We find him supporting his comparison of the three anterior pairs of legs in *Julus* to the three pairs of legs in insects by an argument drawn from embryology; for only the first three pairs of feet are present in *Julus* at birth (Degeer), "an observation, which, together with their position, should cause them to be considered as the representatives of the six thoracic feet of Hexapoda" (p. 44).

His comparison of the Arachnid appendages with those of insects and Crustacea is very curious. As his starting-point he takes *Cyamus*, which has antennæ (two pairs) and mouth parts (four pairs) as in many Crustacea, and then seven pairs of legs; he compares with it *Nymphon*, which has in all seven pairs of appendages. These appendages he homologises with the seven pairs of legs of *Cyamus*, so that the first appendage in *Nymphon* corresponds to the seventh appendage of *Cyamus*. This homology is extended to all Arachnids; their first two pairs of appendages, however they may be modified as "false" mandibles and "false" maxillæ, really correspond to the second and third maxillipedes in Crustacea, and to the second and third pairs of feet in insects. It is interesting to note that he treats *Limulus* as an Arachnid, pointing out that there is as much difference between *Apus* and *Limulus* as between *Cancer* and *Phalangium*. He describes the "gnathobases" in *Phalangium* and *Limulus*. We may note that he had just an inkling of the modern doctrine that all the appendages of Articulates consist of a basal joint bearing an inner and an outer terminal piece, for he observes that the "cirri" of the maxillipedes of Crustacea give the appendage the same bifid appearance as the appendages of the abdomen and the thoracic legs of *Mysis* (p. 50).

V. Audouin, in his memoir, *Recherches anatomiques sur le thorax des animaux articulés*,[1] applied the principle of the unity of plan and composition to the exoskeleton of insects, Crustaceans, and Arachnids. His guiding ideas were, "(1) that the skeleton of articulated animals is formed of a definite number of pieces, which are either distinct or intimately fused with one another; (2) that in many cases, some pieces diminish or altogether disappear, while others reach an excessive development; (3) that the increase of one piece seems to exert on the neighbouring pieces a kind of influence which explains all the differences one finds between the individuals of each order, family and genus" (Sep. copy, p. 16). Geoffroy had already stated, without proof, that the parts of the Arthropod's skeleton, however they might change in shape and size, remained faithful to the principle of connections, at least at their points of insertion.[2] Audouin gave the detailed demonstration of this by his accurate and minute determination of the pieces of the arthropod skeleton. He recognised that the body of Arthropods was made up of a series of similar rings, and that even the compact head of insects consisted of fused segments. In each segment Audouin distinguished a fixed number of hard chitinous parts, the dorsal tergum, the ventral sternum, the lateral "flanc" of three pieces, all to be recognised by their positions

[1] *Ann. Sci. Nat.* (1), i., pp. 97-135, 416-432, 1824.

[2] *Isis*, p. 456, 1820 (2).

relative to one another. Many of the names which he proposed are still in use; it was he who introduced the terms prothorax, mesothorax, and metathorax, for the three segments of the insect's thorax. He used Geoffroy's *Loi de balancement* to explain cases of correlative development, such as the relation between the size of the front wings and the development of the mesothorax. In another paper Audouin compared the three pieces of the dorsal skeleton of Trilobites to the tergum and the upper part of the "flanc."[1] In a third paper of about the same time he tried to establish the homologies of the segments throughout the Articulate series—with less success than Savigny.

Later on, in conjunction with Milne-Edwards, he demonstrated the unity of composition of the nervous system in Crustacea, showing how the concentrated system of the crab was formed by the same series of ganglia as in the Macrura.

The entomologist Latreille also tackled the problem of the homologies of the segments in the different classes of Arthropods (Cuvier, *loc. cit.*, p. cclxxii.). He thought he could find fifteen segments in all Arthropods. He made the retrograde step of likening the head of insects to a single segment. But some of his homologies showed morphological insight, *e.g.*, his comparison of the "first jaws" of Arachnids to antennæ, because they were placed above the upper lip. It was he who first pointed out the resemblance of the leaf-like gills of Ephemerid larvæ to wings, and suggested that wings were "a sort of tracheal feet."

He made also a rather hazy and speculative contribution on Okenian lines to the problem of the relation of Arthropods to Vertebrates, likening the carapace of Crustacea to an enormously developed hyoid, the appendages of the tail to the ventral and anal fins of fish. The masticatory organs of Arthropods were jaws disjointed at their symphysis; antennæ, nostrils turned outside in.

Dugès also made a comparison of Articulates with Vertebrates.[2] He did not accept Geoffroy's vertebral theory

[1] Cuvier, *Mém. Acad. Sci.*, iv., p. cclxx., 1824.

[2] *Acad. Sci.* 18th Oct. 1831. Extract in *Ann. Sci. Nat.*, xxiv., pp. 254-60, 1831.

of the Arthropod skeleton, though he admitted that in Arthropods the dorsal surface was turned towards the ground, basing this assumption on the position of the nervous system, and also, curiously enough, on the inverted position of the embryo on the lower surface of the yolk. He considered that the mandibles and first maxillæ of Arthropods were the homologues of the upper and lower jaws of Vertebrates, adducing as confirmatory evidence the fact that in snakes the rami are separate. The labium was the equivalent of the hyoid, the labial palps and maxillipedes the equivalent of the "hyoid" elements which form the branchial arches.

But Dugès' main contribution to morphological method was his conception of the living organism as a colony of lesser units, which were themselves real "organisms." "By *organism* the author means a complex of organs which taken together suffice to constitute, ideally or actually, a complete animal. An 'organism' is, as it were, an elementary or simple animal; several organisms combined form a complex animal" (p. 255). Dugès hit upon this principle, which was first suggested to him by A. Moquin-Tandon's work on the leech (1827), as a great aid in demonstrating the unity of plan and composition throughout the animal kingdom.[1] According to his view there are three main types of animals—(1) Biserials, including bilaterally symmetrical animals, composed of two parallel series of "organisms"; (2) Radiates, composed of "organisms" arranged like the spokes of a wheel; and (3) Raceme-animals, in which the separate "organisms" were disposed more or less irregularly, in bunches (p. 257). The unitary "organism" is supposed to be the same in all, only the arrangement differing. Dugès of course admitted that the centralisation of the complete organism became greater the higher it stood in the scale, and that this held good also in individual development. The appendages of Articulates and Vertebrates were thought of as the members of as many separate organisms. He went so far as to suggest that the

[1] His views were more fully elaborated in his *Mémoire sur la conformité organique dans l'échelle animale*, Montpellier, 1832.

fingers of a man's hand were the free extremities of as many thoracic members.

Dugès' conception of the organism has often been revived since in a saner form, *e.g.*, by E, Perrier, and it has a certain validity. It has much affinity with the similar conceptions of Goethe and the German transcendentalists.

CHAPTER VII

THE GERMAN TRANSCENDENTALISTS

To complete our historical survey of the morphology of the early 19th century we have now to turn back some way and consider the curious development of morphological thought in Germany under the influence of the *Philosophy of Nature*. We have already seen many of these notions foreshadowed by Goethe, who had considerable affinity with the transcendentalists, but the full development of transcendental habits of thought comes a little later than the bulk of Goethe's scientific work, and owes more to Kielmeyer and Oken than to Goethe himself.

A great wave of transcendentalism seems to have passed over biological thought in the early 19th century, arising mainly in Germany, but powerfully affecting, as we have seen, the thought of Geoffroy and his followers. Many ideas were common to the French and German schools of transcendental anatomy, the fundamental conception that there exists a unique plan of structure, the idea of the scale of beings, the notion of the parallelism between the development of the individual and the evolution of the race. It is difficult to disentangle the part played by each school and to determine which should have the credit for particular theories and discoveries. The philosophy seems to have come chiefly from Germany, the science from France. It must be borne in mind that German comparative anatomy was largely derivative from French, that the Paris Museum was the acknowledged anatomical centre, and that Cuvier was its acknowledged head.

It is probably correct to say that the credit mainly belongs to the German transcendental school for the law of the parallelism between the stages of individual development and the stages of the scale of beings, and the theory of the repetition or multiplication of parts within the individual. The vertebral theory of the skull is a particular application of the second of these generalisations.

The law of parallelism[1] seems to have been expressed first by Kielmeyer (1793),[2] who gave to it a physiological form, saying that the human embryo shows at first a purely vegetative life, then becomes like the lower animals, which move but have no sensation, and finally reaches the level of the animals that both feel and move.

The idea was next taught by Autenrieth in 1797.[3]

Oken (1779-1851) in his early tract *Die Zeugung* (1805), and in his *Lehrbuch der Naturphilosophie* (1809-11) elaborated the thought, and taught that every animal in its development passes through the classes immediately below it. "During its development the animal passes through all stages of the animal kingdom. The fœtus is a representation of all animal classes in time."[4] The Insect, for example, is at first Worm, next Crab, then a perfect volant animal with limbs, a Fly (*ibid.*, p. 542).

As Nature is "the representation of the individual activities of the spirit," so the animal kingdom is the representation of the activities or organs of man. The animal kingdom is therefore "a dismemberment of the highest animal, *i.e.*, of Man" (p. 494). Now "animals are gradually perfected, entirely like the single animal body, by adding organ unto organ"—the way of evolution is the way of development. Hence "animals are only the persistent fœtal stages or conditions of Man," who is the microcosm, and contains within himself all the animal kingdom.

Oken was himself a careful student of embryology; von Baer[5] speaks of his work (published in Oken and Kieser, *Beiträge zur vergleichenden Zoologie, Anatomie und Physi-*

[1] For a full account, see Kohlbrugge, *Zool. Annalen*, xxxviii., 1911.

[2] *Rede über das Verhältnis der organischen Kräfte*, Stuttgart u. Tübingen, 1793 (1814). See Rádl, *loc. cit.*, i., p. 261; ii., p. 57.

[3] *Supplem. ad historiam embryonis*, Tübingen, 1797.

[4] *Lehrbuch der Naturphilosophie*, Eng. trans., p. 491, 1847.

[5] *Ueber Entwickelungsgeschichte der Thiere*, i., p. xvii., 1828.

ologie," 2 pts., 1806-7) as forming the turning-point in our understanding of the mammalian ovum. He had accordingly actually observed a resemblance in certain details of structure between the human fœtus and the lower animals; but the peculiar form which the law took in his hands was a consequence of his hazy philosophy. He saw the relation of teratological to fœtal structure, for he affirmed that "malformations are only persistent fœtal conditions" (p. 492).

The idea of comparing the embryo of higher animals with the adult of lower was widely spread at this time among German zoologists. We find, for example, in Tiedemann's brilliant little textbook[1] the statement that "Every animal, before reaching its full development, passes through the stage of organisation of one or more classes lower in the scale, or, every animal begins its metamorphosis with the simplest organisation" (p. 57).

Thus the higher animals begin life as a kind of fluid animal jelly which resembles the substance of a polyp; the young mammal, like the lower Vertebrates, has only a simple circulation, and, like them, lives in water (the amniotic fluid); the frog is first like a worm, then develops gills and becomes like a fish (p. 57). In his work on the anatomy of the brain,[2] Tiedemann established the homology of the optic lobes in birds by comparing them with fœtal corpora quadrigemina in man (see Serres, *Ann. Sci. nat.*, xii., p. 112).

J. F. Meckel, in 1811, devoted a long essay to a detailed proof of the parallelism between the embryonic states of the higher animals and the permanent states of the lower animals. In a previous memoir in the same collection[3] (i., 1, 1808) he had made some comparisons of this kind in dealing with the development of the human fœtus; in this memoir (ii., 1, 1811) he brings together all the facts which seem to prove the parallelism.

His collection of facts is a very heterogeneous one; he mingles morphological with physiological analogies, and makes the most far-fetched comparisons between organs

[1] *Zoologie*, Landshut, i., 1808.

[2] *Anatomie u. Bildungsgeschichte des Gehirns im Fötus des Menschen*, Nürnberg, 1816.

[3] *Beyträge zur vergleichende Anatomie*, Leipzig, i., 1808-9, ii., 1811-2.

belonging to animals of the most diverse groups. He compares, for instance, the placenta with the gills of fish, of molluscs and of worms, homologising the cotyledons with the separate tufts of gills in *Tethys*, *Scyllæa* and *Arenicola* (p. 26). This is purely a physiological analogy. He compares the closed anus of the early human embryo with the permanent absence of an anus in Cœlentera, and the embryo's lack of teeth with the absence of teeth in many reptiles and fish, in birds, and in many Cetacea (p. 46).[1] These are merely chance resemblances of no morphological importance. He considers bladderworms as animals which have never escaped from their amnion, and *Volvox* as not having developed beyond the level of an egg (p. 7). He lays much stress upon likeness of shape and of relative size, comparing, for instance, the large multilobate liver of the human fœtus with the many-lobed liver of lower Vertebrates and of Invertebrates. In general he shows himself, in his comparisons, lacking in morphological insight.

His treatment of the vascular system affords perhaps the best example of his method (pp. 8-25). The simplest form of heart is the simple tubular organ in insects, and it is under this form that the heart first appears in the developing chick. The bent form of the embryonic heart recalls the heart of spiders; it lies at first free, as in the mollusc *Anomia*. The heart consists at first of one chamber only, recalling the one-chambered heart of Crustacea. A little later three chambers are developed, the auricle, ventricle, and aortic bulb; at this stage there is a resemblance to the heart of fish and amphibia. At the end of the fourth day the auricle becomes divided into two, affording a parallel with the adult heart of many reptiles.

In his large text-book of a somewhat later date, the *System der vergleichenden Anatomie* (i., 1821), he works out the idea again and gives to it a much wider theoretic sweep, hinting that the development of the individual is a repetition of the evolutionary history of the race. Meckel was a timid believer in evolution. He thought it quite possible that much of the variety of animal form was due to a process of

[1] Cetacea were generally considered at this time to be mammals of low organisation.

evolution caused by forces inherent in the organism. "The transformations," he writes, "which have determined the most remarkable changes in the number and development of the instruments of organisation are incontestably much more the consequence of the tendency, inherent in organic matter, which leads it insensibly to rise to higher states of organisation, passing through a series of intermediate states."[1]

His final enunciation of the law of parallelism in this same volume shows that he considered the development of the individual to be due to the same forces that rule evolution. "The development of the individual organism obeys the same laws as the development of the whole animal series; that is to say, the higher animal, in its gradual evolution, essentially passes through the permanent organic stages which lie below it; a circumstance which allows us to assume a close analogy between the differences which exist between the diverse stages of development, and between each of the animal classes" (p. 514).

He was not, of course, able fully to prove his contention that the lower animals are the embryos of the higher, and we gather from the following passage that he could maintain it only in a somewhat modified form. "It is certain," he writes, "that if a given organ shows in the embryo of a higher animal a given form, identical with that shown throughout life by an animal belonging to a lower class, the embryo, in respect of this portion of its economy, belongs to the class in question" (p. 535). The embryo of a Vertebrate might at a certain stage of development, be called a mollusc, if for instance, it had the heart of a mollusc.

He admits, too, that the highest animal of all does not pass through in his development the entire animal series. But the embryo of man always and necessarily passes through many animal stages, at least as regards its single organs and organ-systems, and this is enough in Meckel's eyes to justify the law of parallelism (p. 535).

In his excellent discussion of teratology Meckel points out how the idea of parallelism throws light upon certain

[1] From the French trans., which appeared under the title *Traité gén. d'Anat. comparée*, i., p. 449, 1828.

abnormalities which are found to be normal in other (lower) forms (p. 556).[1]

We may refer to one other statement of the law of parallelism—by K. G. Carus in his *Lehrbuch der vergleichenden Anatomie* (Leipzig, 1834). The standpoint is again that of *Naturphilosophie.* It is a general law of Nature, Carus thinks, that the higher formations include the lower; thus the animal includes the vegetable, for it possesses the "vegetative" as well as the "animal" organs. So it is, too, by a rational necessity that the development of a perfect animal repeats the series of antecedent formations.

As we have said, the main credit for the enunciation of the law of parallelism belongs to the German transcendental school; but the law owes much also to Serres, who, with Meckel, worked out its implications. It might for convenience, and in order to distinguish it from the laws later enunciated by von Baer and Haeckel, be called the law of Meckel-Serres.

Under the "theory of the repetition or multiplication of parts within the organism" may be included, first, generalisations on the serial homology of parts, and second, more or less confused attempts to demonstrate that the whole organisation is repeated in certain of the parts. The recognition of serial homologies constituted a real advance in morphology; the "philosophical" idea of the repetition of the whole in the parts led to many absurdities. It led Oken to assert that in the head the whole trunk is repeated, that the upper jaw corresponds to the arms, the lower to the legs, that in each jaw the same bony divisions exist as in the limbs, the teeth, for instance, corresponding to the claws (*loc. cit.*, p. 408). It led him to distinguish "two animals" in every body—the cephalic and the sexual animal. Each of these has its own organs; thus "in the perfect animal there are two intestinal systems thoroughly distinct from each other, two intestines which belong to two different animals, the sexual and cephalic animal, or the plant and the animal" (p. 382). The intestine of the sexual animal is the large intestine; the lungs of the sexual animal are the kidneys, its glottis is the urethra, its mouth the anus. So, too, the mouth is the stomach of the head. On another line of thought the

[1] *Cf.* Geoffroy (*supra*, p. 70).

sternum is a ventral vertebral column. Limbs are connate ribs, the digits indicating the number of ribs included (*cf.* Dugès, *supra*, p. 88).

J. F. Meckel[1] discusses "homologies" of this kind in the thorough and pedestrian way so characteristic of him. Not only, he says, are the right and left halves of the body comparable with one another, but also the upper and the lower, the dividing line being drawn at the level of the diaphragm. The lumbar complex corresponds to the skull, the anus to the mouth, the urino-genital opening to the nasal opening; in general, the urino-genital system corresponds to the respiratory, the kidneys to the lungs, the ureters to bronchi, the testes and ovaries to the thymus (he had observed the physiological relation between the development of the thymus and the state of the genital organs), the prostate and the uterus to the thyroid gland, and the penis and clitoris to the tongue. The fore-limbs and girdle correspond in detail with the hind limbs and the pelvis—a point already worked out by Vicq d'Azyr; the dorsal and ventral halves of the body are likewise comparable in some respects, the sternum, for example, answering in the arrangement of its bones, muscles and arteries to the vertebral column. The skeleton of each member is in some respects a repetition of the vertebral column.

His brother, D. A. Meckel,[2] worked out an elaborate comparison between the alimentary canal and the genital organs, basing the legitimacy of the comparison upon early embryological relations and upon the state of things in Cœlentera, where genital and digestive organs occupy the same cavity. In his view the uterus corresponded to the stomach, the vagina to the œsophagus, the fallopian tubes to the intestine, and so on.

The vertebral theory of the skull took its origin from the same habit of thought. As part of the wider idea of the metameric repetition of parts it had some scientific worth, but the theory was pushed too far, and the facts were twisted to suit it. Among annulate animals the theory of repetition found ample scope; Oken was able to compare with justice

[1] *Beyträge*, ii., 2, 1812. Also in his *System d. vergl. Anat.*, i., 1821.

[2] In J. F. Meckel's *Beyträge*, ii.

the jaws of crabs and insects with their other limbs, as Savigny did later in a more scientific way. Among Vertebrates the application of the theory of serial repetition was not so obvious, except in the case of the vertebræ. Goethe seems to have been the first to hit upon the idea that the skull is composed of a number of vertebræ, serially homologous with those of the vertebral column. He tells us that the idea flashed into his mind when contemplating in the Jewish cemetery at Venice a dried sheep's skull. The discovery was made in 1790, but not published till 1820.[1]

The idea seems to have been taught by Kielmeyer, one of the earliest of the "philosophers of nature," but it was not published by him.

In a book (*Cours d'Études médicales*), published in 1803, Burdin assimilated the skull to the vertebral column.

Oken, in an inaugural dissertation (Programm) *Ueber die Bedeutung der Schädelknochen*,[2] published in 1807, gave to the theory its necessary development. Autenrieth, also in 1807,[3] distinguishing separate ganglia in the brain, was not far from the hypothesis that each of these ganglia must have its separate vertebra.

In 1808 Duméril read a paper to the Académie des Sciences in which he compared the skull to a gigantic vertebra, basing his hypothesis on the similarity existing between the crests and depressions on the hinder part of the skull and those on the posterior surfaces of the vertebræ.

After Oken's work the vertebral theory was taken up generally by both the German and the French anatomists. Spix published in 1815 a large volume on the skull, entitled *Cephalogenesis*, distinguishing (as Oken did at first) three cranial vertebræ. Bojanus in his *Anatome testudinis europææ* (1819), and in a series of papers in *Isis* (1817-1819, and 1821) established the existence of a fourth cranial vertebra, and this was accepted by Oken in the later editions of his *Lehrbuch*. Meckel and Carus among the Germans, de Blainville and E. Geoffroy among the French, contributed to

[1] *Zur Morphologie*, i., 2, p. 250, 1820; and ii., 2, pp. 122-4, 1824.

[2] See translation, giving the gist of this paper, in Huxley's *Lectures on the Elements of Comparative Anatomy*, pp. 282-6, London, 1864.

[3] Reil's *Archiv. f. Physiol.*, vii., 1807.

the development of the theory. In England the theory was championed particularly by Richard Owen.

It was one thing to assert in a moment of inspiration that the skull was composed of modified vertebræ; it was quite another to demonstrate the relation of the separate bones of the skull to the supposed vertebræ. Upon this much uncertainty reigned; there was not even unanimity as to the number of vertebræ to be distinguished. Goethe found six vertebræ in the skull; Spix, and at first Oken, three only, Geoffroy seven; the accepted orthodox number seems to have been four (Bojanus, Oken, Owen).

As an example of the method of treatment adopted we may take Oken's matured account of the composition of the cranial vertebræ, as given in the English translation of his *Lehrbuch*. "To a perfect vertebra," he says, "belong at least five pieces, namely, the body, in front the two ribs, behind the two arches or spinous processes" (p. 370). In the cervical vertebræ the transverse processes represent the ribs. The skull consists of four vertebræ, the occipital, the parietal, the frontal and the nasal, or, named after the sense with which each is associated, the auditory, the lingual, the ocular and the olfactory. The "bodies" of these vertebræ are the body of the occipital (basioccipital), the two bodies of the sphenoid (basi- and pre-sphenoid), and the vomer. The transverse processes of each are the condyles of the occipitals (exoccipitals), the alæ of the two sphenoids (alisphenoids and orbitosphenoids) and the lateral surfaces of the vomer. The arches or spinous processes are the occipital crest, the parietals, the frontals, and the nasals.

The cranium is thus composed of four rings of bone, each composed of the typical elements of a vertebra.

The arbitrary nature of the comparison is obvious enough. As Cuvier pointed out in the posthumous edition of his *Leçons*, it is only the occipital segment that shows any real analogy with a vertebra—an analogy which Cuvier ascribed to similarity of function. He admitted a faint resemblance of the parietal segment to a vertebra:—"The body of the sphenoid does indeed look like a repetition of the basioccipital, but having a different function it takes on another form, especially above, by reason of its posterior clinoid apophyses."[1] He denied the resemblance of the frontal and nasal "vertebræ" to true vertebræ, pointing out that both parietals and frontals are bones specially developed for the purpose of roofing over and protecting the cerebrum.

A very curious development was given to the vertebral theory by K. G. Carus, who seems to have taken as his text a saying of Oken's, that the whole skeleton is only a repeated vertebra.[2] His system is worthy of some consideration, for he tries to work out a geometry of the skeleton.[3]

His method of deduction is a good example of pure *Naturphilosophie.* Life, he says, is the development of something determinate from something indeterminate. A finite indeterminate thing, that is, a liquid, must take a spherical form if it is to exist as an individual. Hence the sphere is the prototype of every organic body. Development takes place by antagonism, by polarity, typically by the division and multiplication of the sphere. In the course of development the sphere may change, by expansion into an egg-shaped body, or by contraction into a crystalline form, the changes due to expansion being typical of living things, those due to contraction being typical of dead. At the surface of the primitive living sphere is developed the protective *dermatoskeleton*, which naturally takes the shape of a hollow sphere; round the digestive cavity which is formed in the living sphere is developed the *splanchnoskeleton;* round the nervous system (which is, as it were, the animal within the animal) is developed the *neuroskeleton.* All skeletal formations belong to one or other of these systems.

Carus defines his aim to be the discovery of the inner law which presides over the formation of the skeleton throughout the animal kingdom; he desires to know "how such and such a formation is realised in virtue of the eternal laws of reason" (iii., p. 93) Here we touch the kernel of *Naturphilosophie*—the search for rational laws which are active in Nature, the discontent with merely empirical laws.

[1] *Leçons d'anatomie comparée*, 3rd ed., Brussèls reprint, i., p. 414, 1836.

[2] In his Programm, *U. d. Bedeut. d. Schädelknochen*, 1807.

[3] *Traité élémentaire d'anatomie comparée* (French trans.), vol. iii., Paris, 1835. First developed in his volume *Von den Ur-Theilen des Knochen und Schalen-Gerustes*, Leipzig, 1828.

The thesis which Carus sustains is that all forms of skeleton, whether of dermatoskeleton, splanchnoskeleton, or neuroskeleton, can be deduced from the hollow sphere, which is the primary form of any skeleton whatsoever (p. 95). That means, put empirically, that every skeleton can be represented schematically by a number of hollow spheres, suitably modified in shape, and suitably arranged. The chief modification in shape exhibited by bones is one which is intermediate between the organic and the crystalline series of modifications of the sphere. The organic modifications are bounded by curved lines, the crystalline by straight; the intermediate partly by curved and partly by straight lines. They are the dicone (the shape of a diabolo) and the cylinder. These forms must necessarily be of importance for the skeleton, which is intermediate between the organic and the inorganic. "The dicone embodies the real significance of the bone," writes Carus. Each dicone and cylinder composing the skeleton is called by Carus a vertebra.

We may expect then all skeletons to be composed of spheres, cylinders and dicones in diverse arrangements. Nature being infinite, all the possible types of arrangement of these elements must exist in the test or skeleton of some animal, living, fossil, or to come (p. 127). One conceives easily what the main types of skeleton must be. In some animals, *e.g.*, sea-urchins, the skeleton is a simple sphere; in others, *e.g.*, starfish, secondary rows of spheres radiate out from a central sphere or ring; in annulate animals the skeleton consists of a row of partially fused spheres.

In Vertebrates the arrangement is more complex. There are first the protovertebral rings of the dermatoskeleton, these being principally the ribs, limb-girdles, and jaws. Round the central nervous system are developed the deutovertebral rings of the neuroskeleton (vertebræ in the ordinary sense). The apophyses and bodies of the vertebræ, and the bones of the members[1] are composed of columns of tritovertebræ, or vertebræ of the third order. Thus the whole vertebrate skeleton is a particular arrangement of vertebræ, which

[1] Dutrochet in 1821 had tried to prove that the bones of the members belong to the type of the vertebra—the dicone.

in their turn are modifications of the primary hollow sphere.

The German transcendentalists were more or less contemporary with E. Geoffroy, and no doubt influenced him, especially in his later years, as they certainly did his follower Serres. Oken indeed wrote, in a note[1] appended to Geoffroy's paper on the vertebral column of insects, that "Mr Geoffroy [*sic*] is without a doubt the first to introduce in France *Naturphilosophie* into comparative anatomy, that is to say, that philosophy one of whose doctrines it is to seek after the *signification* of organs in the scale of organised beings." This is, however, an exaggeration, for Geoffroy was primarily a morphologist, whereas the morphology of the German transcendentalists was only a side-issue of their *Naturphilosophie.*

Geoffroy, on his part, exercised some influence on the transcendentalists. He asserts[2] indeed that Spix got some of the ideas published in the *Cephalogenesis* (1815) from attending his course of lectures in 1809. It is certainly the case that Spix published before Geoffroy the view that the opercular bones are homologous with the ear-ossicles, adopting, however, a different homology for the separate bones.[3]

Some speculations seem to have been common to both schools—for instance, the law of Meckel-Serres, the vertebral theory of the skull, and the recognition of serial homology in the appendages of Arthropods (Savigny, Oken). Latreille and Dugès, as well as Serres, clearly show in their theoretical views the influence of Oken and the other transcendentalists. Geoffroy's principle of connections and law of compensation were recognised by some at least of the Germans.

But whatever his actual historical relations may have been with the German school, Geoffroy was vastly their superior in the matter of pure morphology. He alone brought to clear consciousness the principles on which a pure morphology could be based: the Germans were transcendental philosophers first, and morphologists after.

[1] *Isis*, pp. 552-9, 1820 (2). [2] *Mém. Mus. d'Hist. nat.*, ix., 1822.

[3] Cuvier and Valenciennes, *Hist. nat. Poissons*, i., p. 311, f.n.

One understands from this how J. F. Meckel, who was in some ways the leading comparative anatomist in Germany at this time, could be at once a transcendentalist and an opponent of Geoffroy. Meckel had a curiously eclectic mind. A disciple of Cuvier, having studied in 1804-6 the rich collections at the Museum in Paris, the translator of Cuvier's *Leçons d'anatomie comparée*, he earned for himself the title of the "German Cuvier," partly through the publication of his comprehensive textbook (*System der vergl. Anatomie*, 5 vols.), partly by his extensive and many-sided research work, partly by his authoritative teaching. His *System* shows in almost every page of its theoretical part the influence of Cuvier; and it is through having assimilated Cuvier's teaching as to the importance of function that Meckel combats Geoffroy's law of connections, at least in its rigorous form. He submits that the connections of bones and muscles must change in relation to functional requirements. He rejects Geoffroy's theory of the vertebrate nature of Articulates. Generally throughout his work the functional point of view is well to the fore.

Yet at heart Meckel was a transcendentalist of the German school. His vagaries on the subject of "homologues" leave no doubt about that, and, in spite of Cuvier, he believed, though not very firmly, in the existence of one single type of structure.

A Cuverian by training, his lack of morphological sense threw him into the ranks of the transcendentalists, to whom perhaps he belonged by nature.

CHAPTER VIII

TRANSCENDENTAL ANATOMY IN ENGLAND—RICHARD OWEN

RICHARD OWEN is the epigonos of transcendental morphology; in him its guiding ideas find clear expression, and in his writings are no half-truths struggling for utterance. But he was, though a staunch transcendentalist, an eclectic of

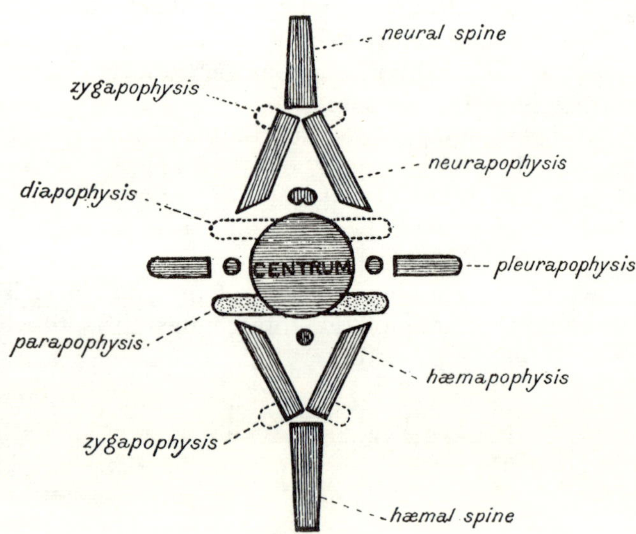

FIG. 4.—Ideal Typical Vertebra. (After Owen.)

the older ideas current in his time; for he picked out what was best in the older systems—Cuvier's teleology, Geoffroy's principle of connections, Oken's idea of the serial repetition of parts. In particular, he assimilated the teaching of Cuvier, the great opponent of the transcendentalists, and reconciled it in part with his own transcendentalism. His main theoretical views are to be found in his volume *On the Archetype and Homologies of the Vertebrate Skeleton* (London, 1848). The master-idea of the book is that the vertebrate skeleton consists of a series of comparable segments, each of

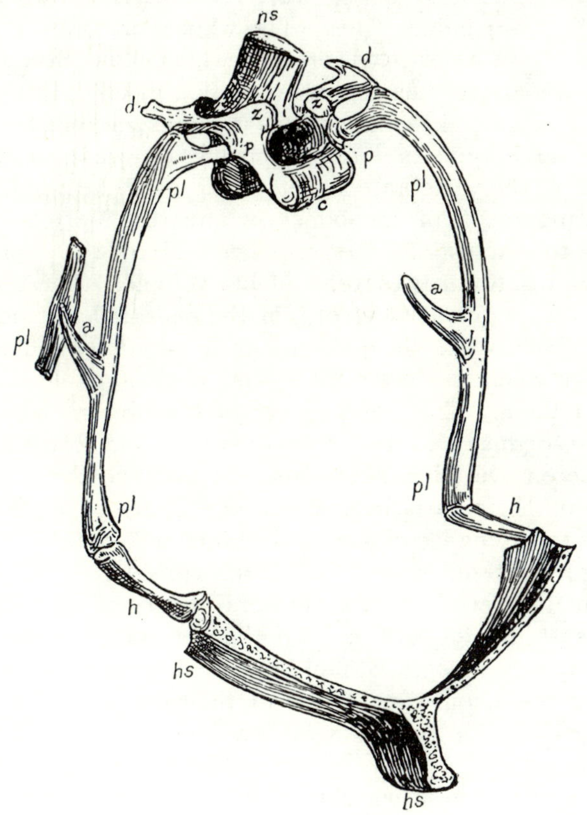

FIG. 5.—Natural Typical Vertebra; Thorax of a Bird. (After Owen.)

which Owen calls a vertebra. His definition of a vertebra is, "one of those segments of the endo-skeleton which constitute the axis of the body, and the protecting canals of the nervous and vascular trunks" (p. 81). The parts of a typical vertebra are shown in Fig. 4, which is copied from Owen's Fig. 14.

In Fig. 5 (page 103) is shown an actual vertebra, as Owen conceives it, the "vertebra" being that of a bird.

A segment of sternum is included as the "hæmal spine" of the vertebra (*hs*); the vertebral rib is the "pleurapophysis" (*pl*); the sternal rib the "hæmapophysis" (*h*); the uncinate process of the vertebral rib is known as the "diverging appendage" (*a*). The whole vertebrate skeleton is composed of a series of vertebræ which show these typical parts. We arrive thus at the conception of an "Archetype" of the vertebrate skeleton, such as is represented in Fig. 6.

The archetype is only a scheme of what is usually constant in the vertebrate skeleton, and both the number and the arrangement of the bones in any real Vertebrate are subject to variation. "It has been abundantly proved," Owen writes, towards the end of his volume, "that the idea of a natural segment (vertebra) of the endoskeleton does not necessarily involve the presence of a particular number of pieces, or even a determinate and unchangeable arrangement of them. The great object of my present labour has been to deduce . . . the relative value and constancy of the different vertebral elements, and to trace the kind and extent of their variations within the limits of a plain and obvious maintenance of a typical character" (p. 146).

It goes without saying that Owen considered the skull to be formed of vertebræ—the vertebral theory of the skull was, in his system, a deduction from the vertebral theory of the skeleton. He recognised four cranial vertebræ; the arrangement of them, and the relation of their constituent bones to the parts of the typical vertebra are shown in the table appearing on page 106. So far as their first three elements are concerned, these vertebræ are practically identical with the vertebræ distinguished in the classical vertebral theory of the skull, as enunciated by Oken. A divergence appears with the determination of the other elements of the vertebræ. The upper and lower jaws are associated with the nasal and frontal vertebræ respectively, not however as limbs of the head, but as constituent elements of these vertebræ. In the same way the hyoid apparatus is part and parcel of the parietal vertebra, and the pectoral girdle and fore-limbs part of the occipital vertebra.

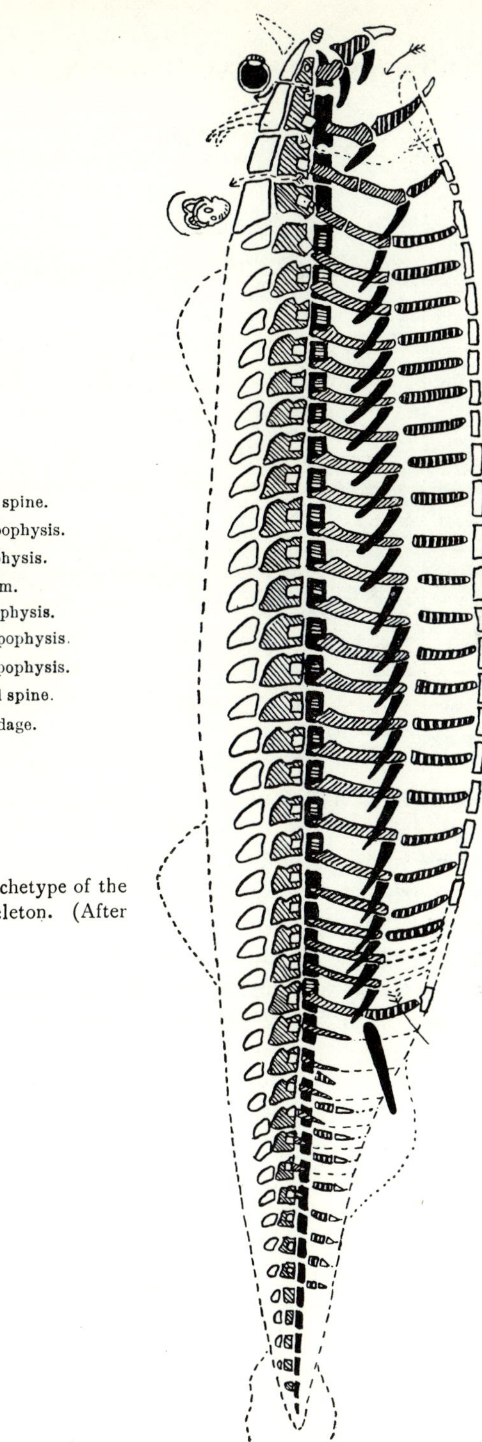

Fig. 6.—The Archetype of the Vertebrate Skeleton. (After Owen.)

Cranial Vertebræ.[1] (After Owen, 1848, p. 165.)

Vertebræ.	Occipital.	Parietal.	Frontal.	Nasal.
Centra.	Basioccipital.	Basisphenoid.	Presphenoid.	Vomer.
Neurapophyses.	Exoccipital.	Alisphenoid.	Orbitosphenoid.	Prefrontal.
Neural Spines.	Supraoccipital.	Parietal.	Frontal.	Nasal.
Parapophyses.	Paroccipital.	Mastoid.	Postfrontal.	None.
Pleurapophyses.	Scapular.	Stylohyal.	Tympanic.	Palatal.
Hæmapophyses.	Coracoid.	Ceratohyal.	Articular.	Maxillary.
Hæmal Spines.	Episternum.	Basihyal.	Dentary.	Premaxillary.
Diverging Appendage.	Fore-limb or Fin.	Branchiostegals.	Operculum.	Pterygoid and Zygoma.

Owen's reasons for considering the pectoral girdle and the fore-limb part of the occipital vertebra are as follows. In fish the pectoral girdle is slung to the skull by means of the post-temporal bone (supra-scapula, according to Owen) which abuts on the occipital arch. In *Lepidosiren*, whose skeleton resembles the archetype in many ways, the pectoral girdle is likewise attached to the occipital segment.

In most other Vertebrates the pectoral girdle has shifted backwards along the vertebral column, by a "metastasis" (Geoffroy) similar to that by which the pelvic fins in many fish have shifted up close to the pectoral girdle. The scapula (with supra-scapula) is the pleurapophysis, the coracoid the hæmapophysis, of the occipital vertebra. The clavicle is homologised with the slender bone in fish now known as the post-clavicle, which shows a connection with the first or atlas vertebra of the vertebral column, forming, according to Owen, the hæmapophysis of the atlas. Owen considers it no objection to this view that in other Vertebrates the

[1] Owen introduced most of the names of bones now current.

clavice is anterior to the coracoid—"its anterior position to the coracoid in the air-breathing Vertebrata is no valid argument against the determination, since in these we have shown that the true scapular arch is displaced backwards" (*On the Nature of Limbs*, p. 63, London, 1849). In the pelvic girdle the ilium corresponds to the scapula, the ischium to the coracoid, the pubis to the clavicle. Hence the ilium is a pleurapophysis, the ischium and pubis are both hæmapophyses. The fore-limb is the developed "appendage" of the occipital vertebra, the hind-limb the developed "appendage" of the pelvic vertebra. They are serially homologous with, for example, the uncinate processes of the ribs in birds (see Figs. 5 and 6). The fore-limb is a simple filament in *Lepidosiren*, and presents few joints in *Proteus* and *Amphiuma*; in other air-breathing Vertebrates it shows a more complete development, the humerus, radius and ulna, and the bones of the wrist and hand becoming differentiated out.

As the fore-limb is equivalent to a single bone of the archetype, it is said to be, in its developed state, "teleologically compound" (p. 103).

Since in the archetype every vertebra has its appendage, more than two pairs of locomotory limbs might have been developed. "Any given appendage might have been the seat of such developments as convert that of the pelvic arch into a locomotive limb; and the true insight into the general homology of limbs leads us to recognise many potential pairs in the typical endoskeleton. The possible and conceivable modifications of the vertebrate archetype are far from having been exhausted in the forms which have hitherto been recognised, from the primæval fishes of the palæozoic ocean of this planet up to the present time" (p. 102). It is not of the essence of the vertebrate type to be tetrapodal.

In determining homologies Owen remained true to Geoffroy's principle of connections. Speaking of an attempt which had been made to determine homologies by the mode of development, he writes, "There exists doubtless a close general resemblance in the mode of development of homologous parts; but this is subject to modification, like the forms, proportions, functions, and very substance of such parts, without their essential homological relationships being thereby obliterated. These relationships are mainly, if not wholly, determined by the relative position and connection of the parts, and may exist independently of form, proportions, substance, function and similarity of development. But the connections must be sought for at every period of development, and the changes of relative position, if any, during growth, must be compared with the connections which the part presents in the classes where vegetative repetition is greatest and adaptive modification least" (p. 6). It is interesting to note that in Owen's opinion comparative anatomy explains embryology. Thus the scapula, which is the pleurapophysis of the occipital vertebra, is vertical on its first appearance in the embryo of tetrapoda, and lies close up to the head (*On the Nature of Limbs*, p. 49)—the embryo shows a greater resemblance to the archetype than the adult. "We perceive a return to it, as it were, in the early phases of development of the highest organised of the actually existing species, or we ought rather to say that development starts from the old point; and thus, in regard to the scapula, we can explain the constancy of its first appearance close to the head, whether in the human embryo or in that of the swan, also its vertical position to the axis of the spinal column, by its general homology as the rib or 'pleurapophysis' of the occipital vertebra" (*Limbs*, p. 56).

We owe to Owen the first clear distinction between "homologous" and "analogous" organs; it was he who first proposed the terms "homologue" and "analogue," which he defined as follows:—"*Analogue.* A part or organ in one animal which has the same function as another part or organ in a different animal." "*Homologue.* The same organ in different animals under every variety of form and function."[1]

He introduced also useful distinctions between Special, General, and Serial Homology. "The relations of homology," he writes, "are of three kinds: the first is that above defined, viz., the correspondency of a part or organ, determined by its relative position and connections, with a part or organ in a different animal; the determination of which homology indicates that such animals are constructed on a common

[1] *Lectures on Invertebrate Animals*, pp. 374, 379, 1843.

type; when, for example, the correspondence of the basilar process of the human occipital bone with the distinct bone called 'basi-occipital' in a fish or crocodile is shown, the *special homology* of that process is determined. A higher relation of homology is that in which a part or series of parts stands to the fundamental or general type, and its enunciation involves and implies a knowledge of the type on which a natural group of animals, the Vertebrate, for example, is constructed. Thus when the basilar process of the human occipital bone is determined to be the 'centrum' or 'body' of the last cranial vertebra, its *general homology* is enunciated.

"If it be admitted that the general type of the vertebrate endoskeleton is rightly represented by the idea of a series of essentially similar segments succeeding each other longitudinally from one end of the body to the other, such segments being for the most part composed of pieces similar in number and arrangement, and though sometimes extremely modified for special functions, yet never so as to wholly mask their typical character—then any given part of one segment may be repeated in the rest of the series, just as one bone may be reproduced in the skeletons of different species, and this kind of repetition or representative relation in the segments of the same skeleton I call 'serial homology'" (p. 7). As an example of serial homology we might take the centra of the vertebræ—the vomer, the presphenoid, the basisphenoid, the basioccipital and the series of centra in the spinal column. Such serially repeated parts are called *homotypes* (p. 8).

Not all the bones of the vertebrate skeleton are included in the archetype as constituents of the vertebræ. Thus the branchial and pharyngeal arches are accounted part of the splanchnoskeleton, as belonging to the same category as the heart bone of some ruminants, and the ossicles of the stomach in the lobster (p. 70). The ossicles of the ear in mammals are "peculiar mammalian productions in relation to the exalted functions of a special organ of sense" (p. 140, f.n.). This recognition of a possible development of new organs to meet new functions shows unmistakably the influence of Cuvier. Owen was indeed well aware of the importance of the functional aspect of living things, and he often adopted the teleological point of view. As a true morphologist, however, he held that the principle of adaptation does not suffice to explain the existence of special homologies. The ossification of the bones of the skull from separate centres may be purposive in Eutheria, in that it prevents injury to the skull at birth; but how explain on teleological principles the similar ossification from separate centres in marsupials, birds and reptiles? How explain above all the fact that the centres are the same in number and relative position in all these groups? Surely we must accept the idea of an archetype "on which it has pleased the divine Architect to build up certain of his diversified living works" (p. 73).

In his study of centres of ossification, Owen made in point of theory a distinct advance on his predecessors. We saw that Geoffroy recognised the importance of studying the ossification of the skeleton, and that Cuvier accepted such embryological evidence as an aid in determining homologies. Owen pointed out that it was necessary to distinguish between centres of ossification which were teleological in import and such as were purely indicative of homological relationships. Many bones, single in the adult, arise from separate centres of ossification, but we must distinguish between "those centres of ossification that have homological relations, and those that have only teleological ones; *i.e.*, between the separate points of ossification of a human bone which typify vertebral elements, often permanently distinct bones in the lower animals; and the separate points which, without such signification, facilitate the progress of osteogeny, and have for their obvious final cause the well-being of the growing animal" (p. 105). There is, for example, a teleological reason why in mammals and leaping Amphibia (*e.g.*, frogs), the long bones should ossify first at their ends, for the brain is thus protected from concussion; in reptiles that creep there is less danger of concussion, and the long bones ossify in the middle (p. 105). But there is no teleological reason why the coracoid process of the scapula should in all mammals develop from a separate centre. The coracoid is however a real vertebral element (hæmapophysis), and in monotremes, birds and reptiles it is in the adult a large and separate bone. Its ossification from a separate centre in mammals has therefore a homological significance. The scapula in mammals is an example of what Owen calls a "homologically compound" bone. All those bones which are formed by a coalescence of parts answering to distinct elements of the typical vertebra are "homologically compound" (p. 105). On the other hand, "All those bones which represent single vertebral elements are 'teleologically compound' when developed from more than one centre, whether such centres subsequently coalesce, or remain distinct, or even become the subject of individual adaptive modifications, with special joints, muscles, etc., for particular offices" (p. 106). The limb-skeleton, corresponding as it does to a single bone of the archetype, is the typical example of a teleologically compound bone. Owen in his definition of teleological compoundness has combined two kinds of adaptation—(1) temporary adaptation of bones to the exigencies of development, birth and growth (*e.g.*, development of long bones from separate centres); (2) definitive adaptation of a skeletal part to the functions which it has to perform (*e.g.*, teleological structure of limbs). Such adaptations are, so to speak, grafted on the archetype.

Owen's general views on the nature of living things merit some attention. Organic forms, according to Owen, result from the antagonistic working of two principles, of which one brings about a vegetative repetition of structure, while the other, a teleological principle, shapes the living thing to its functions. The former principle is illustrated in the archetype of the vertebrate skeleton, in the segmentation of the Articulates, in the almost mathematical symmetry of Echinoderms, and the actually crystalline spicules of sponges. It is the same principle which causes repetition of the forms of crystals in the inorganic world. "The repetition of similar segments in a vertebral column, and of similar elements in a vertebral segment, is analogous to the repetition of similar crystals as the result of polarising force in the growth of an inorganic body" (p. 171). This "general polarising force" it is which mainly produces the similarity of forms, the repetition of parts, and generally the signs of the unity of organisation. The adaptive or "special organising force" or ἰδέα, on the other hand, produces the diversity of organic beings. In every species these two forces are at work, and the extent to which the general polarising or "vegetative-repetition-force" is subdued by the teleological is an index of the grade of the species.

This view is analogous to the Geoffroyan conception that the diversity of form is limited by the unity of plan. Owen thus ranges himself with Geoffroy against Cuvier, who considered that diversity of form is limited only by the principle of the adaptation of parts.

CHAPTER IX

KARL ERNST VON BAER

VON BAER was recognised as the founder of embryology even by his contemporaries. His predecessors, Aristotle,[1] Fabricius,[2] Harvey,[3] Malpighi,[4] Haller,[5] Wolff,[6] had made a beginning with the study of development; von Baer, by the thoroughness of his observation and the strength of his analysis, made embryology a science.

It was to one of the German transcendentalists that von Baer owed the impulse to study development. Ignatius Döllinger, Professor in Würzburg, induced three of his pupils, Pander, d'Alton and von Baer, to devote themselves to embryological research. The development of animals was at this time little known, in spite of recent work by Meckel (1815 and 1817), Tiedemann (*Anatomie u. Bildungsgeschichte des Gehirns*, 1816), by Oken (*loc. cit.*, *supra*, p. 90), and some others.

Pander, with whom apparently Döllinger and d'Alton collaborated, was the first to publish his results;[7] von Baer, who through absence from Würzburg had for a time dropped his embryological studies, started to work in 1819, after the publication of Pander's treatise, and produced in 1828 the first volume of his master-work, *Ueber Entwickelungsgeschichte*

[1] *De generatione Animalium.*

[2] *De formato fœtu*, ? 1600; *De formatione fœtus*, 1604.

[3] *Exercitationes de generatione animalium*, 1651.

[4] *De formatione pulli in ovo*, 1673; *De ovo incubato*, 1686.

[5] *De formatione pulli in ovo*, 1757-8; *Sur la formation du cœur dans le poulet*, 1758.

[6] *Theoria generationis*, 1759; *De formatione intestinorum*, 1768-9.

[7] *Beiträge zur Entwickelung des Hühnchens im Ei.* Würzburg, 1818. Also in Latin in shorter form, 1817.

der Thiere. Beobachtung und Reflexion (Königsberg, 1828). The second volume followed in 1837, but dates really from 1834, and was published in an incomplete form. This second volume is intended as an introduction to embryology for the use of doctors and science students. In it von Baer describes in full detail the development of many vertebrate types—chick, tortoise, snake, lizard, frog, fish, several mammals and man, basing his remarks largely upon his personal observations, but taking account also of all contemporary work. A separate account of the development of a fish (*Cyprinus blicca*) appeared in 1835.[1]

We shall concentrate attention on the first volume. This volume contains the first full and adequate account of the development of the chick, followed by a masterly discussion of the laws of development in general.

When we consider that von Baer worked chiefly with a simple microscope and dissecting needles, the minuteness and accuracy of his observations are astonishing. He described the main facts respecting the development of all the principal organs, and if, through lack of the proper means of observation, he erred in detail, he made up for it by his masterly understanding and profound analysis of the essential nature of development. His account of the development of the chick is a model of what a scientific memoir ought to be; the series of "Scholia" which follow contain the deductions he made from the data, and, in so far as they are direct generalisations from experience, they are valid for all time.

The first Scholion is directed against the theory of preformation, and succeeds in refuting it on the ground of simple observation. The theme of the second Scholion is that the essential nature (*die Wesenheit*) of the animal determines its differentiation, that no stage of development is solely determined by the antecedent stage, but that throughout all stages the *Wesenheit* or idea of the definitive whole exercises guidançe. This guidance is shown most clearly in the regulatory processes of the germ, whereby the large individual variations commonly presented by the

[1] *Untersuchungen ü. die Entwickelungsgeschichte der Fische*; Leipzig, 1835.

early embryo are compensated for or neutralised in the course of further development. Baer in this shows himself a vitalist.

It is, however, the third and subsequent Scholia which must here particularly occupy our attention, for it is in these that von Baer comes to grips with morphological problems. Already in the second Scholion he had definitely enunciated the law which runs as a theme throughout the volume, the observational and the theoretical part alike, the law that development is essentially a process of differentiation by which the germ becomes ever more and more individualised. "The essential result of development," he writes, "when we consider it as a whole, is the increasing independence (*Selbständigkeit*) of the developing animal" (p. 148). In the third Scholion he elaborates this thought and shows that differentiation takes place in triple wise. The three processes of differentiation are "primary differentiation" or layer-formation, "histological differentiation" within the layers, and the "morphological differentiation" of primitive organs.

The first of these differentiations in time is the formation of the germ-layers, which takes place by a splitting or separation of the blastoderm into a series of superimposed lamellæ. Baer's account of the process in the chick is as follows:—

"First of all, the germ separates out into heterogeneous layers, which with advancing development acquire ever greater individuality, but even on their first appearance show rudiments of the structures which will characterise them later. Thus in the germ of the bird, so soon as it acquires consistency at the beginning of incubation, we can distinguish an upper smooth continuous surface and a lower more granular surface. The blastoderm separates thereupon into two distinct layers, of which the lower develops into the plastic body-parts of the embryo, the upper into the animal parts; the lower shows clearly a further division into two closely connected subsidiary layers—the mucous layer and the vessel-layer; the original upper layer also shows a division into two, which form respectively the skin and the parts which I have called the true ventral and dorsal plates—parts which contain in an undifferentiated state the skeletal and muscular systems, the connective tissues, and the nerves belonging to these. In order to have a convenient term for future use, I have named this layer the muscle-layer" (p. 153).

The process of delamination results then in the formation of four layers, of which the upper two (composing the "animal" or "serous" layer) will give origin to the animal (neuromuscular) part of the body, the lower pair to the plastic or vegetative organs. The uppermost layer will form the external covering of the embryo, and also the amniotic folds; from it there differentiates out at a very early stage the rudiment of the central nervous system, forming a more or less independent layer. Below the outermost layer lies the layer from which are formed the muscular and skeletal systems, and beneath this "muscle-layer" comes the "vessel-layer," which gives origin to the main blood-vessels. The innermost layer of the four will form the mucous membrane of the alimentary canal and its dependencies; at the present stage, however, it is, like the other layers, a flat plate.

From all these layers tubes are developed by the simple bending round of their edges. The outermost layer becomes the investing skin-tube of the embryo; the layer for the nervous system forms the tubular rudiment of the brain and spinal cord; the mucous layer curls round to form the alimentary tube; the muscle layer grows upwards and downwards to form the fleshy and osseous tube of the body wall; even the vessel layer forms a tube investing the alimentary canal, but a part of it goes to form the medial "Gekröse," or mesenterial complex, which departs considerably from the tubular form.

When these tubes or "fundamental organs" are formed the process of primary differentiation is complete. The fundamental organs, however, have at no time actually the form of tubes; they exist as tubes only ideally, for morphological and histological differentiation go on concurrently with the process of primary differentiation.

Through morphological differentiation the various parts of the fundamental organs become specialised, through unequal growth, first into the primitive organs and then into the functional organs of the body. "Single sections of the tubes originally formed from the layers develop individual forms, which later acquire special functions: these functions are in the most general way subordinate elements of the function of the whole tube, but yet differ from the functions of other sections. Thus the nerve-tube differentiates into sense-organs, brain and spinal cord, the alimentary tube into mouth cavity, œsophagus, stomach, intestine, respiratory apparatus, liver, bladder, etc. This specialisation in development is bound up with increased or diminished growth" (p. 155). Rapid growth concentrated at one point brings about an evagination; in this manner are formed the sense-organs from the nerve-tube, the liver and lungs from the alimentary tube. Or increased growth over a section of a tube causes it to swell out; in this wise the brain develops from the nerve-tube, the stomach from the alimentary tube. The segmentation which soon becomes so marked, particularly in the muscle layer, is also due to a process of morphological differentiation.

At the same time that the organs of the body are being thus roughly blocked out and moulded from the germ-layers the third process of differentiation is actively going on. "In addition to the differentiation of the layers, there follows later another differentiation in the substance of the layers, whereby cartilage, muscle and nerve separate out, a part also of the mass becoming fluid and entering the blood-stream" (p. 154). Through histological differentiation the texture of the layers and incipient organs becomes individualised. In its earliest appearance the germ consists of an almost homogeneous mass, containing clear or dark globules suspended in its substance (ii., p. 92). This homogeneity gives place to heterogeneity; the structureless mass becomes fibrous to form muscles, hardens to form cartilage or bone, becomes liquid to form the blood, differentiates in a hundred other ways—into absorbing and secreting tissues, into nerves and ganglia, and so forth. It will be noticed that the concept of histological differentiation is independent of the cell-theory; it signifies that textural differentiation which leads to the formation of tissues in

Bichat's sense. The tissues and the germ-layers stand in fairly close relation with one another, for while certain tissues are formed chiefly but not exclusively in one layer, others are formed only in one layer and never elsewhere. For example, peripheral nerves are for the most part formed in the muscle layer, though the bulk of the nervous tissue is formed in the walls of the nerve tube; similarly blood and blood-vessels may arise from almost any layer, though their chief seat of origin is the vessel-layer; on the other hand, bone is formed only in the muscle-layer (i., p. 155, ii., pp. 92-3).

This relation of tissue to germ-layer was more fully discussed and brought into greater prominence by Remak, from the standpoint of the cell-theory, and it will occupy us in a later chapter (Chap. XII.).

The fourth Scholion elaborates the analysis of developmental processes still further, and discusses in particular the scheme of development which is shown by the Vertebrata. The characteristic structure of the vertebrate body is brought about by a "double symmetrical" rolling together of the germ-layers, whereby two main tubes are formed, one above and one below the axis of the body, which is the chorda. The dorsal tube is formed by the two animal layers, the ventral tube by all the layers combined (see Fig. 7).

The process is indicated with sufficient clearness in the diagram. It will be seen that the real foundation and framework of the arrangement is the muscle-layer, with its two tubes, one surrounding the central nervous system and forming the "dorsal plates," the other surrounding the body cavity and forming the "ventral plates." In the dorsal plates, which early show metameric segmentation, the investing skeleton of the neural axis develops; in the ventral plates are formed the ribs, the ventral arches of the vertebræ, the hyoid, the lower jaw and other skeletal structures.

The alimentary or "mucous" tube and the part of the vessel layer which invests it become so closely bound up with one another as to form a single primitive organ—the alimentary canal. The muscles of the alimentary canal are accordingly in all probability developed in the investing part of the vessel layer. From the "Gekröse," or remaining part of the vessel layer develop the Wolffian bodies (*Urnieren*,

Pronephros), the kidneys, the sex glands, and the series of "blood-glands"—suprarenals, thyroid, thymus and spleen. Baer did not attach any special morphological significance to the peritoneal lining of the body cavity, as is done in more modern forms of the germ-layer theory. The gill-slits were largely formed by outgrowths from the alimentary canal.

In his germ-layer theory von Baer was influenced a good deal by Pander, to whom the actual discovery of the process

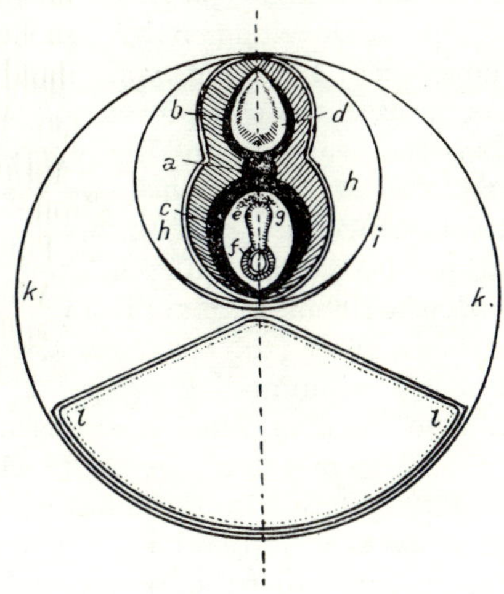

FIG. 7.—Ideal Transverse Section of a Vertebrate Embryo. (After von Baer.)

a. Chorda.
b. Dorsal plates.
c. Ventral plates.
d. Spinal cord.
e. Vessel-layer.
f. Alimentary tube.
g. Pronephros.
h. Skin.
i. Amnion.
k. Serous membrane.
l. Yolk-sac.

of layer-formation is due. Pander, however, had distinguished only three germ-layers, an upper "serous" layer, a lower "mucous" layer and a middle "vessel-layer." He it was who introduced the terms "Keimhaut" (blastoderm) and "Keimblatt" (germ-layer).

The honour of being the founder of the germ-layer theory is sometimes attributed to C. F. Wolff, notably by Kölliker and O. Hertwig. Wolff, it is true, in his memoir *De formatione intestinorum* (1768-9) showed that the alimentary canal was first formed as a flat plate which folded round to form a tube, and in a somewhat vaguely worded passage he hinted that a similar mode of origin might be found to hold good for the other organ-systems. But it seems clear that Wolff had no definite conception of the process of layer-formation as the first and necessary step in all differentiation. This, at any rate, was von Baer's opinion, who assigns to Pander the glory of the discovery of the germ-layers. "You," he writes, "through your clearer recognition of the splitting of the germ—a process which remained dark to Wolff—have shed a light upon all forms of development" (p. xxi.).

We have now seen, following von Baer's exposition, how development is essentially a process of differentiation, a progress from the general to the special, from the homogeneous to the heterogeneous; we have analysed the process into its three subordinate processes—primary, histological and morphological differentiation. So far we have considered development in general and the laws which govern it; we have now to consider the varieties of development which the animal kingdom offers in such profusion, in order to discover what relations exist between them. This is the problem set in the fifth Scholion. Baer at once brings us face to face with the solution of the problem attempted in the Meckel-Serres law. It is a generally received opinion, he writes, that the higher animals repeat in their development the adult stages of the lower, and this is held to be the essential law governing the relation of the variety of development to the variety of adult form. This opinion arose when there was little real knowledge of embryology; it threw light indeed upon certain cases of monstrous development, but it was pushed altogether too far. It complicated itself with a belief in a historical evolution; —"People gradually learnt to think of the different animal forms as developed one from another—and seemed, in some circles at least, determined to forget that this metamorphosis could only be conceptual" (p. 200). At the same time the theory of parallelism led men to rehabilitate the outworn conception of the scale of beings, to maintain that animals form one single series of increasing complexity, a scale which the higher members must mount step by step in their development—from which it followed that evolution, whether conceived as an ideal or as an historical process, could take place only along one line, could be only progressive or regressive. Not all the supporters of the theory of parallelism held these extreme views, but conclusions of this kind were natural and logical enough.

Von Baer had soon found in the course of his embryological studies that the facts did not at all fit in with the doctrine of parallelism ; the developing chick, for example, was at a very early stage demonstrably a Vertebrate, and did not recapitulate in its early stages the organisation of a polyp, a worm or a mollusc. He had published his doubts in 1823, but his final confutation of the theory of parallelism is found in this Scholion.

If it were true, he says, that the essential thing in the development of an animal is this repetition of lower organisations, then certain deductions could be drawn, which one would expect to find confirmed in Nature. The first deduction would be that no structures should appear in the embryo of the higher animals that are not found in the lower animals. But this is not confirmed by the facts—no adult among the lower animals, for instance, has a yolk-sac like that of the chick embryo. Again, if the law of parallelism were true, the mammalian embryo would have to repeat the organisation of, among other groups, insects and birds. But the embryo *in utero* is surrounded by fluid and cannot possibly breathe free air, so it cannot possibly repeat the structure of either insects or birds, which are pre-eminently air-organisms. Generally speaking, indeed, we find in all the higher embryos special structures which adapt them to the very special conditions of their development, and these we never find as permanent structures in the lower animals. The supporters of the theory of parallelism might, however, admit the existence of such special embryonic organs without greatly prejudicing their case, for these temporary organs stand to some extent outside the scope of the theory.

But they would have to face a second and more important deduction from their views, namely, that the higher animals should repeat at every stage of their development the whole organisation of some lower animal, and not merely agree with them in isolated details of structure. The deduction is, however, not borne out by the facts. The embryo of a mammal resembles in many points, at different stages of its development, the adult state of a fish; it has gill-slits and complete aortic arches, a two-chambered heart, and so on. But at no time does it combine all the essential characters of a fish; nor has it ever the tail of a fish, nor the fins, nor the shape. Any recapitulation there may be is a recapitulation of single organs, there is never a repetition of the complete organisation of a fish. This is indeed the fundamental criticism of the theory of parallelism; and if it applies even within the limits of the vertebrate phylum, so much the more does it apply to comparisons between embryonic Vertebrates and adult Invertebrates.

There are also some lesser arguments which might be urged against the theory of parallelism. If the theory were strictly true, no state which is permanent in a higher animal could be passed through by an animal lower in the scale. But birds, which are lower in the scale than mammals, pass through a stage in which they resemble mammals in certain respects much more than they do when adult, for in an embryonic condition they agree with mammals in having no feathers, no air sacs, no pneumatic sacs in the bones, no beak. Their brain also resembles that of mammals more in an earlier stage than it does later. So, too, myriapods and hydrachnids have at birth three pairs of feet, and resemble at this stage adult insects, which form a higher class.

Again, were the analogy between the development of the individual and the evolution of the *Échelle des êtres* complete, organs and organ-systems ought to develop in the individual in the order in which they appear in the scale of beings. But this is not always the case. In fish the hinder extremity develops only its terminal joint, while in the embryos of higher animals the basal joint is the first to appear.

Another consequence one would expect to find realised, were the theory of parallelism correct, is the late appearance in development of parts which are confined to the higher animals. In the development of a Vertebrate accordingly one would not expect the vertebræ to appear before the embryo had passed through many Invertebrate stages. But experience shows the direct contrary, for in the chick the rudiments of the vertebral axis appear sooner than any other part.

The theory of parallelism or recapitulation then is not borne out by the facts, and clearly cannot be the law which we are seeking. But what then is the true relation between the variety of development and the variety of adult structure? Before answering this question we must review the varied forms of adult organisation and consider in what relations they stand to one another. In particular we must enquire whether they belong to one type or to many. One point is here cardinal—we must distinguish between the *type* of organisation and the *grade* of differentiation. By "type" von Baer means the structural plan of the organism. "I call the *type* the spatial relationship of the organic elements and organs" (p. 208). Each type of organisation characterises one of the big groups of animals; the lesser groups represent "grade" modifications of the type. "The product of the degree of differentiation and the type gives the several great groups of animals which are called classes" (p. 208). *Ausbildung* (differentiation) takes place in one or other of several directions, in adaptation, for instance, to life in the water or to life in the air.

There are, von Baer considers, four main types—(1) the peripheral or radiate type, (2) the longitudinal type, (3) the massive or molluscan type, (4) the vertebrate type. The radiate type is shown by discoid infusoria, by medusæ, by starfish and their allies. The longitudinal type characterises such genera as *Vibrio*, *Filaria*, *Gordius*, and all the annulate animals. Mollusca, rotifers, polyzoa, and such infusoria as are not included in types (1) and (2) belong to the massive type, in which the body and its parts form rounded masses. The longitudinal type is predominantly "animal," the massive type predominantly "plastic" (vegetative). The vertebrate type has both the "animal" and the "plastic" organs highly developed. In the symmetrical arrangement of the animal parts it resembles the longitudinal type; its plastic parts with their asymmetrical arrangement and rounded shape belong to the massive type.

These types of von Baer inevitably recall the "Embranchements" of Cuvier, with which they more or less coincide. It seems that von Baer arrived at his types (from the study of adult structure) independently of Cuvier, though the priority of publication rests with Cuvier.[1]

Now it is clear that the development of the individual, which is essentially an *Ausbildung*, a differentiation, is directly comparable with the grade-differentiation of forms within the type. And just as the type rules all its varied modifications, so does the development of the individual take place always within the bounds imposed by type. This is von Baer's chief contribution to the theory of embryonic relationships — the law that "the type of organisation determines the manner of development" (p. xxii.). Development is not merely from the general to the special—there are at least four distinct "general" types, from which the special is developed. The type is fixed in the very earliest stages of development—the embryo of a Vertebrate is from the very beginning a Vertebrate (p. 220), and it shows at no time any agreement in total organisation with any Invertebrate. The types are independent of one another; differentiation and development follow a different course in each of them. Not but what some analogies can be found between the very earliest stages of embryos of different type. Thus vertebrate and annulate embryos agree in certain points at the time of the formation of the primitive streak. And in the earliest stage of all, the egg-stage, there is probably agreement between all the types. In eggs with yolk, whether vertebrate or annulate, there is always a separation into an animal and a plastic layer. It seems, too, as if a hollow sphere were a constant stage in the development of all animals (pp. 224, 258). Apart from these analogies, development takes an entirely independent course in each of the four main types, and no embryo of one of the higher types repeats in its development the peculiar organisation of any adult of the lower types.

[1] Cuvier, in 1812, *Ann. Mus. d'Hist. Nat.*, xix.; von Baer in 1816, *Nova Acta Acad. Nat. Cur.* See *Entwickelungsgeschichte der Thiere*, i., p. vii., f.n.

If we consider now development within the type, which is the only legitimate thing to do, we arrive at certain laws governing the relation of embryos to one another. For instance, at a certain stage vertebrate embryos are uncommonly alike. Von Baer had two in spirit which he was unable to assign to their class among amniotes; they might have been lizard, bird, or mammal, he could not say definitely which.[1] Generally the farther back we go in the development of Vertebrates the more alike we find the embryos. The type-characters are first to appear, then the class characters, then the characters distinguishing the lesser classificatory groups. "From a more general type the special gradually emerges" (p. 221). The chick is first a Vertebrate, then a land-vertebrate, then a bird, then a land-bird, then a gallinaceous bird, and finally *Gallus domesticus.* Development within the type is a progress from the general to the special, a real evolution. The more divergent two adults are, the farther back we must go in their development to find an agreement between their embryos. We can sum up the case in the following laws:—

"(1) *That the general characters of the big group to which the embryo belongs appear in development earlier than the special characters.* In agreement with this is the fact that the vesicular form is the most general form of all; for what is common in a greater degree to all animals than the opposition of an internal and an external surface?

"(2) *The less general structural relations are formed after the more general, and so on until the most special appear.*

"(3) *The embryo of any given form, instead of passing through the state of other definite forms, on the contrary separates itself from them.*

"(4) *Fundamentally the embryo of a higher animal form*

[1] Compare a parallel passage in Prévost et Dumas:—"At the very first sight one will be struck with the resemblance between the forms of the very early embryos of these two classes, a resemblance so extraordinary that one cannot refuse to admit the conclusions resulting from it. The resemblance is so striking that one can defy the most experienced observer to distinguish in any way the embryos of dog or rabbit . . . from those of fowls or ducks of a corresponding age."—*Ann. Sci. nat.*, iii., p. 132, 1824.

never resembles the adult of another animal form, but only its embryo" (p. 224).

These laws relating to development within the limits of type are destructive of even a limited application of the theory of parallelism, for not even within the limits of the type is there a real scale which the higher forms must mount; each embryo develops for itself, and diverges sooner or later from the embryos of other species, the divergence coming earlier the greater the difference between the adult forms. It is only because the lower less-differentiated adult forms happen to be little divergent from the generalised or embryonic type, that they show a certain similarity with the embryos of the higher more differentiated members of the group. Such similarity, however, is due to no necessary law governing the development of the higher animals; it is, on the contrary, merely a consequence of the organisation of these lower animals (p. 224).

Von Baer goes on to show what are the distinguishing embryological characters of the types and classes, working out a dichotomous schema of development, which each embryo must follow, branching off early or late to its terminal point, according to the lower or higher goal it has to reach.

One important consequence for morphology results from von Baer's laws of differentiation within the type. If the embryo develops from the general to the special, then the state in which each organ or organ-system first appears must represent the general or typical state of that organ within the group. Embryology will therefore be of great assistance to comparative anatomy, whose chief aim it is to discover the generalised type, the common plan of structure, upon which the animals of each big group are built. And the surest way to determine the true homologies of parts will be to study their early development. "For since each organ becomes what it is only through the manner of its development, its true value can be recognised only from its method of formation. At present, we form our judgments by an undefined intuition, instead of regarding each organ merely as an isolated product of its fundamental organ, and discerning from this standpoint the correspondences and dissimilarities in the different types" (p. 233). Parts, therefore, which develop from the same "fundamental organ," and in the last resort from the same germ-layer, have a certain kinship, which may even reach the degree of exact homology.

Now since the mode of development in each type is peculiar to that type, organs of the same name in different types must not necessarily be accounted homologous, even if they correspond exactly with one another in their general *functional* relations to the rest of the organs. Thus the central nervous system of Arthropods must not be homologised with the central nervous system of Vertebrates, for it develops in a different manner. So, too, the brain of Arthropods or of Mollusca is not strictly comparable with the brain of Vertebrates. Again, the air-tubes or tracheæ of insects are, like the trachea and bronchi of many Vertebrates, air-breathing organs. But the two organs are not homologous, for the air-tubes of Vertebrates are developed from the alimentary tube ("fundamental organ" of the alimentary system, developed from the vegetative layer), while the air-tubes of insects arise either by histological differentiation, or by invagination of the skin (p. 236). Organs can be homologous only within the limits of the big groups; there can be no question of homology between members of different types.

The development of plants, like the development of animals, is essentially a progress from the general to the special (p. 242). Botanists have not been troubled by any recapitulation theory, and in founding their big groups, Acotyledons, Monocotyledons, and Dicotyledons, upon embryological characters, they were guided by true principles, which ought indeed to be followed in zoology. If we knew the development of all kinds of animals sufficiently well, then the best way to classify them would be according to the characters they show in their early development, for it is in early development that they show the characters of the type in their most generalised form. As it is, we have in our ignorance to establish the big groups by the study of adult structure, but we find, on putting together all we know of comparative embryology, that a classification of animals according to the mode of their development gives, as is only natural, the same four groups as does the study of adult structure. The four types of development are thus:—

(1) The double-symmetrical, which is found in Vertebrates. It is called the double-symmetrical, because in Vertebrates development takes place from a central axis (notochord) in two directions, upwards and downwards, in such a way that two tubes are formed, one above and one below the axis. (2) The second type is the symmetrical, which is shown by Annulates. A primitive streak is formed on the ventral surface of the yolk; development proceeds symmetrically on both sides of the streak. (3) Radiate development is probably typical of the radiate structural type. (4) In the massive type, the development seems to be a spiral one.

Common to most modes is a separation of the germ into animal and plastic layers, a separation which seems to be conditioned largely by the presence of yolk. A classification based upon embryological characters ought to be applied even to the lesser groups and would here prove itself of service. Embryology, for instance, fully supports de Blainville's separation of Batrachia from true reptiles,[1] for reptiles develop an amnion and Batrachia do not.

We come now to the sixth and last Scholion. Development is a true evolution of the special from the general, so runs von Baer's most general law of all. This can be expressed in a slightly different way, and the words which he chooses in the sixth Scholion to express this final and most general result are these:—"The developmental history of the individual is the history of the growing individuality in every respect" (p. 263). The greatest modern treatise on embryology ends on a splendid note. One creative thought rules all the forms of life. And more—"It is this same thought that in cosmic space gathered the scattered masses into spheres and bound them together in the solar system, the same that from the weathered dust on the surface of the metallic planets brought forth the forms of life. And this thought is nought else but life itself, and the words and syllables in which life expresses itself are the varied forms of the living" (p. 264).

Von Baer reminds one greatly of Cuvier. There is

[1] *De l'organisation des Animaux*, i., p. 140, 1822.

the same sheer intellectual power, the same sanity of mind, the same synthetic grip. Von Baer, like Cuvier, never forgot that he was working with living things; he was saturated, like Cuvier, with the sense of their functional adaptedness. In his paper on the external and internal skeleton[1] he gives a masterly analysis of the functional modifications of the limbs in Vertebrates, and the whole paper indeed, with its sober attack on transcendentalism, is a vindication as much of the functional point of view as of the importance of embryology.

Both Cuvier and von Baer, by the very sanity of their views, found themselves in partial opposition to the theories current in their time. Cuvier was the critic of Geoffroy and the transcendentalists, of Lamarck and the believers in the *Échelle des êtres*, evolutionary or ideal. Von Baer also, though influenced greatly by *Naturphilosophie*, turned against the exaggerations of the transcendental school, and by his unanswerable criticism of the theory of parallelism took away the ground from those who too easily believed in an historical evolution.[2]

We have seen what were von Baer's criticisms of the theory of parallelism. If we turn to the later writings of Cuvier we find the essential criticism expressed in similar terms. Speaking of an attempt which had been made to show that fish were molluscs developed to a higher degree, he wrote in 1828,[3] "Let us draw the conclusion that even if these animals can be spoken of as ennobled molluscs, as molluscs raised to a higher power, or if they are embryos of reptiles, the beginnings of reptiles, this can be true of them only in an abstract and metaphysical sense, and that even this abstract statement would be very far from giving an accurate idea of their organisation." From the fact that the respiratory and circulatory organs of fish greatly resemble those of tadpoles the conclusion has been drawn that fish are

[1] "Ueber das äussere und innere Skelet," Meckel's *Archiv für Anat. u. Physiol.*, pp. 327-76, 1826. See, too, his *Entwickelungsgeschichte*, i., pp. 181, ff.

[2] Von Baer wrote an appreciative biography of Cuvier, published posthumously in 1897, *Lebensgeschichte Cuviers*, ed. L. Stieda. French trans. in *Ann. Sci. Nat.* (*Zool.*), ix., 1907.

[3] Cuvier et Valenciennes, *Histoire naturelle des Poissons*, i., p. 550.

in a sense embryos of Amphibia (p. 547). But this manner of viewing things is none the less vicious, "for this reason . . . that it considers only one or two points and neglects all the others" (p. 548), and is directly contrary to common sense. There is never a recapitulation of total organisations, only at the most of single organs.

It will be remembered that Cuvier opposed and demolished the theory of the *Échelle des êtres*, not only by showing that there were in Nature four entirely different plans of animal structure, but also by demonstrating that even the animals of each single *Embranchement* could not readily be arranged in one series, that a serial arrangement was really valid only for their separate organs. Von Baer also held that there are four distinct types of structure; he, too, combated the idea of gradation within the limits of the type. In so far as species represent successive stages in the development, the *Ausbildung*, of the type, so far can the idea of a scale of beings be applied. But the members of a type follow not one line of evolution but several diverging lines, in direct adaptation to different environmental conditions, so that a serial arrangement of them is not as a rule possible. It may be possible to establish a serial arrangement of single organs from the simplest to the most complex. But each organ or organ-system will require a different serial arrangement, for the different systems vary on different lines and an animal may be highly developed in respect of one system and little developed in respect of all the others. Man, for instance, is the highest animal only in respect of his nervous system. The idea of the scale of beings has therefore only a very limited application even within the limits of the type. Applied to the whole animal kingdom it becomes merely absurd.

Another point of resemblance between Cuvier and von Baer was that Cuvier, though essentially a student of adult structure, did recognise the importance of embryology; following up some observations of Dutrochet he studied the fœtal membrane of mammals and tried to establish their homologies.[1] And in his criticism of the vertebral theory of the skull he advanced as an argument against the basi-

[1] *Mém. Mus. d'Hist. Nat.*, iii., pp. 98-119, 1817.

sphenoid being a vertebral centrum the fact (established by Kerkring, 1670), that it develops from two centres.[1]

Von Baer's relation to transcendental anatomy was in some ways a close one, though he was a trenchant critic of the extreme views of the school.[2] He took from Oken the idea that a simple fundamental plan rules the organisation of all Vertebrates; "That jaws and limbs are modifications of one fundamental form is readily apparent, and, after Oken, the fact ought to be accepted by the majority of those naturalists who do not refuse to admit the existence of a general type from which the diversity of structure is developed" (i., p. 192). He accepted the vertebral theory of the skull in its main lines, and used his embryological knowledge to support the idea that jaws correspond to limbs—the latter point as part of the transcendental idea that the hind end of the body repeats the organisation of the anterior part (i., p. 192). The particular form which his theory of the relation of jaws to limbs took is shown in the following passage:—"The maxillary bone has . . . the significance of an extremity and at the same time that of a rib or lower arch of a vertebra, just as the pelvic bones unite in themselves the signification of ribs and proximal members of the hinder extremity" (Meckel's *Archiv*, p. 367, 1826).

He appreciated the morphological idea of the serial repetition of parts, and gave it accurate formulation. The whole vertebrate body, he considered, was composed of a longitudinal series of *morphological elements*, each of which was made up a section from each of the fundamental organs—a vertebra, a section of the nerve-cord, and so on (*Entwickelungsgeschichte*, ii., p. 53). Groups of these morphological elements formed *morphological divisions*, such as the vertebral segments of the head with their highly developed neural arches, or the segments of the neck with their undeveloped hæmal arches. The morphological elements are clearly shown only in the animal parts, but there are indications in the embryo of a segmentation also of the vegetative parts,—the gill-slits, for instance, and the vascular arches.

[1] *Leçons d'Anatomie comparée*, 3rd ed., vol. i., p. 414, Bruxelles, 1836.

[2] In the aforementioned paper in Müller's *Archiv* he criticises Carus vigorously and is sarcastic on Geoffroy.

The vegetative parts, however, develop on the whole unsymmetrically (*cf.* Bichat). These elements which von Baer distinguishes are morphological units, as he himself points out, contrasting them with organs which are not usually units in a morphological sense. " We call organ," he writes, " each part that has by reason of its form or its function a certain distinctiveness, but this concept is very indefinite, and possesses, from a morphological point of view, little value. For this reason it seems necessary to introduce into scientific morphology the concepts of morphological elements and divisions " (ii., p. 84).

Von Baer exercised a very considerable influence upon the subsequent trend of morphological theory. By his criticism of the Meckel-Serres theory, he rid morphology for a time of an idea which was leading it astray ; by his substitution of the law that development is always from the general to the special, he set morphologists looking for the archetype in the embryo, not in the adult alone, and made them realise that homologies could often best be sought in the earliest stages of development ; by formulating the germ-layer theory he supplied morphologists with a new criterion of homology, based upon the special relations of the parts (germ-layers) which are first differentiated in all development. He made the study of development an essential part of morphology.

CHAPTER X

THE EMBRYOLOGICAL CRITERION

PANDER'S work of 1817 was the forerunner of an embryological period in which men's hopes and interest centred round the study of development. "With bewilderment we saw ourselves transported to the strange soil of a new world," wrote Pander, and many shared his hopeful enthusiasm. K. E. von Baer's *Entwickelungsgeschichte* was by far the greatest product of this time, but it stands in a measure apart; we have in this chapter to consider the lesser men who were Baer's contemporaries, friends, followers or critics.

It was largely a German science, this new embryology, and its leaders were all personally acquainted. Pander, von Baer and Rathke were on friendly terms with one another; von Baer dedicated his master-work to Pander; Rathke dedicated the second volume of his *Abhandlungen* to von Baer. Interest in the new science was, however, not confined to Germany. In Italy, Rusconi commenced in 1817 his pioneer researches on the development of the Amphibia with a *Descrizione anatomica degli organi della circolazione delle larve delle Salamandre aquatiche* (Pavia), in which he traced the metamorphoses of the aortic arches. This was followed in 1822 by his *Amours des Salamandres aquatiques* (Milan), and in 1826 by his memoir *Du développement de la grenouille* (Milan). In this last paper he described how the dark upper hemisphere of the frog's egg grows down over the lower white hemisphere and leaves free only the yolk plug; he observed the segmentation cavity and the archenteron, but thought that the former became the alimentary canal; he observed and interpreted rightly the formation of the medullary folds. The circular blastopore in the frog in later years often went by the name of the anus of Rusconi.

In France Dutrochet[1] investigated the fœtal membranes in various vertebrate classes; Prévost and Dumas studied the very earliest stages of development in birds, mammals and amphibia (*Ann. Sci. nat.*, ii., iii., 1824, xii., 1827).

A little later came Dugès' studies of the osteology and myology of developing amphibia (1834),[2] and Coste's careful researches into the early developmental history of mammals.[3]

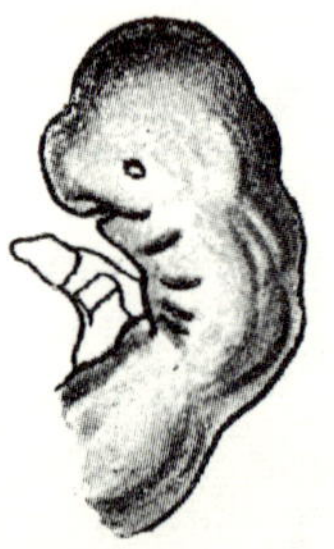

FIG. 8.—Gill-slits of the Pig Embryo. (After Rathke.)

It was in 1825 that Heinrich Rathke (1793-1860), published his famous discovery of gill-slits in the embryo of a mammal,[4] a discovery which aroused considerable interest, and greatly stimulated embryological research. He describes how in a young embryo of a pig he saw four slits in the region of the neck, going right through into the œsophagus. They were separated by partitions which he called *Kiemenbogen* (gill-arches), and immediately in front of the first gill-slit lay the developing lower jaw. He compared these gill-slits with those of a dogfish. We reproduce his drawing of the pig-embryo (*Isis*, Pl. IV., fig. 1).

Later in the same year Rathke discovered gill-slits in the chick,[5] in this case finding only three. He described growing out from in front of the first slit a structure which he compared to the operculum or gill-cover of a fish.

These discoveries were confirmed and extended for the

[1] See review by Cuvier, *Mém. Mus. Hist. nat.*, iii., pp. 82-97, 1817.

[2] *Mém. Savans étrangers*, vi. Extract in *Ann. Sci. nat.* (2) i. (*Zool.*), pp. 366-72, 1834.

[3] *Recherches sur la génération des Mammifères*, 1834. *Embryogénie comparée*, 1837.

[4] "Kiemen bey Säugthieren," *Isis*, pp. 747-9, 1825.

[5] "Kiemen bey Vögeln," *Isis*, pp. 1100-1, 1825.

chick [1] by the embryologist Huschke, a pupil of Oken. Like Rathke, he found only three indubitable gill-slits, but he noticed that the body-wall in front of the first gill-slit was really composed of two arches, which were on the whole similar to the gill-arches. The hinder of these two seemed to him to be a horn of the hyoid, the front one, which was bent at an angle, to be the rudiment of the upper and lower jaws (p. 401). Between these two arches he found an opening, just as between two gill-arches a gill-slit. This opening led into the mouth-cavity, and according to Huschke it became the external ear-passage. He discovered also three pairs of aortic arches in close relation with the gill-arches, so close indeed, that he did not hesitate to call them gill-arteries, and to recognise their resemblance with the aortic arches of fish. He traced, in part at least, the metamorphosis which these aortic arches undergo. This part of his discovery he developed in fuller detail in a paper of 1828,[2] in which he gave some excellent figures.

Shortly after Huschke's first paper, von Baer published his views and observations on this subject in a short memoir in Meckel's *Archiv*.[3] In this paper he confirmed Rathke's discovery, and described the slits and arches in the dog and the chick. Both Rathke and he found gill-slits in the human embryo about this time (p. 557). There were generally present, he found, four gill-slits, and, as Rathke had suggested, the first gill-arch became the lower jaw. Von Baer also confirmed Rathke's discovery of the operculum, assigning it, however, to the second gill-arch. He refused to accept Huschke's derivation of the auditory meatus from the first gill-slit. Von Baer saw what had escaped Rathke and Huschke, that there were, not three nor four, but as many as five aortic arches.

[1] "Ueber die Kiemenbogen und Kiemengefässe beym bebrüteten Hühnchen," *Isis*, xx., pp. 401-3, 1827. (Read in Sept. 1826 to the *Versammlung der deutschen Naturforscher und Aerzte*, then recently founded by Oken).

[2] *Isis*, pp. 160-4, Pl. II., 1828.

[3] "Ueber die Kiemen und Kiemengefässe in den Embryonen der Wirbelthiere," Meckel's *Archiv* for 1827, pp. 556-68. Also in *Ann. Sci. nat.*, xv., pp. 266-80, 280-4, 1828.

In his view of the metamorphosis of the aortic arches in the chick the first two pairs disappeared completely, the third pair gave rise to the arteries of the head and the fore-limbs, the right side of the fourth arch became the aorta, the left half of the fourth and the right half of the fifth arch became the pulmonary arteries, while the left half of the fifth arch disappeared. This schema, which for the last three arches was the same as Huschke's, von Baer upheld for the chick even in the second volume of his *Entwickelungsgeschichte* (p. 116); he rectified it, however, for mammals in the same volume (p. 212), deriving both pulmonary arteries from the fifth arch, and the aorta from the fourth left. He fully recognised the great analogy of the embryonic arrangement of gill-arches and gill-arteries in Tetrapoda with their arrangement in fish (i., pp. 53, 73).

Huschke, in a paper of 1832,[1] chiefly devoted to the development of the eye, figured and described the developing upper and lower jaws, and maintained against von Baer that the first slit turns into the auditory meatus and the Eustachian tube.

These were the first papers of the embryological period. Before going on to discuss the principles which guided embryological research during the next ten or twenty years it is convenient to note what were the main lines of work characterising the period.

The typical figure of the period is Rathke, who produced a great deal of first-class embryological work. He was, even more than von Baer, a comparative embryologist, and there were few groups of animals that he did not study. His first large publication, the *Beiträge zur Geschichte der Thierwelt* (i.-iv., Halle, 1820-27), contained much anatomical work in addition to the purely embryological; he commenced here his series of papers on the development of the genital and urinary organs, continued in the *Abhandlungen zur Bildungs- und Entwickelungs-Geschichte des Menschen und der Thiere* (i., ii., Leipzig, 1832-3). A fellow-worker in this line was Johannes Müller, whose *Bildungsgeschichte der Genitalien* (Düsseldorf) appeared in 1830.

In a memoir on the development of the crayfish which

[1] Meckel's *Archiv*, vi., pp. 1-47, 1832.

appeared in 1829,[1] Rathke found in an Invertebrate confirmation of the germ-layer theory propounded by Pander and von Baer. He was greatly struck by the inverted position of the embryo with respect to the yolk. In following out the development of the appendages he noticed how much alike were jaws and legs in their earliest stage, and how this supported Savigny's contention that the limbs of Arthropods belonged to one single type of structure. In his paper (1832) on the development of the fresh-water Isopod, *Asellus*,[2] Rathke returns to this point. Commenting on the original similarity in development of antennæ, jaws and legs, he writes, "Whatever the doubts one may have reserved as to the intimate relation existing between the jaws and feet of articulate animals after the researches of Savigny on this subject and mine on developing crayfish, they must all fall to the ground when one examines with care the development of the fresh-water Asellus" (p. 147 of French translation).

Further comparative work by Rathke is found in the two volumes of *Abhandlungen* and in a book, *Zur Morphologie, Reisebemerkungen aus Taurien* (1837), which contains embryological studies of many different types, including a study of the uniform plan of arthropod limbs. Later on Rathke devoted himself more to vertebrate embryology, producing among other works his classical papers on the development of the adder (1839), of the tortoise (1848), and of the crocodile (1866). He laid the foundations of all subsequent knowledge of the development of the blood-vascular system in a series of papers of various dates from 1838 to 1856. The diagrams in his paper on the aortic arches of reptiles (1856) were for long copied in every text-book.

Rathke was a foremost worker in another important line of embryological work, the study of the development of the skeleton and particularly of the skull. We shall discuss the

[1] *Untersuchungen über die Bildung und Entwickelung der Fluss-Krebses*, Leipzig, folio, 1829. Preliminary notice in *Isis*, pp. 1093-1100, 1825.

[2] "Untersuchungen über die Bildung und Entwickelung der Wasser-Assel.," *Abh. z. Bild. u. Entwick.-Gesch.*, i., pp. 1-20, 1832. Translated in *Ann. Sci. nat.* (2), ii., (*Zool.*), pp. 139-57, 1834.

history of the embryological study of the skull in some detail below; meantime, we note the two other important lines of research which characterise this period. One is the intensive study of the development of the human embryo, a study pursued by, among others, Pockels, Seiler, Breschet, Velpeau, Bischoff, Weber, Müller, and Wharton Jones.[1] The other important line — the early development of the Mammalia—was worked chiefly by Valentin,[2] Coste,[3] and, above all, by Bischoff, whose series of papers[4] was justly recognised as classical.

What interests us chiefly in the work of this embryological period is, of course, the relation of embryology to comparative anatomy and to pure morphology. The embryologists were not slow to see that their work threw much light upon questions of homology, and upon the problem of the unity of plan. Von Baer, we have seen, recognised this clearly in 1828; Rathke, in one of his most brilliant papers, the *Anatomisch-philosophische Untersuchungen über den Kiemenapparat und das Zungenbein* (Riga and Dorpat, 1832), used the facts of development with great effect to show the homology of the gill-arches and hyoid throughout the vertebrate series; Johannes Müller made great use of embryology in his classical *Vergleichende Anatomie der Myxinoiden* (i. Theil, 1836), and, according to his pupil Reichert, firmly held the opinion that embryology was the final court of appeal in disputed points of comparative anatomy;[5] Reichert himself in a book of 1838 (*Vergleichende*

[1] Kölliker, *Entwickelungsgeschichte*, 2nd ed., p. 17, Leipzig, 1879.

[2] *Handbuch der Entwickelungsgeschichte des Menschen und . . . der Säugethiere und Vögel*, Berlin, 1835.

[3] *Embryogénie comparée*, 1837; *Histoire générale du développement des corps organisés*, 1847-49.

[4] *Entwickelungsgeschichte des Kaninchen-Eies*, Braunschweig, 1842; *Entwickelungsgeschichte des Hunde-Eies*, Braunschweig, 1845; *Entwickelungsgeschichte des Meerschweinchens*, Giessen, 1852; *Entwickelungsgeschichte des Rehes*, Giessen, 1854.

[5] "It is the rôle of embryology, as my great teacher says, to form the court of appeal for comparative anatomy, and it is from embryology particularly, which has in the last decades provided such signal instances of the unravelling of obscure problems, that we have to expect a definite clearing up of the problems relating to the development of the head."—Müller's *Archiv*, p. 121, 1837.

Entwickelungsgeschichte des Kopfes der nackten Amphibien) discussed the two different methods of arriving at the "Type"—the anatomical method of comparing adults, and the embryological method of comparing embryogenies. Of the embryological method, he says, "Its aim is to distinguish during the formation of the organism the originally given, the essence of the type, and to classify and interpret what is added or altered in the further course of development. Embryologists watch the gradual building up of the organism from its foundations, and distinguish the fundament, the primordial form, the type, from the individual developments; they reach thus, following Nature in a certain measure, the essential structure of the organism, and demonstrate the laws that manifest themselves during embryogeny" (p. vi.). The embryologists, influenced in this greatly by von Baer, gradually felt their way to substituting for the "Archetype" of pure morphology what one may perhaps best call the *embryological archetype*. How the transition was made we can best see by following out the course of discovery in one particular line. We choose for this purpose the development of the skull, a subject which excited much interest at this time and upon which much quite fundamental work was done, particularly by Rathke and Reichert.

Following up his discovery of gill-slits and arches in the embryos of birds and mammals, Rathke in two papers of 1832[1] and 1833[2] worked out the detailed homologies of the gill-arches in the higher Vertebrates. He describes how in the embryo of the Blenny there is a short, thick arch between the first gill-slit and the mouth. A furrow appears down the middle of the arch dividing it incompletely into two. In the anterior halves a cartilaginous rod is developed which is connected with the skull; these rods become on either side the lower jaw and "quadrate." In the posterior halves two similar rods are formed which develop into the hyoid. The hyoid is at first connected with the skull,

[1] *Anat.-phil. Unters. ü. d. Kiemenapparat u. d. Zungenbein*, Riga and Dorpat, 1832.

[2] "Bildungs- und Entwickelungs-geschichte des Blennius viviparus," *Abhandl. z. Bild. u. Entwick.-Gesch. des Menschen u. der Thiere*, ii., pp. 1-68, Leipzig, 1833.

but afterwards frees itself and becomes slung to the "quadrate." From the hinder edge of the hyoid arch grows out the membranous operculum, in which develop later the opercular bones and branchiostegal rays. The upper jaw is an independent outgrowth of the serous layer.

The serial homology of the lower jaw and quadrate with the hyoid and with the true gill-arches was thus established in fish, and Rathke had little difficulty in demonstrating a similar origin of lower jaw and hyoid in the embryos of higher Vertebrates. He could even, as we have noted before, find the homologue of the operculum in a flap which grows out from the hyoid arch in the embryo of birds.

But Rathke could not altogether shake himself free from the transcendental notion of the homology of jaws with ribs, and this led him to draw a certain distinction between the first two and the remaining gill-arches, by which the homology of the former with the ribs was asserted and the homology of the latter denied. He thought he could show that the skeletal structures (lower jaw, "quadrate," and hyoid) of the first two arches were formed in the serous layer, just like true ribs, and like them in close connection with the vertebral skeletal axis. The other, "true," gill-arches appeared to him to be formed in the mucous layer, in the lining of the alimentary canal. They had no direct connection with the vertebral column, and seemed therefore to belong to what Carus[1] had called the visceral or splanchno-skeleton. He did not, however, let this distinction hinder him from asserting the substantial homology of all the gill-arches *inter se*, the first two included.

Rathke's discoveries relative to the development of the jaws, the hyoid and the operculum, enabled him to make short work of the homologies proposed for them by the transcendentalists. He could prove from embryology that the jaws were not the equivalent of limbs, as so many Okenians believed. He could reject, with a mere reference to the facts of development, Geoffroy's comparison of the hyoid and the branchiostegal rays in fish with sternum and ribs. He could show the emptiness of the attempts made

[1] *Von den Ur-Theilen des Knochen- und Schalen-Gerustes*, Leipzig, 1828.

by Carus, Treviranus, de Blainville and Geoffroy, to establish by anatomical comparison the homologies of the opercular bones, for he could show that these bones were peculiar to fish, and were scarcely indicated, and that only temporarily, in the development of other Vertebrates.[1] He did not, however, himself realise the relation of the ear-ossicles to the gill-arches, though he knew that Spix and Geoffroy were quite wrong in homologising them with the opercular bones in fish. He described, it is true, the development of the external meatus of the ear and the Eustachian tube from the slit which appears between the first and the second arch, as Huschke had done before him; he described, in confirmation of Meckel, the "Meckelian process" of the hammer running down inside the lower jaw; but the discovery of the true homologies of the ear-ossicles was not made until a year or two later by Reichert.

In his further study of the development of *Blennius viviparus*, Rathke observed some important facts about the development of the vertebral column and skull. He found that the vertebral centra were first formed as rings in the chorda-sheath, which give off neural and hæmal processes. The vertebra later ossifies from four centres. The chorda (notochord) is prolonged some little way into the head, and the base of the cranium is formed by the expanded sheath, which reaches forward in front of the end of the notochord. This cranial basis shows a division into three segments, in which Rathke was inclined to see an indication of three cranial vertebræ. (It turned out that this division into three segments did not really exist, and Rathke later acknowledged that he had made an error of observation.) The side walls of the skull grow out from this base and form a fibrous capsule for the brain. The cranial section of the chorda itself shows no sign of segmentation; but later on the cranial portion of the chorda-sheath ossifies, like the vertebræ, from several centres. The vomer, which, in the classical form of the vertebral theory of the skull, was the centrum of the fourth, or foremost, cranial vertebra, does not, according to Rathke, develop in continuity with the cranial basis and the chorda sheath, but develops separately in the facial region.

[1] *Kiemenapparat*, pp. 107-118.

Von Baer, like Rathke at this time, was also to some extent a believer in the vertebral theory of the skull. In his second volume (1834, pub. 1837) he holds that the development of the skull, as the sum of the anterior vertebral arches, is in general the same as that of the other neural arches, and is modified only by the great bulk of the brain (*Entwickelungsgeschichte*, ii., p. 99). He had, however, some doubts as to the entire correctness of the vertebral theory, doubts suggested by a study of the developing skull. "In the course of the formation of the head in the higher animals, something additional is introduced which does not originally belong to the cranial vertebræ. At first we see the vertebration in the hinder region of the skull very clearly. Afterwards it becomes suddenly indistinct, as if some new formation overlaid it" (i., p. 194).

Even more clearly is his doubt expressed in his paper on *Cyprinus*. "Upon the formation of the vertebral column only this need be said, that at this stage the notochord is very clearly seen, and the upper and lower arches and spinous processes are visible right to the end of the tail, but the separation into vertebræ ceases abruptly where the back passes into the head. I do not hesitate to assert *that bony fish, too, have at this stage an unsegmented cartilaginous cranium* (as cartilaginous fish have all their life), the prominences and hollows of which constitute its only resemblance with the vertebral type" (1835, p. 19).

A convinced supporter of the vertebral theory was Johannes Müller, who, in his classical memoir on the Myxinoids,[1] discussed at some length the relation between the development of the vertebræ and the development of the skull. His memoir is principally devoted to comparative anatomy, but in treating of the skeleton he pays much attention to development. He describes the formation of the vertebræ in elasmobranch embryos; for the facts regarding other Vertebrates he relies largely on work by Rathke (*Blennius*, 1833) and Dugès (1834). He recognises as the basis of his comparisons the homology of the notochord

[1] *Vergleichende Anatomie der Myxinoiden.* Part I. (Osteology and Myology). (*Abh. königl. Akad. Wiss. Berlin*, for 1834, pp. 65-340, 9 pls., 1836.) Also separately.

in all vertebrate embryos with the persistent notochord which forms the chief part or the whole of the vertebral column in the Cyclostomes. The notochord possesses an inner and an outer sheath and the outer sheath is continuous with the *basis cranii* (p. 92). It is in the outer sheath that the vertebræ develop—from four separate pieces, in fish at least, plus an additional element which helps to form the centrum. The skull of Vertebrates consists, according to Müller, of three vertebræ, whose centra are the basioccipital, the basisphenoid and the presphenoid. Other bones besides those belonging to the vertebræ are present, but this formation out of three vertebræ gives the essential schema for the skull. Now the brain capsule, like the sheath of the spinal cord, is a development from the outer sheath of the notochord. If the skull consists of vertebræ we should expect the centra of the skull-vertebræ to develop in the outer sheath at the sides of the cranial section of the notochord as two separate halves, just as do the bodies of the vertebræ; we should expect further the cartilaginous side-walls of the cranium to develop in the membranous brain-sheath just as the neural arches develop in the membranous sheath of the spinal column. In Rathke's discovery (!) of a segmentation of the *basis cranii* into three parts, and of the isolated formation of the vomer, Müller sees a confirmation of his view that the skull is composed of three and not four vertebræ. But there is nothing in Rathke's observations to support the idea that the centra of the cranial vertebræ are formed from separate halves. Müller has to be content with a reference to the state of things in *Ammocoetes* (which, by the way, he did not know to be the young of *Petromyzon*). In the simple skull of *Ammocoetes* the base is formed chiefly by two cartilaginous bars lying more or less parallel with the longitudinal axis of the skull and embracing with their hinder ends the cranial portion of the notochord.

These bars, declares Müller, are clearly the still separate halves of the *pars basilaris cranii*, and represent the divided centra of the two hinder cranial vertebræ. To complete the parallel between the development of the skull and of the vertebræ, it would have been necessary to show that the side walls of the cranium developed in a similar manner from separate pieces. Müller could not prove this point from the available embryological data, and indeed the facts which he did use had to be twisted to suit his theory. A curious apparent confirmation of his idea that the centra of the cranial vertebræ are formed from separate halves was supplied in 1839 by Rathke's discovery of the trabeculæ in the embryonic skull of the adder.

The next big step in the study of the development of the skull was taken by a pupil of Müller, C. B. Reichert, who showed in his work very distinct traces of his master's influence. Reichert's first and most important contribution to the subject was his paper on the metamorphosis of the gill, or, as he called them, the visceral arches in Vertebrates,[1] particularly in the two higher classes. Reichert describes the similar origin in embryo of bird and mammal (pig) of three "visceral" arches. These arches stand in close relation to the three cranial vertebræ which Reichert, like Müller, distinguishes. He makes the retrograde step of admitting only three aortic arches, and he is not inclined to consider the three visceral arches as equivalent to the gill-arches of fish—in his opinion they have more analogy with ribs, though differing somewhat from ribs in their later modifications. The visceral arches are processes of the visceral plates (von Baer), which grow downwards and meet in the middle line, leaving between one another and the undivided body wall three visceral slits opening into the pharynx. The first visceral process is different in shape from the others, for it sends forward, parallel with the head and at right angles to its downward portion, an upper portion in which later the upper jaw is formed. The other two processes are straight. From the hinder edge of the second visceral arch there develops, as Rathke had seen, a fold which is comparable with the operculum of fish. The first slit develops externally into the ear-passage, internally into the Eustachian tube, and in the middle a partition forms the tympanic ring and tympanum. Inside each of the visceral processes on either side a cartilaginous rod develops. In

[1] "Ueber die Visceralbogen der Wirbelthiere in Allgemeinen und deren Metamorphosen bei den Vögeln und Säugethiere," Müller's *Archiv*, pp. 120-222, 1837.

the first process this rod shows three segments, of which the first lies inside that portion of the process which is parallel with the head. This upper segment forms the foundation for the bones of the upper jaw. The lowest segment of the cartilaginous rod becomes Meckel's cartilage, and on the outer side of this the bones of the lower jaw are formed. The middle segment becomes in mammals the incus (one of the ear-ossicles), and in birds the quadrate. Meckel's cartilage, which was discovered by Meckel[1] in fish, amphibians and birds, is a long strip of cartilage which runs from the ear-ossicle known as the hammer in mammals,[2] to the inside of the mandible. Reichert shows how this relation comes

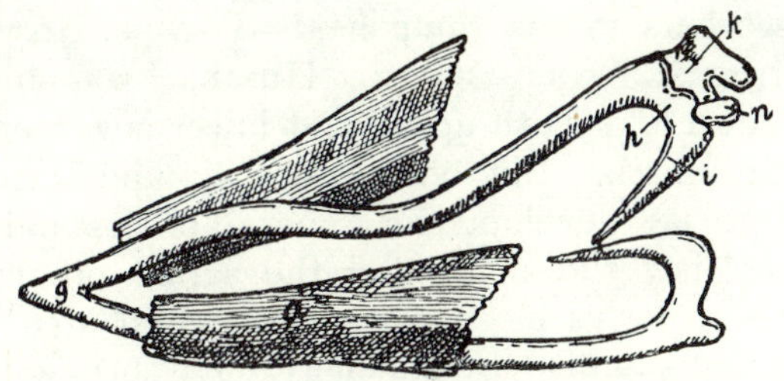

FIG. 9.—Meckel's Cartilage and Ear-ossicles in Embryo of Pig. (After Reichert.)

a. Mandible. *g*. Meckel's cartilage. *h*. Hammer. *i*. Handle of Hammer. *k*. Incus. *n*. Stapes.

about. The hammer, according to his observations on the embryo of the pig, is simply the proximal end of Meckel's cartilage, which later becomes separated off from the long distal portion (see Fig. 9). The third ear-ossicle of mammals, the stapes, comes not from the first arch but from the second. The cartilaginous rod of the second arch segments like the first into three pieces. Of these the uppermost disappears, the middle one, which lies close up to the labyrinth of the ear, becomes the stapes, and the lowest becomes the anterior

[1] *Handbuch d. menschl. Anatomie*, iv., p. 47.

[2] This was shown by Serres (*Ann. Sci. nat.*, xi., p. 54 f.n., 1827), who found in a human embryo a long cartilaginous piece extending from the ear-ossicles to the inside of the lower jaw, and suggested that it was the foundation of the permanent mandible.

horn of the hyoid. The stapes forms a close connection with the hammer and the incus. In birds, where there is a single ear-ossicle, the columella, the middle piece of arch I forms, as we have seen, the quadrate, by means of which the lower jaw is joined to the skull. The proximal end of Meckel's cartilage, which in mammals forms the hammer, here gives the articular surface between the lower jaw and the quadrate. The columella is formed from the middle piece of the three into which the cartilage of the second arch segments. It is, therefore, the homologue of the stapes in mammals. The third arch takes a varying share, together with the second, in the formation of the hyoid apparatus.

In this paper Reichert made a distinct advance on the previous workers in the same field—Rathke, Huschke, von Baer, Martin St Ange, Dugès. Huschke was indeed the first to suggest that both upper and lower jaws were formed in the first gill-arch. But both von Baer and Rathke[1] held that the upper jaw developed as a special process independent of the lower jaw rudiment, and the actual proof that the upper jaw is a derivative of the first visceral arch seems to have been first supplied by Reichert. His brilliant work on the development of the ear-ossicles founded what we may justly call the classical theory of their homologies. His views were attacked and in some points rectified, but the main homologies he established are even now accepted by many, perhaps the majority of morphologists.

In a paper of 1838 on the comparative embryology of the skull in Amphibia,[2] Reichert added to his results for mammals and birds an account of the fate of the first and second visceral arches in Anura and Urodela.

The first visceral arch, he found, gave in Amphibia practically the same structures as in the higher Vertebrates. Its skeleton segmented, as in mammals and birds, into three parts; the upper part gave rise to the palatine and pterygoid in Anura, but seemed to disappear in Urodeles, where the so-called palatine and pterygoid developed in the mucous membrane of the mouth; the middle part gave, as in birds,

[1] *Abhandl.*, i., p. 102, 1832; ii., p. 25, 1833. (*Blennius* paper).

[2] *Vergleichende Entwickelungsgeschichte des Kopfes der nackten Amphibien*, Königsberg, quarto, 276 pp., 1838.

the quadrate, which formed a suspensorium for both arches; the lower part, as Meckel's cartilage, formed a foundation for the bones of the lower jaw. Of arch II., the lower part became the horn of the hyoid, the upper part had a varying fate. In some Anura it formed the ossicle of the ear (homologue of the columella of birds and the stapes of mammals), in others it disappeared. In reptiles the upper segment of the second arch formed, as in birds, the columella.

The account of the metamorphoses of the visceral arches in Amphibia forms only a small part of Reichert's memoir of 1838, the chief object of which was to discover the general "typus" of the vertebrate skull, and to follow out its modifications in the different classes. Von Baer had shown that the generalised type appeared most clearly in the early embryo; Reichert therefore sought the archetype of the skull in the developing embryo. He brought to his task the preconceived notion that the skull could be reduced to an assemblage of vertebræ, but he saw that comparative anatomy alone could not effect this reduction; he had recourse, therefore, to embryology, hoping to find in the simplified structure of the embryo clear indications of three primitive cranial vertebræ (p. 121, 1837).

In the head he distinguished two tubes, the upper formed by the dorsal plates, the lower by the ventral or visceral plates. Both of these tubes were derived from the serous or animal layer (*cf.* von Baer, *supra*, p. 118). The walls of the lower tube were formed by the visceral processes, within which later the skeleton of the visceral arches developed. The walls of the upper tube formed the bones and muscles of the cranium proper. The facial part of the head was formed by elements from both upper and lower tubes. The dorsal tube showed signs of a division into three cranial vertebræ (*Urwirbeln*, primitive vertebræ). In mammals and birds, as Reichert had shown in his 1837 paper, the three cranial vertebræ were indicated by transverse furrows on the ventral surface of the still membranous skull (see Fig. 10, p. 148)

Even in mammals and birds, however, the positions of the eye, the ear-labyrinth, and the three visceral arches were the safest guides to the delimitation of the cranial vertebræ (pp. 134-138, 1837). In Amphibia generally there were no definite lines of separation on the skull itself. "At this stage," he writes of the cartilaginous cranium of the frog, "we find no trace of a veritable division into vertebræ in the cartilaginous trough formed by the *basis cranii* and the side parts. On the contrary, it is quite continuous, as it is also in the higher Vertebrates during the process of chondrification" (p. 44, 1838). The vertebræ in the membranous or cartilaginous skull could be delimited in Amphibia by the help of the eye and the ear-labyrinth, which lie more or less between the first and second, and the second and third vertebræ, but, above all, by the vesicles of the brain.

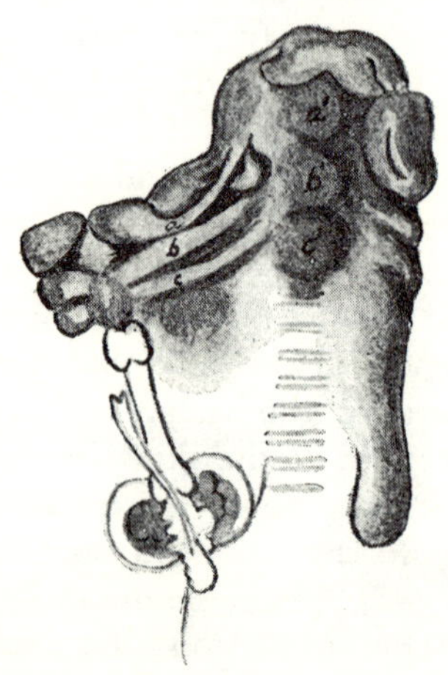

FIG. 10. — Cranial Vertebræ and Visceral Arches in Embryo of Pig. Ventral Aspect. (After Reichert.)

As in the higher Vertebrates, the visceral arches are associated with the cranial vertebræ as their ventral extensions, being equivalent to the visceral plates which form the ventral portion of the "primitive vertebræ" or primitive segments of the trunk.

If the three cranial vertebræ are not very distinct in the early stages of development when the skull is still membranous or cartilaginous, they become clearly delimited when ossification sets in. Three rings of bone forming three more or less complete vertebræ are the final result of ossification.

The composition of these rings is as follows :—

	Base.	Sides.	Top.
First vertebra . .	Presphenoid	Orbitosphenoids	Frontals
Second vertebra . .	Basisphenoid	Alisphenoids	Parietals
Third vertebra . .	Basioccipital	Exoccipitals	Supraoccipital

The other bones of the skull are not included in the vertebræ, and this is in large part due to the fact that the sense capsules are formed separately from the cranium (p. 29, 1838). The ear-labyrinth, it is true, fuses indissolubly with the cranium at a later period, but the bones which develop in its capsule are not for all that integral parts of the primitive cranial vertebræ. This point, it is interesting to note, had already been made by Oken in his *Programm* (1807). But many of the bones developed in relation to the sense organs can find their place in the generalised embryonic schema or archetype of the vertebrate skull, for they are of very constant occurrence during early development.

Having arrived at a generalised embryonic type for the vertebrate skull, of which the fundamental elements are the three cranial vertebræ and their arches, Reichert goes on to discuss the particular forms under which the skull appears in adult Vertebrates. He accepts in general von Baer's law that the characters of the large groups appear earlier in embryogeny than the characters of the lesser classificatory divisions. "When we observe new and not originally present rudiments in very early embryonic stages, as, for instance, that for the lacrymals, the probability is that they belong to the distinctive development of one of the *larger* vertebrate groups. From these are to be carefully distinguished such rudiments as arise later during ossification, mostly as *ossa intercalaria*, in order to give greater strength to the skull in view of the greater development of the brain, etc.; the latter give their individual character to the *smaller* vertebrate groups, and comprise such bones as the *vomer*, the *Wormian bones*, the lowermost turbinal, etc." (p. 63, 1838).

He did not accept the Meckel-Serres law of parallelism. He recognised the great similarity between the unsegmented cartilaginous cranium of Elasmobranchs, and the primordial cranium of the embryos of the higher Vertebrates, but he did not think that the cranium of Elasmobranchs was simply an undeveloped or embryonic stage of the skulls of the higher forms. Rather "do the *Holocephala*, *Plagiostomata*, and *Cyclostomata* appear to us to be lower developmental stages individually differentiated, so that the other fully differentiated Vertebrates cannot easily be referred directly to their type" (p. 152, 1838). The skull of these lower fishes is itself a specialised one; it is an individualised modification of a simple type of skull. And this holds good in general of the skulls of the lower Vertebrates—they are individualised exemplars of a simple general type, not merely unmodified embryonic stages of the greatly differentiated skulls of the higher Vertebrates (p. 250, 1838). Differentiation within the vertebrate phylum is therefore not uniserial, but takes place in several directions. Reichert describes two sorts of modifications of the typical skull—class modifications and functional modifications. The causes of the modifications which characterise classificatory groups are unknown; the second class of modifications occur in response to adaptational requirements.

Reichert's two papers are of considerable importance, and Müller's remark in his review[1] of them is on the whole justified. "These praiseworthy investigations supply from the realm of embryology new and welcome foundations for comparative anatomy" (p. clxxxvii.).

The development of the skull was, however, more thoroughly worked out by Rathke, and with less theoretical bias, in his classical paper on the adder.[2] This memoir of Rathke's is an exhaustive one and deals with the development of all the principal organ-systems, but particularly of the skeletal and vascular. He confirmed in its essentials Reichert's account of the metamorphoses of the first two visceral arches, describing how the rudiment of the skeleton of the first arch appears as a forked process of the cranial basis, the upper prong developing into the palatine and pterygoid, the lower forming Meckel's cartilage, while the quadrate develops from the angle of the fork. The actual bone of the upper jaw (maxillary) develops outside and separate from the palato-pterygoid bar. The cartilaginous rod supporting the second visceral arch divides into three pieces on each side, of which the lower two form the hyoid, the uppermost the columella. Like Reichert he held the visceral arches to be parts of the visceral plates, containing, however, elements from all three germ-layers—the serous, mucous, and vessel layers.

[1] Müller's *Archiv* for 1838.

[2] *Entwickelungsgeschichte der Natter*, Königsberg, 1839.

The first gill-slit, or, as Rathke here prefers to call it, pharyngeal slit, closes completely in snakes and in Urodeles. It forms the Eustachian tube in all other Tetrapoda. As regards the vertebræ, Rathke describes them as being formed in the sheath of the chorda from paired rudiments, each of which sends two branches upwards, and two branches downwards. The two inner pairs of processes coalesce round the chorda, and later form the centrum; the upper outer pair meet above the spinal column; the lower outer pair form ribs. The odontoid process of the axis vertebra is the centrum of the atlas (p. 120). The formation of vertebral rudiments begins close behind the ear-labyrinth, but in front of this the chorda-sheath gives origin to a flat membranous plate which afterwards becomes cartilaginous. This plate reaches forward below the third cerebral vesicle as far as the infundibulum. The notochord ends in this plate, which is the *basis cranii*, just at the level of the ear-labyrinth. In no Vertebrate does the notochord extend farther forward (p. 122). The *basis cranii* gives off three trabeculæ. The middle one is small and sticks up behind the infundibulum; it is absent in fish and Amphibia, and soon disappears during the development of the higher forms. The lateral trabeculæ are long bars which curve round the infundibulum and reach nearly to the front end of the head. Together they are lyre-shaped. The cranial basis and the trabeculæ are formed, like the vertebræ, in the sheath of the notochord, and the only differences between the two in the early stage of their development are that the formative mass for the cranial basis is much greater in amount than that for the vertebræ, and that the cranial basis by means of its processes, the trabeculæ, reaches well in front of the terminal portion of the notochord (p. 36). The capsule for the ear-labyrinth develops quite independently of the cranial basis and the notochord. It resembles on its first appearance, in form, position, composition, and connections, the ear-capsule of Cyclostomes, and so do the ear-capsules of all embryonic Vertebrates (p. 39). It manifests clearly the embryonic archetype, . . . "there exists one single and original plan of formation, as we may suppose, upon which is built the labyrinth of Vertebrates in general"

(p. 40). When ossification sets in, the ear-capsule forms three bones, of which two fuse with the supraoccipital and exoccipitals.

During the formation of the ear-capsule the cranial basis develops from a plate to a trench, for in its hinder section the side parts grow up to form the side walls of the brain, in exactly the same way as the processes of the vertebral rudiments grow up to enclose the spinal column (pp. 122, 192). The foundations of the skull are now complete, and ossification gradually sets in. The basioccipital is formed

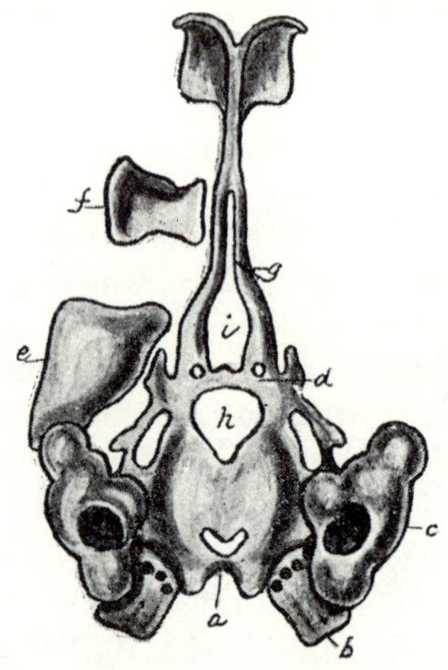

FIG. 11.—Embryonic Cranium of the Adder. Ventral Aspect. (After Rathke.)

a. Basioccipital.
b. Exoccipital.
c. Ear capsule.
d. Basisphenoid.
e. Alisphenoid.
f. Orbitosphenoid.
g. Trabeculæ.
h. Foramen.
i. Pituitary space.

in the posterior part of the *basis cranii*, and the exoccipitals in the side walls of the trench in continuity with the fundament of the basioccipital (see Fig. 11). The supra-occipital is formed in cartilage above the exoccipitals. The basisphenoid develops, like the basioccipital, in the flat *basis cranii*, but towards its anterior edge, between the large foramen (*h*) and the pituitary space (*i*). It is formed from two centres, each of which is originally a ring round the carotid foramen. The presphenoid develops in isolation between the lateral trabeculæ, just behind the point where they fuse. The side parts of the basisphenoid and presphenoid (forming the alisphenoids and the orbitosphenoids respectively) develop in cartilage separately from the cranial basis, not like the exoccipitals in continuity with it. The hinder parts of the trabeculæ become enclosed by two processes of the basisphenoid; their front parts remain in a vestigial and cartilaginous state alongside the presphenoid. The frontals and parietals show a peculiar mode of origin in the adder, differing from their origin in other Vertebrates. The frontals develop in continuity with the orbitosphenoids, the parietals in continuity with the alisphenoids, and so have much resemblance with the vertebral neural arches which surround the spinal column (p. 195)

Through Rathke's work the real embryonic archetype of the vertebrate skull was for the first time disclosed. Rathke discussed this archetype and its relation to the vertebral theory of the skull in another paper of the same year (1839), but before going on to this paper, we shall quote from the paper on the adder the following passage, remarkable for the clear way in which the idea of the embryological archetype is expressed. "Whatever differences may appear in the development of Vertebrates, there yet exists for the different classes and orders a universally valid idea (plan, schema, or type) ruling the first formation of their separate parts. This idea must first be worked out, though possibly with modifications, before more special ideas can find play. The result of the latter process, however, is that what was formed by the first idea is not so much hidden as partially or wholly destroyed" (p. 135).

Rathke's general paper on the development of the skull in Vertebrates[1] treats the matter on a broader comparative basis than his paper on the adder, and takes into account all the vertebrate classes, in so far as their development was then known. He here makes the interesting suggestion, later entirely confirmed, that the *basis cranii* or basilar plate is first laid down as two strips, one on each side of the chorda —the structures now known as parachordals (pp. 6, 27). For this supposition, he thinks, speaks the structure of the

[1] *Bemerkungen über die Entwickelung des Schädels der Wirbelthiere*, Königsberg, 1839.

skull in *Ammocoetes*, which in this respect is the simplest of all Vertebrates (pp. 6, 22). In *Ammocoetes*, as Johannes Müller had shown, the foundation of the skull is formed by two long cartilaginous bars, between the hinder portions of which the notochord ends. In these Rathke was inclined to see the homologues of his trabeculæ, and of the parachordals which he was ready to assume from his embryological observations.

Müller was, of course, very ready to accept Rathke's opinions on this subject, for he considered that they supported his own theory of the vertebral nature of the skull. After describing in his *Handbüch der Physiologie* the cartilaginous bands in *Ammocoetes* and their highly differentiated homologues in the Myxinoids, he writes in the later editions, "Hence we see that in the cranium, as in the spinal column, there are at first developed at the sides of the chorda dorsalis two symmetrical elements, which subsequently coalesce, and may wholly enclose the chorda. Rathke has recently observed, in the embryos of serpents and other animals, before the formation of the proper cranial vertebræ, two symmetrical bands of cartilage, similar to those which I discovered as a persistent structure in *Ammocoetes*. . . . At a later period the *basis cranii* of vertebrate animals contains three parts analogous to the bodies of vertebræ, the most anterior of which, in the majority of animals, is generally small, and its development frequently abortive, whilst in man and mammiferous animals the three are very distinct. These parts are developed by the formation of three distinct points of ossification, one behind the other, in the basilar cartilage."[1]

Rathke was very cautious about accepting the vertebral theory of the skull; he saw that the facts of development were not altogether favourable to the theory, and he gave his adherence with many reservations and saving clauses. His general attitude may be summed up as follows.[2]

[1] *Handbüch der Physiologie des Menschen*, Koblenz, 1835; Eng. trans. by W. Baly, ii., p. 1615, 1838.

[2] For a full statement of Rathke's conclusions, see the translation given by Huxley in *Lectures on the Elements of Comparative Anatomy*, London, 1864.

The chorda sheath is the common matrix of the vertebræ and of a large part of the skull. The basilar plate and the trabeculæ, which are developed from the chorda sheath, give origin to three bones, which might possibly be considered equivalent to vertebral centra—the basioccipital, the basisphenoid, and the *Riechbein* (ethmoid). The *Riechbein* develops from the fused ends of the trabeculæ. The presphenoid might also be considered as a vertebral body, but it develops independently of the basilar plate and trabeculæ.

Now of these bones, the basioccipital is in every way equivalent to a vertebral centrum, for it develops in the basilar plate round the notochord. With the exoccipitals, which arise just like neural arches, it forms a true vertebra. The supraoccipital is an accessory bone developed in relation to bigger brains. The basisphenoid appears in the basilar plate, but in front of the notochord, nor does it arise in exactly the same way as the centrum of a vertebra. The basisphenoid with the alisphenoids, which develop independently in the side walls of the brain, may, however, still be considered as forming a vertebra, though the resemblance is not so great as in the case of the occipital ring. The presphenoid, being long and pointed, is very unlike a vertebral body. The orbitosphenoids develop separately from it. The ethmoid also differs from a vertebra, for it surrounds not the whole nervous axis as the two hinder "vertebræ" do, but only two prolongations of it, the olfactory lobes. In its development and final form it shows no particular resemblance to a vertebra. Its body, the *pars perpendicularis* (mesethmoid) shows no similarity with a vertebral centrum. Completing the three hinder cranial "vertebræ" and roofing in the brain are the supraoccipital, the parietals and the frontals. The premaxillaries, vomer, and nasals do not belong to the cranial scheme; they are covering bones connected with the ethmoid. So, too, the ear-capsule is not part of the cranial vertebræ, but is rather to be compared to the intercalary bones in the vertebral column of certain fish. Summing up as regards the cranial vertebræ Rathke writes, "We find that the four different groups of bones, consisting of the basioccipital with its intercalary (the supraoccipital), the basisphenoid with its intercalaries (parietals), the presphenoid with its intercalaries (frontals), and the ethmoid with its outgrowths (turbinals and cribriform plate), taking them in order from behind forwards, show an increasing divergence from the plan according to which vertebræ as commonly understood develop, so that the basioccipital shows the greatest resemblance to a vertebra, the ethmoid the least" (p. 30).

In a posthumous volume published in 1861 the same opinion is put forward. "In the head, too," he writes, "some vertebræ can be recognised, although in a more or less modified form. Yet at most only four cranial vertebræ can be assumed, and these differ from ordinary well-developed vertebræ in their manner of formation the more the farther forward they lie."[1]

Rathke was an able and careful critic of the vertebral theory of the skull, but he accepted it in the main. Actual attack on the theory upon embryological grounds was begun by C. Vogt, in his work on the development of *Coregonus*,[2] and in his paper on the development of *Alytes*."[3] He described for *Coregonus* an origin of the skull in the main similar to that established by Rathke for the adder. There was a "nuchal plate" in which the front end of the notochord was imbedded; the notochord ended at the level of the labyrinth; there were two lateral bands, comparable to Rathke's lateral trabeculæ; a "facial plate" was also formed, which seems on the whole equivalent to the plate formed by the fused anterior ends of the trabeculæ. A little later the cranium formed a complete cartilaginous box surrounding the brain, very similar to the adult cranium of a shark.

In his criticism of the vertebral theory of the skull, Vogt started by defining the vertebra as a ring formed round the chorda. Now since only the occipital segment of the skull is formed actually round the notochord, the parts of the skull

[1] *Entwickelungsgeschichte der Wirbelthiere*, p. 142, 1861.

[2] *Embryologie des Salmones.* A separate volume of L. Agassiz's *Histoire naturelle des Poissons d'Eau douce de l'Europe centrale*, Neuchâtel, 1842.

[3] *Untersuchungen über die Entwickelungsgeschichte der Gebürtshelferkröte*, Solothurn, 1842.

lying in front of this cannot themselves be vertebræ, though they may be considered as prolongations of the occipital or nuchal vertebra. "We must regard the nuchal plate as a true vertebra, modified, it is true, in its formation and development by its particular functions. Now, since the notochord ends with the nuchal plate we can no longer regard as vertebræ the parts of the skull that lie beyond, such as the lateral processes of the cranium and the facial plate, for they have no relation with the notochord" (p. 123).

To support this view he adduced the fact that the vertebral divisions (primitive vertebræ) visible in the trunk do not extend into the head. He used precisely the same arguments in his paper on *Alytes* to destroy the vertebral theory of the skull. We quote the following passage translated by Huxley (1864, p. 295) from this paper. "It has therefore become my distinct persuasion that the occipital vertebra is indeed a true vertebra, but that everything which lies before it is not fashioned upon the vertebrate type at all, and that efforts to interpret it in such a way are vain; that, therefore, if we except that vertebra (occipital) which ends the spinal column anteriorly, there are no cranial vertebræ at all."

L. Agassiz, himself a pupil of Döllinger, in the general part (1844) of his *Recherches sur les Poissons fossiles* (Neuchâtel, 1833-43), repeats in the main his pupil Vogt's criticism of the vertebral theory (vol. i., pp. 125-9).

These arguments of Vogt and Agassiz were not considered by Müller to dispose of the theory,[1] which maintained a firm hold even upon embryologists. It was still upheld by Reichert, and Kölliker in 1849 showed himself convinced of its general validity.

A useful step in the analysis of the concept "vertebra" was taken by Remak,[2] who showed what a complex affair the formation of vertebræ really is, involving as it does a complete resegmentation (*Neugliederung*) of the vertebral column, whereby the original vertebral bodies were replaced by the secondary definitive bodies (p. 143). Remak showed, as he thought, that the protovertebral segmentation of the dorsal

[1] Müller's *Archiv* for 1843, p. ccxlviii.

[2] *Untersuchungen über die Entwickelung der Wirbelthiere*, Berlin, 1850-55.

muscle-plates did not extend into the head, and he denied Reichert's assertion (1837) that the cranial basis in mammals showed transverse grooves delimiting three cranial vertebræ (p. 36). The gill-slits, he considered, could not possibly be regarded as marking the limits of head vertebræ.

In 1858 appeared Huxley's well-known Croonian Lecture, *On the Theory of the Vertebrate Skull*,[1] in which he stated with great clearness and force the case for the embryological method of determining homologies, and criticised with vigour the vertebral theory of the skull. By this time the two rival methods in morphology had become clearly differentiated, and Huxley was able to contrast them, or at least to show how necessary the new embryological method was as a corrective and a supplement to the older anatomical, or, as he calls it, "gradation" method. Applied to the "Theory of the Skull," the gradation method consists in comparing the parts of the skull and vertebral column in adult animals with respect to their form and connections. "Using the other method, the investigator traces back skull and vertebral column to their earliest embryonic states and determines the identity of parts by their developmental relations" (p. 541). This second method is the final and ultimate. "The study of the gradations of structure presented by a series of living beings may have the utmost value in suggesting homologies, but the study of development alone can finally demonstrate them" (p. 541). As an example of the utility and, indeed, the necessity of applying the embryological method Huxley takes the case of the quadrate bone in birds. This bone had been generally regarded by anatomists as the equivalent of the tympanic of mammals, on account of its connection with the tympanum; but Reichert showed (1837) that the same segment of the first visceral arch developed into the incus in mammals, and into the quadrate in birds, and that therefore the quadrate was homologous with the incus. Similarly, on developmental grounds, the malleus or hammer of mammals is the homologue of the articular of birds, since both are developed from a portion

[1] Delivered 17th June 1858. Reprinted in *The Scientific Memoirs of T. H. Huxley*, edited by M. Foster and E. Ray Lankester, vol. i., pp. 538-606 (1898).

of Meckel's cartilage identical in form and connections in the two groups. The homologies of the bones connected with the jaws in bony fishes had long been a subject of contention among comparative anatomists; Huxley shows from his personal observations how the development of the visceral arches throws light upon these difficulties. The mandibular arch in the developing fish is abruptly angled, as in the embryo of Tetrapoda; the upper prong of it ossifies into the palatine and pterygoid; at the angle is formed the quadrate (jugal, Cuvier), and to the quadrate is articulated the lower jaw, which ossifies round the lower prong or Meckel's cartilage. The scheme of development of the jaws is accordingly similar in fish to what it is in other Vertebrates, and this similarity of development enables Huxley to recognise what are the true homologues of the quadrate, the palatine and the pterygoid in adult bony fish, and to prove that the symplectic and the metapterygoid (tympanal, Cuvier) are bones peculiar to fish. In developing Amphibia Huxley found a suspensorium of hyoid and mandibular arches similar to the hyomandibular of fish.

Tackling his main problem of the unity of plan of the vertebrate skull, Huxley shows, by a careful discussion of the anatomical relationships of the chief bones in typical examples of all vertebrate classes, that there is on the whole unity of plan as regards the osseous skull. This unity of composition can be established, on the gradation method, by considering the connections of the bones of the skull with one another, their relations to the parts of the brain and to the foramina of the principal cranial nerves. The assistance of the embryological method is, however, necessary in determining many points with regard to the bones developed in relation to the visceral arches. But there is a further step to be taken. "Admitting . . . that a general unity of plan pervades the organisation of the ossified skull, the important fact remains that many vertebrated animals—all those fishes, in fact, which are known as *Elasmobranchii*, *Marsipobranchii*, *Pharyngobranchii* and *Dipnoi* have no bony skull at all, at least in the sense in which the words have hitherto been used" (p. 571). The membranous or cartilaginous skull of these fishes shows a general resemblance in its main features to the ossified skull of other Vertebrates; the relations of the ear to the vagus and trigeminal nerves are, for instance, the same in both; the main regions of the cartilaginous skull can be homologised with definite bones or groups of bones in the bony skull; but discrepancies occur. It is again to development that we must turn to discover the true relationship of the cartilaginous to the ossified skull. "The study of the development of the ossified vertebrate skull . . . satisfactorily proves that the adult crania of the lower *Vertebrata* are but special developments[1] of conditions through which the embryonic crania of the highest members of the sub-kingdom pass" (p. 573). It is with the embryonic cranium of higher Vertebrates that the adult skull of the lower fishes must be compared, and the comparison will show a substantial though not a complete agreement between them. Thus, speaking of the development of the frog's skull, Huxley writes:—"If, bearing in mind the changes which are undergone by the palatosuspensorial apparatus, . . . we now compare the stages of development of the frog's skull with the persistent conditions of the skull in the *Amphioxus*, the lamprey, and the shark, we shall discover the model and type of the latter in the former. The skull of the *Amphioxus* presents a modification of that plan which is exhibited by the frog's skull when its walls are still membranous and the notochord is not yet embedded in cartilage. The skull of the lamprey is readily reducible to the same plan of structure as that which is exhibited by the tadpole when its gills are still external and its blood colourless. And finally, the skull of the shark is at once intelligible when we have studied the cranium in further advanced larvæ, or its cartilaginous basis in the adult frog" (p. 577). Development, therefore, proves what comparative anatomy could only foreshadow—the unity of plan of all vertebrate skulls, ossified and unossified alike. "We have thus attained to a theory or general expression of the laws of structure of the skull. All vertebrate skulls are originally alike; in all (save *Amphioxus?*) the base of the primitive cranium undergoes the mesocephalic flexure, behind which the notochord terminates, while immediately in front of it

[1] *Cf.* Reichert, *supra*, p. 149.

the pituitary body is developed;[1] in all, the cartilaginous cranium has primarily the same structure—a basal plate enveloping the end of the notochord and sending forth three processes, of which one is short and median, while the other two, the lateral trabeculæ, pass on each side of the space on which the pituitary body rests, and unite in front of it; in all, the mandibular arch is primarily attached behind the level of the pituitary space, and the auditory capsules are enveloped by a cartilaginous mass, continuous with the basal plate between them. The amount of further development to which the primary skull may attain varies, and no distinct ossifications at all may take place in it; but when such ossification does occur, the same bones are developed in similar relations to the primitive cartilaginous skull" (p. 578).

In a word, there is a general plan or primordial type which is manifested in the higher forms most clearly in their earliest development—an embryological archetype therefore.

Huxley now goes on to consider the relation of this general plan or type of the skull to the structure and development of the vertebral column. Does the skull in its development show any signs of a composition out of several vertebræ? The vertebral column develops as a segmented structure round the notochord; the skull develops first as an unsegmented plate extending far beyond the notochord. The processes of this basilar plate, the trabeculæ, are quite unlike anything in the vertebral column. It is true that when the process of ossification begins, separate bones are differentiated in the basilar plate one in front of the other, giving an appearance of segmentation. The hindmost of these bones, the basioccipital, ossifies round the notochord, quite like a vertebral centrum, and its side parts which form the occipital arch develop in a "remotely similar" way to the neural arches of the vertebræ. The next bone, however, the basisphenoid, develops in front of the notochord, and shows very little analogy with a vertebral body. The analogy is even more far-fetched when applied to the axial

[1] The origin of the pituitary body from the roof of the mouth was first described by Rathke (1839).

bones in front of the basisphenoid. The cranium might indeed be divided upon ossification into a series of segments bearing a more or less remote analogy with vertebræ. "In the process of ossification there is a certain analogy between the spinal column and the cranium, but that analogy becomes weaker and weaker as we proceed towards the anterior end of the skull" (p. 585). The best way to state the facts is to say that both skull and vertebral column start in their development from the same point, but immediately begin to diverge. The clear indications of segmentation which fully ossified adult skulls undoubtedly show are, therefore, secondary, and the vertebral theory of the skull, which was originally based upon the appearance of such fully ossified crania, is on the whole negatived by embryology.

We have now to turn back a few years in order to follow up another line of discovery which had an important bearing upon the theory of the vertebrate skull—the working out of the distinction between membrane and cartilage bones.

As early as 1731, R. Nesbitt,[1] in two lectures delivered to the Royal College of Surgeons, demonstrated that in the human fœtus some bones were formed not in cartilage but directly in fibrous tissue, and this observation was confirmed by other human anatomists, particularly by Sharpey at a considerably later date. In 1822 Arendt[2] focussed attention upon the remarkable structure of the skull of the Pike, with its cartilaginous brain-box studded all over with bony plaques, an arrangement which had already attracted the interest of Cuvier and Meckel. K. E. von Baer[3] in 1826 discussed at some length the relation between the bony and the cartilaginous skull in fishes, with particular reference to the sturgeon, coming to the following just conclusion :—"If we consider the fibrous skeleton of *Ammocoetes* as the first foundation of the skeleton of Vertebrates, we can form a

[1] *Human Osteogeny explained in two Lectures*, London, 1736.

[2] *De capitis ossei Esocis lucii structura singulari. Dissert. inaug.* Regiomonti, 1822.

[3] "Ueber das äussere und innere Skelet," Meckel's *Archiv*, pp. 327-76, 1826.

series among the cartilaginous fishes, according as a cartilaginous skeleton penetrates more and more into this fibrous foundation. In the same way the process of ossification supplants the cartilaginous skeleton. So long as the ossifications lie in the skin, as in the sturgeon, they form corneous bones (*Hornknochen*), but when they lie under the skin, they form true bones, *e.g.*, the bones of the skull in the pike" (p. 374).

Embryologists soon become aware that a similar distinction between a primitive cartilaginous foundation and a secondary overlying ossification of the skull showed itself in the development of all Vertebrates. Dugès, in his *Recherches sur l'ostéologie et la myologie des Batraciens* (1834), distinguished between such bones as are formed by direct ossification of the cartilaginous groundwork of the skull, and such as are developed in the periosteal fibrous tissue.

Reichert in 1838[1] noted that several of the skull bones in Amphibia are formed without the intermediary of cartilage, such as the nasals, the maxillaries and the lacrymals. So, too, the frontals and parietals of Teleosts developed independently of the cartilaginous skull, and belonged to the skeletal system of the skin, not to the true vertebral axial skeleton (pp. 215-6). Even more interesting was his discovery, afterwards confirmed by Hertwig,[2] that in the newt several bones connected with the palate were formed in the mucous membrane of the mouth by the fusion of a number of little conical teeth (p. 97). Certain of these bones he considered to be the substitutes, not the equivalents, of the palatine and pterygoid of other Vertebrates, which are formed from the upper part of the first visceral arch, a part missing in the newt (p. 100). Owing to the difference of development he would not homologise these bones in the newt with the palatine and pterygoid of other Vertebrates. He recognised also that the bone now known as the parasphenoid was developed in the frog in the mucous membrane of the mouth, and had originally no connection with the cranial basis (p. 34). Rathke in 1839 also allowed the distinction between cartilage and membrane bone, but laid no stress upon it (*Entw. d. Natter.*, p. 197).

[1] *Vergl. Entwick. d. Kopfes d. nackten Amphibien* (p. 186).

[2] *Arch. f. mikr. Anat.*, xi., Suppl., 1874.

Jacobson in 1842[1] introduced the useful term, "primordial cranium," for the primitive cartilaginous foundation of the skull, and drew a sharp distinction between cartilage bones and membrane bones.

In his *Recherches sur les Poissons fossiles*,[2] L. Agassiz used Vogt's work on the development of *Coregonus* to establish a classification of the bones of the skull in fish, a classification which had the merit of drawing a sharp distinction between the cartilaginous groundwork and the "protective plates" of the fish's skull. He recognised that the protective plates developed in a different way from the other bones of the skull. "We must distinguish," he writes, "two kinds of ossification; one which tends to transform the primitive parts of the embryonic cranium directly into bone, and another which leads to the deposition of protective plates round this core, which develop not only upon the upper surface, as has hitherto been supposed, but also on the lateral walls and on the lower surface of the cranium" (p. 112). In the skull of all fish there are three elements—(1) the cartilaginous base, including the nuchal plate, the trabeculæ and the facial plate, together with the auditory capsules; (2) the cartilaginous cerebral envelope; (3) the bony protective plates (absent in Elasmobranchs). The bones developed in relation to these cranial elements can be classified as follows:—(1) the basioccipital, exoccipitals (paroccipitals?), supraoccipital and "petrous" (*rocher*), developed from the nuchal plate; the ali- and orbito-sphenoids developed from the trabeculæ; the "cranial ethmoid"[3] developed from the facial plate; (2) the parietals, frontals and nasals formed from the "superior" protective plate; the "anterior" and "posterior" frontals and the temporal, from the "lateral" plates; the body of the sphenoid and the vomer from the "inferior" plates. The other element, the cartilaginous brain-box, does not ossify, and tends to become absorbed (p. 124).

In 1849 Kölliker published a paper[4] dealing with the

[1] "Om Primordial-Craniet," *Förhandlingar Skand. Naturf. Möle*, Stockholm, 1842.

[2] Vol. I., General part, pub. 1844.

[3] *Entosphenoid*, Owen.

[4] *Zweiter Bericht zootom. Anstalt zu Würzburg*, 1849.

morphological significance of the distinction between membrane and cartilage bones, and in 1850[1] he defended his views against the criticisms of Reichert[2] in a further note entitled *Die Theorie des Primordialschädels festgehalten.* It is convenient to consider these papers together. Kölliker held that there was (1) a histological and (2) a morphological difference between the two categories of bones. The histological development of the two kinds was different, but this difference was not sufficient to establish a morphological distinction between them, a distinction in their anatomical *Bedeutung*. The true morphological distinction between them was their development in different skeleton-forming layers. Membrane bones were developed in fibrous tissue lying between the skin and the deep layer which formed the primordial cranium, and it was this formation in a separate layer that gave them a different morphological significance from the bones formed directly in the deep layer. Kölliker's distinction, therefore, was between the bones formed in the primordial cartilaginous cranium on the one hand, and the superficial ossifications in fibrous tissue on the other hand. The cartilaginous cranium in Kölliker's opinion was formed upon the vertebral type, and the membrane bones were accessory. This, at least, was his opinion in 1849. In 1850, after Stannius had shown that membrane bones occurred as integral parts of the vertebræ in certain fish, he modified his view of the membrane bones, and admitted them, at least in some cases, as constituents of the cranial vertebræ.

On this morphological distinction of membrane and cartilage bones future comparative osteology was to be based :—

"My sole aim is to state again the principle upon which comparative osteology is to be based and extended, and this is that first place should be assigned to anatomical considerations, and among these to the manner of origin of the whole bone in relation to the skeleton-forming layers" (1850, p. 290).

The homologies established by this new principle might

[1] *Zeits. f. wiss. Zool.*, ii., pp. 281-91.

[2] Müller's *Archiv* for 1849, pp. 443-515.

run counter to the homologies indicated by the study of adult structure. "Thus, for instance, although the lower jaw in position, function, form and shape, appears to be the same bone throughout, yet it must be admitted that it shows a difference in the different classes. In Mammals and Man it is an entirely secondary bone (an extremity according to Reichert), in Birds, Amphibia and Fishes only partially so, for its articular belongs to Meckel's cartilage and is accordingly analogous to a rib; indeed, in the Plagiostomes, etc., the whole lower jaw along with the articular is a persistent Meckel's cartilage" (p. 290, 1850).

So, too, the supraoccipital in man cannot be fully homologised with the supraoccipital of many mammals, for its upper half arises at first in isolation as a secondary bone (p. 290).

Reichert objected to the distinction drawn by Kölliker, and denied that there was either a histological or a morphological difference between membrane and cartilage bones. It was shown a few years later by H. Müller[1] that there was in truth no essential difference in histological development between the two categories of bone, that the cartilage cells were replaced by bone cells identical with those taking part in the formation of membrane bones. The morphological distinction continued however to be recognised, particularly by the embryologists. Rathke in his volume of 1861[2] classified the bones of the skull according to their origin from the primordial cranium or from the overlying fibrous layer, distinguishing as membrane bones, the parietals, frontals, nasals, lachrymals, maxillaries and premaxillaries, jugals, tympanic, parts of the "temporal," vomer, part of the supraoccipitals in some mammals, and the mandible (with the exception of the articular in such as have a quadrate bone). Huxley was also inclined in 1864[3] to recognise the distinction, but he writes with some reserve:—"Is there a clear line of demarcation between membrane bones and cartilage bones? Are certain bones always developed primarily from cartilage, while certain others as constantly originate in membrane? And further,

[1] *Zeits. f. wiss Zool.*, ix., 1858.

[2] *Entw. d. Wirbelthiere*, pp. 139-40, 1861.

[3] *Lectures on the Elements of Comparative Anatomy.*

if a membrane bone is found in the position ordinarily occupied by a cartilage bone, is it to be regarded merely as the analogue and not as the homologue of the latter?" (p. 296).

We may note here that many comparative anatomists of the period were quite ready to decide Huxley's last question in a sense favourable to the older, purely anatomical, view of homology. Owen, for instance, held that difference of development did not disturb homologies established by form and connections. "Parts are homologous," he writes, "in the sense in which the term is used in this work, which are not always similarly developed: thus the 'pars occipitalis stricte dicta,' etc., of Soemmering is the special homologue of the supraoccipital bone of the cod, although it is developed out of pre-existing cartilage in the fish and out of aponeurotic membrane in the human subject."[1] Similarly he pointed to the diversities of development of the vertebral centrum in the different vertebrate classes as proof that development could not always be relied upon in deciding homologies (p. 89). But he could not deny that the archetype was better shown in the embryo than in the adult (*supra*, p. 108).

J. V. Carus[2] likewise stood firm for the older method of determining homologies by comparison of adult structure. "We can regard as homologous," he writes, "only those parts which in the fully formed animal possess a like position and show the same topographical relations to the neighbouring parts" (p. 389). Parts homologous in this sense might develop in different ways, but no great importance was to be attached to such a circumstance. Membrane and cartilage bones developed in practically the same way, from the same skeleton-forming layer, and no morphological significance attached to their distinction (pp. 227, 457). Embryology was of considerable value in helping to determine homologies, but the evidence that it supplied was contributory, not conclusive. Perhaps the greatest service which the study of development rendered was to disentangle, by a comparison of the earliest embryos, the generalised type (p. 389).

[1] *On the Archetype of the Vertebrate Skeleton*, p. 5, 1848.

[2] *System der thierischen Morphologie*, Leipzig, 1853.

We have now traced, by our historical study of the theory of the skull, the gradual evolution of the tendency to find in development the surest guide to determining homologies. We have seen how the embryological "type" came to be substituted, in whole or in part, for the anatomical "type" derived from the study of adult structure. But we have had to do only with a modification, not with a transformation, of the criterion of homology recognised by the anatomists. Homology is still determined by position, by connections, in the embryo as in the adult. "Similarity of development" has become the criterion of homology in the eyes of the embryologist, but "similarity of development" means, not identity of histological differentiation, but similarity of connections throughout the course of development. For the purposes of morphology, development has to be considered as an orderly sequence of successive forms, not in its real nature as a process essentially continuous. Morphology has to replace the living continuity by a kinematographic succession of stages. Since it is the earliest of these stages that manifest the simplest and most generalised structural relations of the parts, it is in the earlier stages that homologies can be most easily determined. But these homologies are still determined solely by the relative positions and connections of the parts, just as homologies are determined in the last of all the stages of development, the adult state. And since the generalised type is shown most clearly in the earliest stages and tends to become obscured by later differentiation, homologies observed in embryonic life are to be upheld even if the relations in adult life seem to indicate different interpretations.

CHAPTER XI

THE CELL-THEORY.

WITH the founding of the cell-theory by Schwann in 1839 an important step was taken in the analysis of the degrees of composition of the animal body. Aristotle had distinguished three—the unorganised material, itself compounded of the four primitive elements, earth and water, air and fire, the homogeneous parts or tissues and the heterogeneous parts or organs, and this conception was retained with little change even to the days of Cuvier and von Baer. Those of the old anatomists who speculated on the relations of organic elements to one another were dominated by Aristotle's simple and profound classification, and proposed schemes which differed from his only in detail. Bichat enlarged and deepened the concept of tissue, but the degree of composition below this was for him, as for all anatomists of his time, a fibrous or pulpy "cellulosity," living, indeed, but showing no uniform and elemental structure. It was Schwann's merit to interpose between the tissue and the mere unorganised material a new element of structure, the cell. And, as it happened, a few years before Schwann published his cell-theory, Dujardin hinted at another degree of composition which was later to take its place between the cell and the chemical elements—sarcode or protoplasm.

As is well known, the concept of the cell arose first in botany. Robert Hooke discovered cells in cork and pith in 1667, and his discovery was followed up by Grew and Malpighi in 1671, and by Leeuenhoek in 1695. But they did not conceive the cell as a living, independent, structural unit. They were interested in the physiology of the plant as a whole, how it lived and nourished itself, and they studied cells and sieve-tubes, wood fibres and tracheæ with a view rather to finding out their functions and their significance for the life of the plant than to discovering the minutiæ of their structure. The same attitude was taken up by the few botanists who in the 18th century paid any heed to the microscopical anatomy of plants. For C. F. Wolff,[1] the formation of cells was a result of the secretion of drops of sap in the fundamental substance of the plant, this substance remaining as cell-walls when cell-formation was completed—no idea here of cells as units of structure.

In the early 19th century, interest in plant anatomy revived somewhat, and much work was done by Treviranus, Mirbel, Moldenhawer, Meyen and von Mohl.[2] As a result of their work the fact was established that the tissues of plants are composed of elements which can, with few exceptions, be reduced to one simple fundamental form—the spherical closed cell. Thus the vessels of plants are formed by coalescence of cells, fibres by the elongation of cells and the thickening and toughening of their walls. At this time, interest was concentrated on the cell-wall, to the almost total neglect of the cell-contents; the "matured framework" of plant cells, to use Sach's convenient phrase, was the chief, almost the sole, object of study. And it was natural enough that the mere architecture of the plant should monopolise interest, that the composition of the tissues out of the cells, and the fitting together of the tissues to form the plant should awaken and hold the curiosity of the investigator; even the modifications of the cell-walls themselves, their rings and spiral thickenings and pits, offered a fascinating field of enquiry.

The idea that the cell-contents might show a characteristic and individual structure had hardly dawned upon botanists when Schleiden published his famous paper, *Beiträge zur Phytogenesis*.[3] Schleiden's theme in this paper is the origin

[1] *Theoria generationis*, Halae, 1759.

[2] See J. v. Sachs, *Geschichte der Botanik*, book ii., Eng. Trans., 2nd impr., 1906.

[3] Müller's *Archiv*, pp. 137-76, 1838.

and development of the plant cell, a subject then very obscure, in spite of pioneer work by Mirbel. A few years before, Robert Brown had called attention to the presence in the epidermal cells of orchids and other plants of a characteristic spot which he called the areola or nucleus.[1] Schleiden saw the importance of this discovery, confirmed the constant presence of the nucleus in young cells, and held it to be an elementary organ of the cell. He named it the cytoblast because, in his opinion, it formed the cell. It was embedded in a peculiar gummy substance, the cytoblastem, which formed a lining to the cellulose cell-wall. Within the nucleus there was often a small dark spot or sphere—the nucleolus. The nucleus, Schleiden thought, originated as a minute granule in the cytoblastem which gradually increased in size, becoming first a nucleolus (*Kernchen*), and then, by further condensation of matter round it, a nucleus. Several nuclei might be formed in this way in a single cell. New cells took their origin directly from a full-grown nucleus, in a peculiar way which Schleiden describes as follows:—"As soon as the cytoblasts have reached their full size a delicate transparent vesicle arises on their surface; this is the young cell, which at first takes the shape of a very flat segment of a sphere, of which the plane surface is formed by the cytoblast, the convex side by the young cell itself, which lies upon the cytoblast like a watch-glass on a watch" (p. 145). The young cells increase in size and fill up the cavity of the old cell, which is in time resorbed. Cell-development always takes place within existing cells, and either one or many new cells may be formed within the mother-cell. Schleiden's views on cell-formation were drawn from some rather imperfect observations on the embryo-sac and pollen-tube, but he extended his theory to cell-formation in general. Though wrong in almost all respects the theory had at least the merit of fixing attention upon the really important constituents of the cell, the nucleus and the cell-plasma. To Schleiden, too, we owe the conception of the cell as a more or less independent living unity, whose life is not entirely identified with the life of the plant as a whole. "Each cell," he writes, "carries on a double life; one a quite independent and self-contained

[1] *Trans. Linnean Soc.*, xvi., p. 710, 1833.

life, the other a dependent life in so far as the cell has become an integral part of the plant" (p. 138).

So long as the definition of the plant cell embraced little more than the hardened cell-wall it was little wonder that "cells" in this sense were not recognised in animal tissues, except in a few exceptional cases—as in the notochord by Johannes Müller.[1] Careful observation of animal tissues discovered in some cases the existence of discontinuous units of structure, but these were not, as a rule, recognised before 1838 as analogous to plant cells. Von Baer, for example, observed that the young chick embryo was composed partly of an albuminous mass and partly of *Kügelchen* or little globules suspended in it (*Entwickelungsgeschichte*, i., pp. 19, 144). Since such *Kügelchen* disposed in a row formed the notochord (i., p. 145) it seems probable that his *Kügelchen* were really cells. Similarly A. de Quatrefages[2] in 1834 saw and figured segmentation spheres in the developing egg of *Limnæa*, but he called them globules and did not recognise their analogy with the cells of plants. According to M'Kendrick,[3] Fontana, so far back as 1781,[4] described cells with nuclei in various tissues, and used acids and alkalis to bring out their structure more clearly. But it was not till 1836-7-8 that a fairly widespread occurrence of cells in animal tissues was recognised. The pioneer in this seems to have been Purkinje, who described cells in the choroidal plexus in 1836,[5] and compared gland cells with the cells of plants in 1837.[6] Henle in 1837[7] and 1838[8] described various kinds of epithelial tissue, distinguishing them according to the kind of cell composing them; he also discovered the mode of growth of stratified epithelium.

[1] *Myxinoiden*, i. Theil., p. 89, 1835.

[2] *Ann. Sci. nat.* (2) (*Zool.*) ii., pp. 107-18, pl. 11, 1834.

[3] *Proc. Phil. Soc. Glasgow*, xix., pp. 71-125, 1887-8.

[4] *Traité sur le venin de la vipère*, 1781.

[5] Müller's *Archiv*, 1836.

[6] J. Müller, *Jahresbericht ü. d. Fortschritte der anat.-physiol. Wissenschaften im jahre* 1838. Müller's *Archiv*, 1838.

[7] *Symbolæ ad anatomiam villorum imprimis eorum epithelii*, Berlin, 1837.

[8] *U. d. Ausbreitung des Epitheliums im menschlichen Körper.* Müller's *Archiv*, 1838.

Valentin[1] appears to have seen cells in cartilage and epithelium even before Henle, and to have observed cells in the blastoderm of the chick. In his report on the progress of anatomy during 1838 Johannes Müller was able to refer to quite a number of papers dealing with the occurrence of cells in animal tissues. In addition to those already noted, he mentions work by Breschet and Gluge on the cells of the umbilical cord, by Dumortier on the cells in the liver of molluscs, by Remak and by Purkinje on nerve cells, by Donné on the cells of the conjuctiva, cornea and lens. He reports, too, that Turpin had compared the epithelial cells of the vagina with the cell-tissue of plants. Müller himself had not only recognised the cellular nature of the notochord, but had observed the cells of the vitreous humour, fat cells and pigment cells, and even the nuclei of cartilage cells. From Schwann (1839) we learn that C. H. Schults had followed back the corpuscles of the blood to their original state of nucleated cells, and that Werneck had recognised cells in the embryonic lens. A preliminary notice of Schwann's own work appeared in 1838 (Froriep's *Notizen*, No. 91, 1838), the full memoir in 1839, under the title *Mikroskopische Untersuchungen über die Uebereinstimmung in der Struktur und dem Wachstume der Tiere und Pflanzen.*[2]

Theodor Schwann was a pupil of Johannes Müller, and we know that Müller took much interest in the new histology. It is probably to his influence that we owe Schwann's brilliant work on the cell, which appeared just after Schwann left Berlin for Löwen. Schwann was himself, as his later work showed, more a physiologist than a morphologist; he did quite fundamental work on enzymes, discovering and isolating the pepsin of the gastric juice; he proved that yeast was not an inorganic precipitate but a mass of living cells; he carried out experiments directed to show that spontaneous generation does not occur. We shall see in his treatment of the cell-theory clear indications of his physio-

[1] See Schwann's *Bemerkungen* at the end of his *Mikroskopische Untersuchungen.*

[2] Republished in Ostwald's *Klassiker der exakten Wissenschaften*, No. 176, Leipzig, 1910 References in the text are to the original pagination.

logical turn of mind. Schwann was only twenty-nine when his master-work appeared, and the book is clearly the work of a young man. It has the clear structure, the logical finish, which the energy of youth imparts to its chosen work. So the work of Rathke's prime, the *Anatomische-philosophische Untersuchungen* of 1832 shows more vigour and a more reasoned structure than his later papers. Schwann's book is indeed a model of construction and cumulative argument, and even for this reason alone justly deserves to rank as a classic.

The first section of his book is devoted to a detailed study of the structure and development of cartilage cells and of the cells of the notochord, and to a comparison of these with plant cells. He accepts Schleiden's account of the origin and development of nuclei and cells as a standard of comparison; and he seeks to show that nucleus and nucleolus, cell-wall and cell-contents, show the same relations and behave in the same manner in these two types of animal cells as in the plant-cells studied by Schleiden. The types of cell which he chose for this comparison are the most plant-like of all animal cells, and he was even able to point to a thickening of the cell-wall in certain cartilage cells, analogous to the thickening which plays so important a part in the outward modification of plant-cells. The analogy indeed in structure and development between chorda and cartilage cells and the cells of plants seemed to him complete. The substance of the notochord consisted of polyhedral cells having attached to their wall an oval disc similar in all respects to the nucleus of the plant-cell, and like it containing one or more nucleoli. Inside the mother-cell were to be found young developing cells of spherical shape, lacking however a nucleus. Cartilage was even more like plant tissue. It was composed of cells, each with its cell membrane. The cells lay close to one another, separated only by their thickened cell-wall and the intercellular matrix, showing thus even the general appearance of the cellular tissue of plants. They contained a nucleus with one or two nucleoli, and the nucleus was often resorbed, as in plants, when the cell reached its full development. Other nuclei were in many cases present in the cell, round which young cells could be seen to develop, in exactly the same manner as in plants. These nuclei had accordingly the same significance as the nuclei of plants, and deserved the same name of cytoblasts or cell-generators. The true nucleus of the cartilage cell was probably in the same way the original generator of the mother-cell.

Having proved the identity in structure and function of the cells of these selected tissues with the cells of plants, as conceived by Schleiden, Schwann had still to show that the generality of animal tissues consisted either in their adult or in their embryonic state of similar cells. This demonstration occupies the second and longest section of his book.

His method is throughout genetic; he seeks to show, not so much that all animal tissues are actually in their finished state composed of cells and modifications of cells, as that all tissues, even the most complex, are developed from cells analogous in structure and growth with the cells of plants.

All animals develop from an ovum; it was his first task to discover whether the ovum was or was not a cell. It happened that, some years before Schwann wrote, a good deal of work had been done on the minute structure of the ovum, particularly by Purkinje and von Baer. Purkinje in 1825[1] discovered and described in the unfertilised egg of the fowl a small vesicle containing granular matter, which he named the *Keimbläschen* or germinal vesicle. It disappeared in the fertilised egg. As early as 1791 Poli had seen the germinal vesicle in the eggs of molluscs, but the first adequate account was given by Purkinje. In 1827[2] von Baer discovered the true ova of mammals and cleared up a point which had been a stumbling block ever since the days of von Graaf, who had described as the ova the follicles now bearing his name.[3] Even von Graaf had noticed that the early uterine eggs were smaller than the supposed ovarian eggs; Prévost and Dumas[4] had observed the presence in the Graafian follicle of a minute spherical body, which, however, they hesitated to call the ovum; it was left to von Baer to elucidate the structure of the follicle and to prove

[1] *Symbolae ad ovi avium historiam.*

[2] *De ovi mammalium et hominis genesi.*

[3] *De mulierum organis*, 1672.

[4] *Ann. Sci. nat.*, iii., p. 135, 1842.

that this small sphere was indeed the mammalian ovum. His discovery was confirmed by Sharpey and by Allen Thomson. Von Baer found the germinal vesicle in the eggs of frogs, snakes, molluscs, and worms, but not in the mammalian ovum; he considered the whole mammalian ovum to be the equivalent of the germinal vesicle of birds—a comparison rightly questioned by Purkinje (1834). In 1834 Coste[1] discovered in the ovum of the rabbit a vesicle which he considered to be the germinal vesicle of Purkinje; he observed that it disappeared after fertilisation. Independently of Coste, and very little time after him, Wharton Jones[2] found the germinal vesicle in the mammalian ovum. Valentin in 1835,[3] Wagner in 1836,[4] and Krause in 1837,[5] added considerably to the existing knowledge of the structure of the ovum. Wagner in his *Prodromus* called attention to the widespread occurrence, within the germinal vesicle of a darker speck which he called the *Keimfleck* or germinal spot, known sometimes as Wagner's spot. He recognised the *Keimfleck* in the ova of many classes of animals from mammals to polyps. Frequently more than one *Keimfleck* occurred.

Schwann had therefore a good deal of exact knowledge to go upon in discussing the significance of the ovum for the cell-theory. There were two possible interpretations. Either the ovum was a cell and the germinal vesicle its nucleus, or else the germinal vesicle was itself a cell within the larger cell of the ovum and the germinal spot was its nucleus. Schwann had some difficulty in deciding which of these views to adopt, but he finally inclined to the view that the ovum is a cell and the germinal vesicle its nucleus, basing his opinion largely upon observations by Wagner which tended to prove that the germinal vesicle was formed

[1] *Recherches sur la génération des Mammifères.* Report by Academy Committee. *Ann. Sci. nat.* (2) (*Zool.*) ii., pp. 1-18, 1834; also *Embryogénie comparée*, 1837.

[2] *Lond. and Edin. Phil. Mag.* (3) vii., 1835; *Phil. Trans.* 1837.

[3] *Handbuch der Entwickelungsgeschichte*, 1835, and Müller's *Archiv*, 1836.

[4] *Prodromus historiæ generationis hominis atque animalium*, Lipsiæ, 1836.

[5] Müller's *Archiv*, 1837.

first and the ovum subsequently formed round it. But the ovum was not, in Schwann's view, a simple cell, for within it were contained yolk-granules, one set apparently containing a nucleus, the others not. Even the second set, those composing the yellow yolk, were considered by Schwann to deserve the name of cells, because, although a nucleus could not be observed in them, they had a definite membrane, distinct from their contents—a conception of the cell obviously dating from the earliest botanical notions of cells as little sacs. The yolk cells were not mere dead food material but living units which took part in the subsequent development of the egg. The relation between the unfertilised egg and the blastoderm which arises from it is not made altogether clear by Schwann. According to his account the cells of the blastoderm are formed actually in the ovum. Round the nucleus of the egg appears a *Niederschlag* or precipitate which is the rudiment of the blastoderm (p. 68). When the egg leaves the ovary the nucleus disappears, leaving behind it this rudiment of the blastoderm, which rapidly grows and increases in size. The blastoderm of the chick before incubation is found to be composed of spherical anucleate bodies which Schwann considers to be cells, because they almost certainly develop into the cells of the incubated blastoderm, which are clearly recognisable as such after eight hours' incubation. The serous and mucous layers can be distinguished after sixteen hours' incubation, and it is found that the cells of the serous layer contain definite nuclei, though such seem to be absent in the cells of the mucous layer. Between the two layers other cells are formed belonging to the vessel layer, which is, however, in Schwann's opinion not a very definitely individualised layer.

Schwann's next step is a detailed demonstration of the origin of each tissue from simple cells such as those composing the incubated blastoderm.

"The foregoing investigation has taught us that the whole ovum shows nothing but a continual formation and differentiation of cells, from the moment of its appearance up to the time when, through the development of the serous and mucous layers of the blastoderm, the foundation is given for all the tissues subsequently appearing: we have found this common parent of all tissues itself to consist of cells; our next task must be to demonstrate not only in this general way that tissues originate from cells, but also that the special formative mass of each tissue is composed of cells, and that all tissues are either constituted by simple cells or by one or other of the manifold kinds of modified cells" (p. 71). Five classes of tissue can be distinguished, according to the extent and manner of the modifications which the cells composing them have undergone. There are first of all independent and isolated cells, such as the corpuscles of the blood and lymph, not forming a coherent tissue in the ordinary sense. Next there are the assemblages of cells lying in contiguity with one another, but not in any way fused; examples of this class are the epidermal tissues and the lens of the eye. In the third class come tissues the cells of which have fused by their walls, but whose cell-cavities are not in continuity, such as osseous tissue and cartilage. In the tissues of the fourth class, comprising the most highly specialised of all, not only are the cell-walls continuous but also the cell-cavities; to this class belong muscle, nerve and capillary vessels. A fifth class, of rather a special nature, includes the fibrous tissues of all kinds. This is the first classification of tissues upon a cellular basis, and it marks the foundation of a new histology which took the place of the "general anatomy" of Bichat. The exhaustive account which Schwann gives of the structure and development of the tissues in this section of his book constitutes the first systematic treatise on histology in the modern sense, and it is still worth reading, in spite of many errors in detail.

Schwann found it easy to demonstrate the cellular nature of the tissues of his first three classes. With the other two classes he had more difficulty. Fibres of all kinds, he considered, arose by an elongation of cells, which afterwards split longitudinally into long strips, forming as the case might be white or elastic fibrous tissue. Muscle-fibres and nerve-fibres were formed in a totally different way, by coalescence of cells; each separate muscle-fibre and nerve-fibre was thus a compound cell. Capillaries, Schwann held, were formed by cells hollowed out like drain-pipes, and set end to end—a mistaken view soon corrected by Vogt (*Embryologie des Salmones*, p. 206, 1842).

In this detail part of his book Schwann accumulates material for a general theory of the cell which he develops in the third and last section. Taking up the physiological or dynamical standpoint, he points out that one process is common to all growth and development of tissues both in animals and plants, namely, the formation of cells, a process which he conceives to take place in the following manner. There is, first of all, a structureless substance, the cytoblastem, the matrix in which all cells originate. The cytoblastem may be either inside the cells, or, more usually, in the spaces between them. It is not a substance of definite chemical and physical properties, for the matrix of cartilage and the plasma of the blood alike come within the definition. It has largely the significance of food material for the developing cells. In plants, according to Schleiden, cells are never formed in the intercellular substance—the cytoblastem is within the cells; but extra-cellular cell formation seems to be the general rule in animals. An intracellular formation of cells occurs only in the ovum, in cartilage cells and chorda cells and in a few others, and even there it is not the exclusive method of formation; a formation of cells within cells never occurs in muscles and nerves, nor in fibrous tissue (p. 204). In the cytoblastem granules appear, which gradually increase in size and take on the characteristic shape of nuclei; round each of these a young cell is formed. Sometimes the young cells appear to have no nuclei, as in the intracellular brood of chorda cells, but, as a rule, a nucleus is clearly visible. The nucleus is indeed the most characteristic constituent of the cell. "The most important and most constant criterion of the existence of a cell is the presence or absence of the nucleus," writes Schwann near the beginning of his book (p. 43).

As a general rule the nucleolus is formed first, and round it by a sort of condensation or concretion the nucleus, which is frequently hollow, and round this again, by a somewhat similar process, the cell. "The whole process of the formation of a cell consists in the precipitation round a small previously formed corpuscle (the nucleolus) of first one layer (the nucleus) and then later round this a second layer (the cell substance)" (p. 213). The outermost layer of the cell usually thickens to form the membrane, but this membrane formation does not always occur, and the membrane is not present in all cells. The nucleus is formed in exactly the same manner as the cell, and it might with much truth itself be called a cell—a cell of the first order, while ordinary nucleated cells might be designated cells of the second order (p. 212). In anucleate cells there is probably only a single process of layer formation round an infinitely small nucleolus. In almost all nucleate cells the nucleus is resorbed when the cell reaches its full development, and it is larger and more important the younger the cell is.

The cell was for Schwann not a morphological concept at all, but a physiological; the cell was a dynamical, not a statical unit. Cell-formation was the process at the back of all production of life, and cells were the centres of all vital activity. Each cell was itself an organism, and its life and activities were to some extent independent of the lives and activities of all the other cells. The multicellular organism was a colony of unicellular organisms, and its life was a sum of the lives of its constituent elements. This "theory of the organism," which holds so important a place in biology even at the present day, is developed by Schwann in the concluding pages of his book.

He begins by contrasting the teleological with the materialistic conception of living things. In the teleological view, a special force works in the living organism, guiding and directing its activities towards a purposeful end. According to the materialistic view there are no other forces at work in the living organism than those which act in the inorganic realm, or at least there are none but forces at one with these in their blindness and necessity. True, the purposiveness of living processes cannot be denied; but its ground lies, according to this view, not in a vital force which guides and rules the individual life, but in the original creation and collocation of matter according to a rational plan. The purposiveness of life is part of the purposiveness of the universe; just as the stars circle for ever in harmoniously adjusted paths, so do the processes of life work together towards a common end. Both are the inevitable result of the original distribution of matter in the primitive chaos, a distribution fixed by a rational and foreknowing Being (p. 222).

Which of the two conceptions is to be adopted in biology? Teleological explanations have long been banished from the physical sciences, and in biology they are only a last resort when physical explanations have proved incomplete (p. 223). And if the ground of the purposiveness of living Nature is the same as the ground of the purposiveness of the universe, is it not reasonable to suppose that explanations which have proved satisfactory for inorganic things will in time with sufficient knowledge prove adequate also for organic things?

The teleological conception, again, leads to difficulties particularly when it is applied to the facts of reproduction. If we suppose that a vital force unifies and co-ordinates the organism and is its very essence, we must also suppose that this force is divisible and that a part of it — separated in reproduction — can bring about the same results as the whole. If on the contrary the forces having play in the organism are the mere result of the particular combination of the matter composing it, the reconstruction of a particular combination of molecules in the ovum is all that is necessary to set development a-going along exactly the course taken by the ovum of the parent. Another argument against the teleological view is derived from the facts of the cell-theory. The cell-theory tells us that the molecules of the living body are not immediately built up in manifold combinations to form the organism, but are formed first into unit-constructions or cells, and that these units of composition are invariably formed in all development, of plants and animals alike, however diverse the goal of development may be. If there were a vital principle would we not expect to find that, scorning this roundabout way of reaching its goal, it went straight to the mark, taking a different and distinctive course for each individual development, building up the organism direct without the intermediary of cells? But since there is a universal principle of development, namely, the formation of cells, does it not seem that the cells must be the true organisms, that the whole "individual" organism must be an aggregate of cells, and that the concept of individuality applied to the organism is accordingly a logical fiction? And it is just upon this notion of the individuality of the organism that the teleological concept is based. The teleological view can perhaps not be completely refuted until the adequacy of materialistic explanations has been finally shown; but it is certain that the most promising method for research is the materialistic (p. 226).

"We start out then from the assumption that the basis of the organism is not a force acting according to a definite plan; on the contrary, the organism arises through the action of blind and necessary laws, of forces which are as much implicit in matter as those of the inorganic world. Since the chemical elements in organic Nature differ in no way from those of inorganic Nature, the ground or cause of organic phenomena can consist only in a different mode of combination of matter, either in a peculiar mode of combination of the elementary atoms to form atoms of the second order, or in the particular arrangement of these compound molecules to form the separate morphological units of the organism or the whole organism itself" (p. 226). Accepting then the materialistic conception of the organism, we have to consider this further problem. Does the ground of organic processes lie in the whole organism or in its elementary parts? Translated into terms of metabolism — note the physiological point of view—the question runs, are metabolic processes the result of the molecular construction of the organism as a whole, or does the centre of metabolic activity lie in the cell? Is it the cell rather than the organism that is the immediate agent of assimilatory processes? In the first alternative the cause of the growth of the constituent parts lies in the totality of the organism; in the other alternative:—"Growth is not the result of a force having its ground in the organism as a whole, but each of the elementary parts possesses a force of its own, a life of its own, if you will; that is to say, in each elementary part the molecules are so combined as to set free a force whereby the cell is enabled to attract new molecules and so to grow, and the whole organism exists only through the reciprocal action of the single elementary parts. . . . In this eventuality it is the elementary parts that form the active element in nutrition, and the totality of the organism can be indeed a condition, but on this view it cannot be a cause" (p. 227).

To help in the decision of this question, appeal must be made to the facts established as to the cellular nature of the organism and of its reproductive elements. We know that every organism is composed of cells, which are formed and grow according to the same laws wherever they are found, whose formation therefore is everywhere due to the same forces. If we find that certain of these cells—all of which we know to be essentially identical one with another—have the power when separated from the others of growing and developing into new organisms, we can infer that not only such cells but also all other cells have this assimilatory power. The ova of animals, the spores of plants, the isolated cells of lower organisms in general, all show the power of separate assimilation and development. "We must therefore, in general, ascribe to the cell an individual life, that is to say, the combination of the molecules in the single cell does suffice to produce the force whereby the cell is enabled to draw to itself new molecules. The ground of nutrition and growth lies not in the organism as a whole, but in the separate elementary parts, the cells. The fact that it is not every cell that can continue to grow when separated from the organism is not in itself an objection to this theory, any more than it is an objection to the individual life of a bee that it cannot continue to exist apart from the swarm. The activation of the forces existing within the cell depends on conditions which the cell encounters only in connection with the whole" (pp. 228-9).

Schwann's next step is to discover what are the essential forces active in the cell, and here he enters the realm of hypothesis. He finds they can be reduced to two—an attractive force and a metabolic force. The attractive force is seen in the process of cell-formation, where first of all the nucleolus is formed by a concentration and precipitation of substances found free in the cytoblastem, and in the same way the nucleus and later the cell are laid down as concentric precipitates from the cytoblastem. Cell-formation also involves the second or metabolic force, by means of which the cell alters the chemical composition of the medium surrounding it so as to prepare it for assimilation. Schwann's attractive force brings about the actual taking up of the prepared substance; his metabolic force is the cause of the digestion of food substances, and is nearly identical with enzyme action. With what inorganic process, he now asks (p. 239), can the process of cell-formation be most nearly compared, and the answer obviously is, with the process of crystallisation. Cells are, it is true, quite different in shape and consistency from crystals, and they grow by intussusception, not by apposition—their plastic or attractive forces seem therefore to be different. A still more important difference is that the metabolic force is peculiar to the cell. Yet there are important analogies between crystals and cells. They agree in the important respect that they both grow in solutions at the cost of the dissolved substance, according to definite laws, and develop a definite and characteristic shape. It might even be maintained, Schwann thinks, that the attractive force of crystals is really identical with that of cells, and that the difference in result is due merely to the difference between the substance of the cell and the substance of the crystal. He points out how organic bodies are remarkable for their powers of imbibition, and he seeks to show that the cell is the form under which a body capable of imbibition must necessarily crystallise, and that the organism is an aggregate of such imbibition-crystals. The analogy between crystallisation and cell-formation he works out in the following manner:—"The substance of which cells are composed possesses the power of chemically transforming the substance with which it is in immediate contact, in somewhat the same way as the well-known preparation of platinum changes alcohol into acetic acid. Each part of the cell possesses this property. If now the cytoblastem is altered by an already formed cell in such a way that a substance is formed that cannot become part of the cell, it crystallises out first as the nucleolus of a new cell. This in its turn alters the composition of the cytoblastem. A part of the transformed substance may remain in solution in the cytoblastem or may crystallise out as the beginning of a new cell; another part, the cell-substance, crystallises round the nucleolus. The cell-substance is either soluble in the cytoblastem and crystallises out only when the latter is saturated with it, or it is insoluble and crystallises as soon as it is formed, according to the aforementioned laws of the crystallisation of imbibition-bodies; it forms thus one or more layers round the nucleolus, etc. If one imagines cell-formation to take place in this way, one is led to think of the plastic force of the cell as identical with the force by means of which a crystal grows" (pp. 249-50).

Two difficulties have to be faced by this theory—(1) the origin of the metabolic power of the cells, (2) the reason why the cells arrange themselves so as to form an organism of complex and definite structure. Schwann tries to explain the origin of the "metabolic" action, the analogy of which with the contact-action of colloidal platinum he recognises, by attributing it to the peculiar structural arrangements of molecules. In attempting to account for the harmonious structure of the organism he points to the analogy of ordinary crystals, which often form complex and regular tree-like arrangements; plants in particular resemble these regularly shaped crystal-aggregates.

The whole ingenious theory is offered merely as an hypothesis and a guide to research. It is interesting as one of the most carefully thought-out attempts ever made to give a thorough-going materialistic account of the origin and development of organic form, and it arose directly out of the cell-theory.

Schleiden and Schwann started out from an erroneous theory of the origin and development of cells, which impaired to some extent the value of their results. It was not long, however, before their theory of the origin of cells by "crystallisation" from an intra- or extra-cellular cytoblastem was challenged and overthrown, and the generalisation that cells originate by division from pre-existing cells put in its place.

This was established for plant cells by Meyen, Unger, von Mohl, Naegeli and Hofmeister in or about the forties.[1] Criticism of the Schwann-Schleiden theory from the zoological side was suggested by the study of the segmentation of the ovum—the developmental process in which the multiplication of cells is most easily observed. The segmentation of the ovum was well known to Schwann, for the process had been described in the frog by Prévost and Dumas in 1824,[2] in the frog and newt by Rusconi,[3] and an elaborate study of the process in the frog had been made by von Baer.[4] Schwann indeed suspected that there must be some connection between the segmentation of the ovum and the formation of cells, but he did not realise that the cellular blastoderm of the chick was formed by the division or segmentation of the egg-cell.

Segmentation was soon found to be of widespread occurrence. Von Siebold in 1837 described the process in Entozoa,[5] and in the same year Lovén saw segmentation in *Campanularia*,[6] and Sars in the starfish and in Nudibranchs.[7]

In 1838 Bischoff[8] observed segmentation in the mammalian ovum, and the whole course of segmentation in the ovum of the rabbit from the 2-celled to the morula stage was carefully described and figured by Barry[9] in 1839. C. Vogt[10] in 1842 described segmentation in *Coregonus* and *Alytes.* The discovery of segmentation in the ovum of

[1] Sachs, *History of Botany*, Book ii.

[2] *Ann. Sci. nat.*, i., pp. 110-14, 1824. Swammerdam is said to have observed the 2-celled stage in the egg of the frog (*Bibl. Nat.*, 1752), and Rösel v. Rosenhof the same stage in the tree-frog (*Hist. nat. ranarum nostratium*, 1758).

[3] *Développement de la grenouille commune*, Milan, 1826. *Biblioteca italiana*, lxxix., 1836, and Müller's *Archiv*, 1836. Agassiz is said by Vogt (1842) to have seen segmentation in the Perch as early as 1831.

[4] Müller's *Archiv*, 1836.

[5] In Burdach, *Die Physiologie als Erfahrungswissenschaft*, 2nd Ed., vol. ii.

[6] Wiegmann's *Archiv*, 1837.

[7] *Bericht Versamml. deutsch. Naturf. in Prag*, 1837.

[8] *Bericht Versamm. deutsch. Naturf. in Freiburg*, 1838. Later in his *Entw. d. Wirbelth.*, and in his papers on the development of the rabbit.

[9] *Phil. Trans.*, 1839. See particularly Pl. vi., figs. 105-12.

[10] *Embryologie des Salmones* 1842.

birds was not made until 1847, by Bergmann,[1] confirmed independently by Coste[2] in 1850. By 1848 segmentation had been noted in *Hydra* and various hydroids, in acalephs, in starfish, polyzoa, nematodes, rotifers, leeches, oligochætes, polychætes, in most groups of molluscs and arthropods, and in all the vertebrate classes.[3]

The process was at first held to be merely one of yolk-division, or *Dotterfurchung*, and its details were by most interpreted in the light of the Schleiden-Schwann theory of cell-formation.

The first steps towards a truer conception of the process seem to have been taken by Bergmann, who in 1841[4] called attention to the presence of nuclei in the segmentation-spheres of the frog's egg, and by Bagge in the same year, who observed that division of the nuclei preceded the multiplication of the segmentation spheres.[5] He considered the nuclei to be anucleate cells, and the same view was taken by Kölliker in 1843.[6] Next year, however, in his classical paper on Cephalopod development[7] Kölliker came to the opinion that they were really nuclei. He showed that segmentation was brought about by cell-division, that between "total" and "partial" segmentation there was a difference of degree and not of kind, and that the cells of the body were formed by division of the segmentation spheres. He held, however, that the nuclei multiplied endogenously and not by division. The division of nuclei was observed by Coste in 1846.[8] Leydig in 1848[9] took the necessary step in advance and maintained that the nuclei as well as the cells increased always by division. He was supported by Remak, who in a paper of 1852,[10] and more fully in his monumental

1 Müller's *Archiv*, 1847.

2 *C.R. Acad. Sci.*, xxx., p. 638.

3 See review by Leydig in *Isis*, 1848, pp. 161-193.

4 Müller's *Archiv*, pp. 89-102, 1841.

5 *De evolutione Strongyli auric. et Ascaridis acum.*, Erlangen, 1841.

6 Müller's *Archiv*, pp. 66-141, 1843.

7 *Entwickelungsgeschichte der Cephalopoden*, Zurich, 1844.

8 *Froriep's Notizen*, No. 800, 1846.

9 *Isis*, 1848.

10 Müller's *Archiv*, p. 47, 1852, also 1854 and 1858.

Untersuchungen über die Entwickelung der Wirbelthiere (Berlin, 1850-55), proved that in the frog's egg at least segmentation was a simple process of cell-division, initiated always by division of the nucleus.[1]

One point Remak left undecided—the fate of the *Keimbläschen* or egg-nucleus. It was generally held, even so late as the 'fifties, that the egg-nucleus disappeared just before segmentation began — Bischoff clung to this belief even in 1877.[2] Though Barry had held in 1839 that the egg-nucleus does not disappear in segmentation, J. Müller seems to have been the first actually to prove that it forms by division the nuclei of the first two segmentation spheres. He furnished the demonstration in the egg of *Entoconcha mirabilis*,[3] and his paper was known to Remak, who could not, however, observe a similar division of the egg-nucleus in the frog. Müller's discovery was confirmed for *Oceania armata* by Gegenbaur,[4] and for *Notommata sieboldii* by Leydig.[5]

In 1854 Virchow,[6] previously a supporter of Schwann, crystallised the new views in the famous phrase—*Omnis cellula e cellula*—and gave wide publicity to them in his classical lectures on Cellular Pathology, delivered in 1858.[7] The new doctrine of cell-formation was also taught by Leydig[8] in his text-book of histology, published in 1857.

The Schleiden-Schwann theory of the origin of cells by generation in a cytoblastem was now definitely overthrown.

The importance of the protoplasmic content of the cell was brought into prominence through the work of Dujardin,[9]

[1] See particularly Plate IX., figs. 3-7.

[2] *Hist.-krit. Bemerkungen zu den neuesten Mittheilungen ü. d. erste Entwickelung d. Säugethiereier*, München, 1877.

[3] *Monatsber. Akad. Wiss. Berlin*, 1851.

[4] *Zur Lehre von Generationswechsel u. d. Fortpflanzen d. Medusen u. Polypen.*

[5] *U. d. Bau u. d. system. Stellung d. Räderthiere*, 1854.

[6] *Arch f. path. Anat. Phys.*, vii., pp. 1-39, 1854. Also in his *Beiträge z. spec. Path. u. Therapie.*

[7] *Die Cellularpathologie*, Berlin, 1858.

[8] *Lehrbuch der Histologie*, 1857.

[9] *Ann. Sci. nat.* (2) iii., pp. 108-9 and pp. 312-4, 1835. Also iv, pp 343-77.

Purkinje,[1] Cohen[2] and Max Schultze.[3] The last-named in 1861 proposed a definition of the cell which might be accepted at the present day. "A cell," he wrote, "is a little blob of protoplasm containing a nucleus" (p. 11).

[1] 1839 or 1840.

[2] *Nova Acta Acad. Leop.*, xxii., 1850. Trans. in 1853 for Ray Society.

[3] *Arch. f. Anat. u. Physiol.*, pp. 1-27, 1861.

CHAPTER XII

THE CLOSE OF THE PRE-EVOLUTIONARY PERIOD

THE influence of the cell-theory on morphology was not altogether happy. The cell-theory was from the first physiological; cells were looked upon as centres of force rather than elements of form, and the explanation of all the activities of the organism was sought in the action of these separate dynamic centres. There resulted a certain loss of feeling for the problems of form. The organism was seen no longer as a cunningly constructed complex of organs, tissues and cells; it had become a mere cell-aggregate; the higher elements of form were disregarded and ignored.

We have seen this physiological attitude expressed with the utmost clearness by the founder of the cell-theory himself; we shall see the same attitude taken up by most of his successors. Thus Vogt, who was later to become one of the protagonists of materialism in Germany, developed in his memoir on the embryology of *Coregonus*[1] the theory of the independent or individual life of the cell. "Each cell," he wrote, "represents in some measure a separate organism, and while their development necessarily conforms to the general plan and the particular tendencies of the parent organism, they nevertheless each follow their own particular tendency and do not lose their independence until, by reason of the metamorphoses which they undergo, they lose their cellular nature" (p. 275).

And again, ". . . we are obliged to admit the existence in the cell of an independent life, which makes its development self-sufficient. . . . Each cell consequently represents a little independent organism, which assimilates foreign substances, builds them up, and rejects those that are useless;

[1] *Embryologie des Salmones*, 1842.

from this point of view the embryo can be compared up to a certain point with a zoophyte stock, of which each polyp, while living its own independent life, is yet incorporated in the common corm, which impresses its distinctive character upon every polyp" (p. 293).

Classical expression was given to the "colonial theory" of the organism by Virchow in his lectures on "Cellular Pathology."[1] For Virchow the organism resolves itself into an assemblage of living centres, the cells; the organism has no real existence as a unity, for there is no one single centre from which its activities are ruled. Even the nervous system, which appears to act as a co-ordinating centre, is itself an aggregate of discrete cells. "A tree is a body of definite and orderly composition, the ultimate elements of which, in every part of it, in leaf and root, in stem and flower, are cellular elements—so also are animal forms. *Every animal is a sum of vital units*, each of which possesses the full characteristics of life. The character and the unity of life cannot be found in one definite point of a higher organisation, for example in the brain of man, but only in the definite, constantly recurring disposition shown individually by each single element. It follows that the composition of the major organism, the so-called individual, must be likened to a kind of social arrangement or society, in which a number of separate existences are dependent upon one another, in such a way, however, that each element possesses its own particular activity, and, although receiving the stimulus to activity from the other elements, carries out its own task by its own powers" (2nd ed., pp. 12-13).

Analysis, decomposition, or disintegration of the organism is here pushed to its extreme point, and the problem of recomposition, synthesis and co-ordination shirked or forgotten.

The harmful influence of the cell-theory upon morphology did not pass unnoticed by the broader-minded zoologists of the day. Virchow's earlier paper[2] on the application of the

[1] *Die Cellularpathologie in ihrer Begründung auf physiologische und pathologische Gewebelehre*, Berlin, 2nd ed. 1859; Eng. trans., by Chance, 1860.

[2] *Arch. path. Anat. Phys.*, vii., pp. 1-39 (1854).

cell-theory to physiology and pathology called forth a vigorous protest from Reichert,[1] who discussed in a very instructive way the contrast between the older "systematic" and the newer "atomistic" attitude to living Nature.

Is it really true, he asks, that the cell is the dominant element in all organisation; is the cell comparable in importance to the atom of the chemists; or is it not rather the servant of a higher regulatory power? Johannes Müller, who was Reichert's master, had in his *Physiology*[2] argued splendidly for the existence of a creative force which guides and rules development, and brings to pass that unity and harmony of composition which distinguish living things from inorganic products. Reichert sought in vain in the writings of the biological "atomists" for any smallest recognition of these broader characteristics of living things upon which Müller had rightly laid stress. For the atomists the cell was the only element of form; they ignored the combination of cells to form tissues, of tissues to form organs, of organs to form an organism. For the morphologists the cell was one element among many, and the lowest of all.

The difference of attitude is clearly shown if we consider from the two points of view a complicated organ-system such as the central nervous system. The atomist sees in this a mere aggregate of cells or at the most of groups of cells. "The morphologist," on the other hand, "sees in the central nervous system a *proximate* element in the composition of the body—a primitive organ. From this point of view he apprehends and judges its morphological relations with, in the first place, the other co-ordinated primitive organs in the system as a whole; in all this the cells remain in the background, and have nothing to do directly with the determination of these morphological relations" (p. 6). Within the nervous system there are separate organs which stand to one another in a definite morphological and functional relationship. These organs are, it is true, composed of cells; but between the form and connections of

[1] *Bericht über die Fortschritte der mikroskopischen Anatomie im jahre* 1854. Müller's *Archiv*, 1855. See also 1856.

[2] *Hndb. d. Physiol.*, i., 1835.

these organs and the cells which compose them there is no direct and necessary relation (p. 6). It is true that the cell is the ultimate element of organic form, and that all development takes place by multiplication and form-change of cells. Yet is the cell in all this not independent of the unity of the developing embryo, and what the cells produce, they produce, so to speak, not of their own free will, nor by chance, but under the guiding influence of the unity of the whole, and in a certain measure as its agents (p. 7). The atomists will not admit the truth of this; they see in development nothing more than a process of the form-change and multiplication of cells. The full meaning of development escapes them, for they take no cognisance of the increasing complexity of the embryo, of the separating-out of tissues, of the moulding of organs, of the harmonious adaptation and adjustment of the parts to form a working whole.

In general, the fault of the atomists is that they do not respect the limits which Nature herself has prescribed to the process of logical analysis and disintegration of the organism; they do not recognise the existence of natural and rational units or unities; they forget the one great principle of rational analysis, "that, by universally valid, inductive, logical method, natural objects must in all cases be accepted and dealt with in the combination and concatenation in which they are given" (p. 10).

The atomists at least recognised one natural organic element, the cell; the materialistic physiologists of the time resolved even this unity into an aggregate of inorganic compounds, and regarded the organism itself as nothing but a vastly complicated physico-chemical mechanism. From this point of view morphology had no right of existence, and we find Ludwig, one of the foremost of the materialistic school, maintaining that morphology was of no scientific importance, that it was nothing more than an artistic game, interesting enough, but completely superseded and robbed of all value by the advance of materialistic physiology.[1]

Naturally enough, morphologists did not accept this rather contemptuous estimate of their science, but held

[1] See Leuckart's reply to Ludwig's criticism, in *Zeit. f. wiss. Zool.*. ii., p. 271, 1850.

firmly to the morphological attitude. So Leuckart in his reply to Ludwig, so Rathke in a letter to Leuckart published in that reply, so Reichert in his *Bericht*, so J. V. Carus in his *System der thierischen Morphologie*,[1] upheld the validity, the independence, of morphological methods. Leuckart and Rathke called attention to the absolute impossibility of explaining by materialistic physiology the unity of plan underlying the diversity of animal form. J. V. Carus, who was convinced of the validity of physiological methods within their proper sphere, drew a sharp distinction between systematics and morphology on the one hand, and physiology on the other. Physiology had nothing to do with the problems of form at all; its business was to study the physical and chemical processes which lay at the base of all vital activities. Morphology, on its part, had to accept form as something given, and to study the abstract relations of forms to one another. "On this point," he writes, "stress is to be laid, that morphology has to do with animal form as something *given* by Nature, that though it follows out the changes taking place during the development of an animal and tries to explain them, it does not enquire after the conditions whose necessary and physical consequence this form actually is" (p. 24). He expressed indeed a pious hope (p. 25) that physiology might one day be so far advanced that it could attempt with some hope of success to discover the physico-chemical determinism of form, but this remained with him merely a pious hope. Reichert, in his *Bericht*, applied to the rather wild theorisings of the physiologist Ludwig the same clear commonsense criticism that he bestowed on the other "atomists."

It would take too long to describe the great development that materialistic physiology took at this time, and to show how the separation of morphology from physiology, which originally took place away back in the 17th century, had by this time become almost absolute. The years towards the end of the first half of the century marked indeed the beginning of the classical period as well of physiology as of dogmatic materialism. Moleschott and Buchner popularised materialism in Germany in the 'fifties, while Ludwig, du Bois

[1] Leipzig, 1853.

Reymond and von Helmholtz began to apply the methods of physics to physiology. In France, Claude Bernard was at the height of his activity, rivalled by workers almost as great. The doctrine of the conservation of energy was established about this same time.

Between the cell-theory on the one side, and physiology on the other, it was a wonder that morphology kept alive at all. The only thing that preserved it was the return to the sound Cuvierian tradition which had been made by many zoologists in the 'thirties and 'forties. It is a significant fact that this return to the functional attitude coincided in the main with the rise of marine zoology, and that the man who most typically preserved the Cuvierian attitude, H. Milne-Edwards, was also one of the first and most consistent of marine biologists. Milne-Edwards describes in his interesting *Rapport sur les Progrès récents des Sciences zoologiques en France*" (Paris) 1867, how "About the year 1826, two young naturalists, formed in the schools of Cuvier, Geoffroy and Majendie, considered that zoology, after having been purely descriptive or systematic and then anatomical, ought to take on a more physiological character; they considered that it was not enough to observe living objects in the repose of death, and that it was desirable to get to understand the organism in action, especially when the structure of these animals was so different from that of man that the notions acquired as to the special physiology of man could not properly be applied to them" (p. 17). The two young naturalists were H. Milne-Edwards and V. Audouin. In pursuance of these excellent ideas they set to work to study the animals of the seashore, producing in 1832-4 two volumes of *Recherches pour servir à l'histoire naturelle du littoral de la France.* After Audouin's early death A. de Quatrefages was associated with Milne-Edwards in this pioneer work, and their valiant struggles with insufficient equipment and lack of all laboratory accommodation, and the rich harvest they reaped, may be read of in Quatrefage's fascinating account of their journeyings.[1] Note that though they called themselves

[1] *Souvenirs d'un Naturaliste*, 2 vols., Paris, 1854. Eng. Trans. as *Rambles of a Naturalist on the Coasts of France, Spain, and Italy*, 2 vols., 1857.

physiologists they meant by physiology something very different from the mere physical and chemical study of living things. They were interested, as Cuvier was, primarily in the problems of form; they sought to penetrate the relation between form and function; their chief aim was, therefore, the study not of physiology[1] in the restricted sense, but physiological morphology. As a matter of fact they produced more taxanomic and anatomical work than work on physiological morphology, but this was only natural, since such a wealth of new forms was disclosed to their gaze. Milne-Edwards' masterly *Histoire Naturelle des Crustacés*[2] and A. de Quatrefage's *Histoire Naturelle des Annelés marins et d'eau douce*[3] were typical products of their activity.

In the North, men like Sars and Lovén were starting to work on the littoral fauna of the fjords; in Britain, Edward Forbes was opening up new worlds by the use of the dredge; Johannes Müller was using the tow-net to gather material for his masterly papers on the metamorphoses of Echinoderms.[4] Work on the taxonomy and anatomy of marine animals was in general in full swing by the 'fifties and 'sixties.

This return to Nature and to the sea had a very beneficial effect upon morphology, bringing it out from the laboratory to the open air and the seashore. It saved morphology from formalism and aridity, and in particular from a certain narrowness of outlook born of too close attention paid to the details of microscopical anatomy. It brought morphologists face to face again with the wonderful diversity of organic forms, with the unity of plan underlying that diversity, with the admirable adjustment of organ to function and of both to the life of the whole.

Milne-Edwards' theoretical views, as expounded in his *Introduction à la zoologie générale* (1851), well reflect this Cuvierian attitude.[5] He acknowledges himself the debt he

[1] Milne-Edwards later published a classical textbook on comparative anatomy and physiology—*Leçons sur la Physiologie et l'Anatomie comparées*, 14 vols., Paris, 1857-80.

[2] Paris, 1834-40. Three volumes of the *Suites à Buffon.*

[3] Paris, 1865. Two volumes of the *Suites à Buffon.*

[4] *U. d. Metamorphose der Ophiuren u. Seeigel.*, Berlin, 1848. *U. d. Metamorphose der Holothurien u. Asterien.*, Berlin, 1851.

[5] As I have been unable to obtain a copy of the *Introduction*,

owes to Cuvier; "the further I advance in the study of the sciences which he cultivated with so sure a hand," he writes in 1867, "the more I venerate him."

Milne-Edwards frankly takes up the teleological standpoint, and interprets organic forms on the assumption that they are purposive and rationally constructed. "To arrive at an understanding of the harmony of the organic creation," he writes, "it seemed to me that it would be well to accept the hypothesis that Nature has gone about her work as we would do ourselves according to the light of our own intelligence, if it were given us to produce a similar result. Comparing and studying living things as if they were machines created by the industry of man, I have tried to grasp the manner in which they might have been invented, and the principles whose application would have led to the production of such an assemblage of diversified instruments" (p. 435). The problem is to discover the laws which rule the diversity of organic forms. The first and most obvious of these laws is the "law of economy," or the law of unity of type. Nature, as Cuvier pointed out, has not had recourse to all the possible forms and combinations of organs; she appears to work with a limited number of types and to get the greatest possible diversity out of these by varying the proportions of the constitutive materials of structure. Within the limits of each type Nature has brought about diversity by raising her creatures to different degrees of perfection. This is the second law of organic form, and it is this law that Milne-Edwards chiefly elaborates. Degrees of perfection mean for him, as for Aristotle, primarily degrees of perfection of function, but since structure is necessarily in close relation with function, perfection of function brings in its train increased perfection of organisation. This can only be attained by a division of labour[1]

the passages which follow are taken from the *Rapport* of 1867, where Milne-Edwards gives a complete exposition of his doctrine, sometimes in the words of the original.

[1] This principle was first developed by Milne-Edwards in 1827, in the *Dictionnaire classique d'Hist. naturelle.* It was probably suggested to him by his studies on the Crustacea, among which the principle is so beautifully exemplified in the concentration and specialisation of the appendages and the ganglionic chain.

among the organs and by their consequent differentiation. An animal is like a workshop where some complicated product is manufactured, and the organs are like the workmen. Each workman has his own special piece of work to do, at which he becomes thoroughly expert; and the finished product is manufactured more rapidly and efficiently by the co-operation of workers each skilled in one department than it would be if each workman had to produce the whole. Applied to the organism this principle of the division of labour means the differentiating out of the separate functions, their localisation in different parts of the organism, and their co-ordination to produce a combined result.

This differentiation of functions implies a corresponding differentiation of organs, but it is functional differentiation which always takes the lead. "Where division of labour has not been introduced into the organism there must exist a great simplicity of structure. But just as uniformity in the functions of the different parts of the body implies a uniformity in their mode of constitution, so diversity in function must be accompanied by particularities in structure; and, in consequence also, the number of dissimilar parts must be augmented and the complication of the machine increased" (p. 463). Since function comes before form there is not always a special organ for every function. "It is a grave error to believe that a particular function can be performed only by one and the same organ. Nature can arrive at the desired result by various ways, and when we look down through the animal kingdom from the highest to the lowest forms we see that the function does not disappear even when the special instrument provided for the purpose in the higher types ceases to exist" (p. 470).

Nature, holding fast to the law of economy, does not even always create a new organ for a new function; she may simply adapt an undifferentiated part to special functions, or she may even convert to other uses an organ already specialised (p. 464). So, for example, the function of respiration is in the lowest animals diffused indifferently over the whole surface of the body, and only as organisation advances is it localised in special organs, such as gills. Now suppose that Nature wishes to adapt a fish, which breathes by gills, to life in the air; she does not create an organ specially for this purpose, but utilises the moist gill-chamber (*e.g.*, in *Anabas scandens*), modifying it in certain ways so that the fish can take advantage of the oxygen it contains. But this gill-chamber lung is at best a makeshift, and when she comes to the more definitely terrestrial Amphibia Nature gives up the attempt to use the gill-chamber as a lung, and creates a new organ, the true vertebrate lung, specially adapted for breathing air (p. 475).

But whatever means Nature adopts, her aim is always the same—to specialise, to differentiate, to produce diversity from uniformity.

Differentiation not only raises the level of organisation; it usually also takes the direction of adaptation to particular habits of life, and this is perhaps the most fruitful cause of diversity. Everywhere we find animals specialised in adaptation to their environment—to life in air or water, or on land—and many of their most striking differences are due to this cause. But adaptation may also act in reducing diversity, for there necessarily occur many instances of parallel adaptation or convergence. So we get the extraordinary parallelism between the families of marsupials and the orders of placentals,[1] the remarkable similarity between the respiratory organs of land-crabs and air-breathing fish—to mention only two out of an immense range of analogous facts.

The last cause of diversity that Milne-Edwards adduces is what he calls a "borrowing" of peculiarities of structure from another systematic group. Thus, "among reptiles, the tortoises seem to have borrowed from birds some of their characteristic features of organisation; and among the sauroid fishes the piscine type seems to have been influenced by the type from which reptiles are derived" (p. 479). So many riddles that, a little later on, stimulated the ingenuity of the evolutionists!

Such, then, were the factors which Milne-Edwards

[1] Studied by Isidore Geoffroy St Hilaire in his paper *Classification parallélique des Mammifères*, *C. R. Acad. Sci.*, xx., 1845. Remarked upon by Cuvier, *Règne animal.*, i., p. 171, 1817, also by de Blainville.

considered adequate to explain the rich variety of animal forms. We cannot do better than quote his own summary of his doctrine:—"To sum up, then, the great differences introduced by Nature into the constitution of animals seem to depend essentially upon the existence of a certain number of general plans or distinct types, upon the perfecting in various degrees either of the whole or of parts of each of these structural plans, upon the adaptation of each type to varied conditions of existence, and upon the secondary imitation of foreign types by certain derivatives of each particular type" (p. 480).

We have laid stress on the fact that Milne-Edwards put function before form, for this is the mark of the true Cuvierian. With it goes the belief that Nature forms new parts to meet new requirements, that she is not limited, as Geoffroy thought, to a definite number of "materials of organisation," but can produce others at need. Cuvier held, for example, that many of the muscles and even the bones of fish were peculiar to them, and without homologues in the other Vertebrates, having been created by Nature for special ends.[1] So, too, Johannes Müller, who in many ways and not least in his sane vitalism was a follower of the Cuvierian tradition, recognised that many of the complicated cartilages in the skull of Cyclostomes were specially formed for the important function of sucking, and had no equivalent in other fish.[2]

So, too, the embryologists after Cuvier often came across instances of the special formation of parts to meet temporary needs. Thus Reichert interpreted the "palatine" and "pterygoid," which are formed in the mouth of the newt larva by a fusion of conical teeth, as special adaptations to enable the little larva to lead a carnivorous life.[3]

Not many years after the publication of Milne-Edwards' *Introduction à la zoologie générale* (1851) there appeared a book by H. G. Bronn in which was offered a very similar analysis of organic diversity. The curious thing was that

[1] Cuvier et Valenciennes, *Hist. nat. des Poissons*, i. p. 550, 1828.

[2] *Myxinoiden*, Th. I. *Abh. k. Akad. Wiss. Berlin* for 1834, pp. 100, 110, 179, etc.

[3] *Vergl. Entw. Kopf. nackt. Amphibien*, p. 101, 1838.

Bronn approached the problem from quite a different standpoint, from the standpoint, indeed, of *Naturphilosophie.* Of this the title of the book is itself sufficient proof—*Morphologische Studien über die Gestaltungs-gesetze der Naturkörper überhaupt und der organischen insbesondere* (Leipzig and Heidelberg, 1858).[1] The linking up of organic with inorganic form is characteristic; there is much talk, too, in the book of *Urstoffe* and *Urkräfte*, but underlying the *Naturphilosophie* we can trace the same Cuvierian treatment of form, and see crystallise out laws of progressive development that bear no small analogy with the laws established by Milne-Edwards.

According to Bronn, the ideal fundamental form of the plant is an ovoid or strobiloid [2] body, for a plant reaches out in two directions in search of food—towards the sun and towards the earth. Animals differ from plants in being endowed with sensation and mobility (*cf.* Aristotle and Cuvier), and it is this characteristic that gives them their distinctive form. The main types of animal form—the Amorphozoa, Actinozoa, and Hemisphenozoa — are essentially adaptations to particular modes of locomotion. Animals either are fixed, or they move in all directions without reference to any definite axis, or they move in one main direction.

The Amorphozoa or shapeless animals include many of the Protozoa and sponges; they have no typical form, and most of them are sessile. The Actinozoa include such animals as the Cœlentera, which are fixed, and the Echinoderms, which have a central point and move indifferently along any radial axis; their form differs from the strobiloid mainly in having radiate rather than spiral symmetry. The Hemisphenozoa, or bilaterally symmetrical animals, include all those that habitually move forward; they have a front end and a hind end, a dorsal surface and a ventral, and the mouth, sense-organs and "brain" are concentrated

[1] I have not seen the companion volume on palæontological progression, *Unters. ü. d. Entwickelungsgesetze der organischen Welt während der Bildungszeit unserer Erdoberfläche*, Stuttgart, 1858.

[2] "Strobiloid" because of its spiral development. The theory of the spiral growth of plants played an important part in botanical morphology about this time.

in the front end to form a head—all in direct adaptation to this forward movement; they make up the vast majority of animals.

The fundamental forms of living things are, however, merely so many themes on which a multitude of further variations are woven, through the action of the laws which rule the detail of organic diversities. These further laws may be set down under four main heads. Under the first comes the law of the existence of certain fundamentally distinct structural types, which are distinguished from one another by their ground-form, by the number of organ-systems, and by the number of homotypic organs they possess, but principally by the relative position of the organs to one another (principle of connections). The form and connections of the nervous system are of particular importance in distinguishing the types (*cf.* Cuvier). The second factor in the diversity of organic form is the action of certain laws of progressive development[1] (*Entwickelungsgesetze*), which bear the same relation to the development of the animal kingdom as the laws of individual development bear to the development of the embryo, for organs appear in the different animal series in much the same order and manner as they develop in the individual. These laws are (1) progressive differentiation of functions and organs; (2) numerical reduction of serially repeated parts; (3) concentration of functions and their organs in particular parts of the body; (4) centralisation of organ-systems and parts of such, so that they come to depend upon one central organ; (5) internalisation of the "noblest" organs, unless these are necessarily external, and (6) increase in size of the whole or of parts. Of these the law of differentiation is by far the most important, and most of the others are in a sense merely special cases of this fundamental law. To this law of differentiation is due the increase in complexity or perfection of organisation which is shown by all the animal series. Bronn himself recognised the great similarity of this law of progressive differentiation to Milne-Edwards' principle of the division of labour; he seems, however, to have arrived at it independently.

[1] *Cf.* Meckel's Principle of progressive Evolution, *supra*, p. 93.

Bronn's third factor in the production of variety of form is adaptation to environment, or better, functional response to environment. Bronn gives an excellent account of adaptational modifications and calls attention, just as Milne-Edwards did, to the numerous analogies of structure which adaptation brings about. He works out the interesting view that there is some connection between classificatory groups and adaptational forms, especially such as are connected with the function of locomotion:—"Based upon a common characteristic method of locomotion are whole or nearly whole sub-phyla (Hexapoda), classes (mammals and reptiles, birds, fishes, gastropods, pteropods, brachiopods, Bryozoa, Rotifera, jelly-fish, polypes, sponges), sub-classes (mobile and immobile lamellibranchs, echinoderms, walking and swimming Crustacea, parasitic and free-living worms, and so on), often, however, only orders and quite small groups (snakes, eels, bats, sepias, medusæ, etc.)" (p. 141).

It was characteristic of the 'forties and 'fifties that transcendental anatomy, along with Nature-philosophy, went rather out of fashion, its false simplicities and premature generalisations being overwhelmed by the flood of new discoveries. A few stalwarts indeed upheld transcendental views. We have already discussed the morphological system built up by Richard Owen in the late 'forties, a system transcendental in its main lines. We have seen the vertebral theory of the skull still maintained in the 'fifties by such men as Reichert and Kölliker, and we find J. V. Carus in 1853[1] taking it as almost conclusively proved.[2]

We may mention, too, as showing clear marks of the influence of transcendental ideas, L. Agassiz's work on the principles of classification.[3] And Serres, who was Geoffroy's

[1] *System der thierischen Morphologie*, pp. 33, 457. Also C. Bruch, *Die Wirbeltheorie des Schädels, am Skelette des Lachses geprüft*, Frankfort-on-Main, 1862.

[2] In France the vertebral theory was advocated by Lavocat in his *Nouvelle Ostéologie comparée de la tête des animaux domestiques*, Toulouse, 1864. It seems also that Lacaze-Duthiers held fast to it even in 1872—*Arch. zool. exp. gén.*, i., p. 51, 1872.

[3] *An Essay on Classification*, Boston, 1857, London, 1859. He considered the classificatory categories to be the categories of the Creator's thought, and hence natural, and in no sense mere conventions.

chief disciple, recanted not a whit of his doctrine of recapitulation, but re-affirmed and expanded it from time to time, and particularly in a lengthy memoir published in 1860.[1] But in general we may say that pure morphology in the Geoffroyan or Okenian sense was becoming gradually discredited. A curious indication of this is seen in the fact that not only the idea but the very word "Archetype" came to be regarded with suspicion. Thus even J. V. Carus, who had much affinity with the transcendentalists, wrote of the vertebrate archetype (which he took over almost bodily from Owen)—"It may here be observed that this schema may be used as a methodological help, but it is not to be placed in the foreground" (*loc. cit.*, p. 395). Huxley, who was definitely a follower of von Baer, was much more outspoken with regard to ideal types. In an important memoir on the general anatomy of the Gastropoda and Cephalopoda,[2] he set himself the task of reducing all their complex forms to one type. In summing up, he writes :—"From all that has been stated, I think that it is now possible to form a notion of the archetype of the Cephalous Mollusca, and I beg it to be understood that in using this term, I make no reference to any real or imaginary 'ideas' upon which animal forms are modelled. All that I mean is the conception of a form embodying the most general propositions that can be affirmed respecting the Cephalous Mollusca, standing in the same relation to them as the diagram to a geometrical theorem, and like it, at once imaginary and true" (i., p. 176). Again, in his Croonian lecture on the theory of the vertebrate skull, he remarks that a general diagram of the skull could easily be given. "There is no harm," he continues, "in calling such a convenient diagram the 'Archetype' of the skull, but I prefer to avoid a word whose connotation is so fundamentally opposed to the spirit of modern science" (*Sci. Memoirs*, vol. i., p. 571).

It is instructive to find that between Serres and Milne-Edwards there existed the same antagonism as between von

[1] "Principes d'Embryogénie, de Zoogénie et de Teratogénie," *Mém. Acad. Sci.*, xxv., pp. 1-943, pls. xxv., 1860.

[2] "On the Morphology of the Cephalous Mollusca," *Phil. Trans.*, 1853, *Sci. Memoirs*, i., pp. 152-92.

Baer and the German transcendentalists. Milne-Edwards was a constant critic of the law of parallelism which Serres continued to uphold with little modification for over thirty years, just as von Baer was a critic of that form of the doctrine which was current in the early part of the century. As early as 1833, Milne-Edwards, through his studies of crustacean development,[1] had come to the conclusion, independently of von Baer, that development always proceeded from the general to the special; that class characters appeared before family characters, generic characters before specific. In an interesting paper published in 1844,[2] he discussed the relation of this law of development to the problems of classification, and arrived at results almost identical with those set forth by von Baer in his Fifth Scholion.

Like von Baer he rejected completely the theory of parallelism and the doctrine of the scale of beings; like von Baer he held that the type of organisation—of which there are several—is manifested in the very earliest stages and becomes increasingly specialised throughout the course of further development; like von Baer, too, he sketched a classification based upon embryological characters.

These views were further developed in his volume of 1851, and also in his *Rapport* of 1867.

They brought him into conflict with his confrere in the Academy of Sciences, Étienne Serres, who in a number of papers published in the 'thirties and 'forties,[3] and particularly in his comprehensive memoir of 1860, still maintained the theory of parallelism and the doctrine of the absolute unity of type. His memoir of 1860 shows how completely Serres was under the domination of transcendental ideas. Much of it indeed goes back to Oken. "The animal kingdom," he writes, "may be considered in its entirety as a single ideal and complex being" (p. 141). His views have become a little more complicated since his first exposition of them in 1827,

[1] "Observations sur les changements de forme que les divers Crustacés éprouvent," *Ann. Sci. nat.* (1) xxx., p. 360, 1833.

[2] "Considérations sur quelques principes relatifs à la classification naturelle des animaux," *Ann. Sci. nat.* (3) i., p. 65, 1844.

[3] *Supra*, pp. 79-83. Also *Précis d'anatomie transcendante, principes d'organogénie*, Paris, 1842.

and he has been forced to modify in some respects the rigour of his doctrine. But he still holds fast to the main thesis of transcendentalism—the absolute unity of plan of all animals, vertebrate and invertebrate alike,[1] the gradual perfecting of organisation from monad to man, the repetition in the embryogeny of the higher animals of the "zoogeny" of the lower.

He recognised, however, that the idea of a simple scale of beings is only an abstraction, and that the true repetition is of organs rather than of organisms. He was willing even to admit, at least in the later pages of his memoir, that there might be not one animal series but several parallel series, as had been suggested by Isidore Geoffroy St Hilaire (p. 749). In general, his views are now less dogmatic than they were in his earlier writings, but they are not for all that changed in any essential. For, in summing up his main results, he writes, "The whole animal kingdom can in some measure be regarded ideally as a single animal, which, in the course of formation and metamorphosis in its diverse manifestations, here and there arrests its own development, and thus determines at each point of interruption, by the very state it has reached, the distinctive characters of the phyla, the classes, families, genera, and species" (p. 833).[2]

To settle the dispute pending between two of its most illustrious members, the Academy proposed in 1853, as the subject of one of its prizes, "the positive determination of the resemblances and differences in the comparative development of Vertebrates and Invertebrates." A memoir was presented the next year by Lereboullet[3] which met with the approval of the Academy in so far as its statements of fact were concerned, but seemed to them to require amplifica-

[1] The inversion of the organs shown by Vertebrates as compared with Invertebrates is due to the reversed position of the embryo relatively to the yolk! (pp. 821-6).

[2] It is worth while recording that Serres enunciated a "law of symmetry" according to which the embryo is formed by the union of its two symmetrical halves—a law which recalls the "concrescence theory" of His and some modern embryologists.

[3] "Embryologie comparée du Brochet, de la Perche, et de l'Ecrévisse," *Ann. Sci. nat.* (4), i., p. 237, 1854; ii., p. 39, 1854. *Mém. Savans etrangers*, xvii.

tion in its theoretical part. But even in this memoir Lereboullet was able to show that the balance of evidence was greatly in favour of Milne-Edwards' views, and his general conclusions in 1854 were that "in the presence of such fundamental differences, one is obliged to give up the idea of one single plan in the formation of animals; while, on the contrary, the existence of diverse plans or types is clearly demonstrated by all the facts" (p. 79). To fulfil the Academy's requirements, Lereboullet continued his work, and in 1861-63 he published a series of elaborate monographs[1] on the embryology of the trout, the lizard and the pond-snail *Lymnæa*, and rounded off his work with a full discussion[2] of the theoretical questions involved. In this considered and authoritative judgment he completely disposed of Serres' theories of the unity of plan and the unity of genetic formation. Except in the very earliest stages of oogenesis there is no real similarity between the development of a Zoophyte, a Mollusc, an Articulate and a Vertebrate, but each is stamped from the beginning with the characteristics of its type. The lower animals are not, and cannot possibly be the permanent embryos of the higher animals. "The results which I have obtained," he writes, "are diametrically opposed to the theory of the zoological series constituted by stages of increasing perfection, a theory which tries to demonstrate in the embryonic phases of the higher animals a repetition of the forms which characterise the lower animals, and which has led to the assertion that the latter are permanent embryos of the former. The embryo of a Vertebrate shows the vertebrate type from the very beginning, and retains this type throughout the whole course of its development; it never is, and never can be, either a Mollusc or an Articulate" (xx., p. 54).

"We are led to establish . . . as the general result of our researches, the existence of several types, and, consequently, of different plans, in the development of animals. These different types are manifested from the very beginning of embryonic life; the characters distinguishing them are there-

[1] *Ann. Sci. nat.* (4) xvi., p. 113, 1861; xvii., p. 88, 1862; xviii., p. 5, 1862; xix., p. 5, 1863.

[2] xx., p. 5, 1863.

fore primordial, and we can say with M. Milne-Edwards that *everything goes to prove that the distinction established by Nature between animals belonging to different phyla is a primordial distinction*" (p. 58).

In other directions also von Baer's work was confirmed and extended by later observers—those parts of it particularly that had reference to the germ-layer theory, and to the concept of histological differentiation. His germ-layer theory was accepted in its main lines by Rathke, Bischoff and Lereboullet, and applied by them to the multitude of new facts they discovered. Rathke, in particular, was a firm upholder of the doctrine, and made considerable use of it in his writings.[1] Even before the publication of von Baer's book he had interpreted in terms of the germ-layer theory sketched by his friend Pander the splitting of the blastoderm which occurs in the early development of *Astacus*, whereby there are formed a serous and a mucous layer, one inside the other—like the coats of an onion, to use his own expressive phrase.[2]

An ingenious application of the Pander-Baer theory was made by Huxley, who compared the outer and inner cell-layers which form the groundwork of the Cœlentera with the serous and mucous layers of the vertebrate germ.[3] He laid stress, it is true, rather on the physiological than on the morphological resemblance. "A complete identity of structure," he writes, "connects the 'foundation membranes' of the Medusæ with the corresponding organs in the rest of the series; and it is curious to remark, that throughout, the outer and inner membranes appear to bear the same physiological relation to one another as do the serous and mucous layers of the germ; the outer becoming developed into the muscular system, and giving rise to the organs of offence and defence; the inner, on the other hand, appearing to be more closely subservient to the purposes of nutrition and generation" (p. 24). Von Baer had already hinted at this homology

[1] Particularly in his *Blennius* (1833) and *Natter* (1839).

[2] In the "preliminary notice" of his Crayfish paper—*Isis*, pp 1093-1100, 1825.

[3] "On the Anatomy and the Affinities of the Family of the Medusæ," *Phil. Trans.*, 1849; *Sci. Memoirs*, i., pp. 9-32.

in the second volume of his *Entwickelungsgeschichte* (1837), where he says with reference to the separation of the blastoderm of the chick into two layers. "Yet originally there are not two distinct or even separable layers, it is rather the two surfaces of the germ which show this differentiation, just as polyps show the same contrast of an external surface and an internal digestive surface. In between the two layers there is in our germ as in the polyp an indifferent mass" (p. 67). The terms ectoderm and entoderm were introduced by Allman[1] in 1853 for the two cell-layers in the Hydrozoa.

Remak is the second great name in the history of the germ-layer theory. He had the great advantage over von Baer of being able to make use of the cell-theory in interpreting the formation of the germ-layers. Microscopical technique also had been greatly improved since 1828.[2]

Remak's greatest service was that he put the germ-layer theory in direct relation with the cell-theory by demonstrating the cellular continuity from egg-cell to tissue, and by showing that each germ-layer possessed distinctive histological characteristics. Hardly less important was his clear marking-off of the "middle layer" as a separate and distinct layer of the germ. He it was who introduced the modern conception of the mesoderm, and cleared up the confusion in which Pander and von Baer had left the organs formed between the serous and the mucous layer. Remak's middle layer was a different thing from Pander's ill-defined "vessel-layer"; it included and unified from a new point of view the "vessel" and "muscle" layers of von Baer.

There are in the unincubated blastoderm of the chick, according to Remak,[3] two cell-layers, of which the undermost

[1] *Phil. Trans.*, cxliii., p. 368, 1853.

[2] The principle of achromatism was discovered (by Fraunhofer) and achromatic microscopes introduced in the early part of the 19th century. The use of chemical reagents, such as acetic acid, and various hardening fluids, came into fashion not long after. J. Müller seems to have been one of the first to realise their importance. Remak himself invented one or two fixing and hardening mixtures (pp. 87, 127, 1855), which enabled him to cut excellent hand sections. Section-cutting machines were not invented till later (V. Hensen, 1866, His, 1870).

[3] *Untersuchungen über die Entwickelung der Wirbelthiere*, folio, pp. xxxvii + 195, 12 plates, Berlin, 1850-1855.

subsequently splits into two. Three layers are thus formed—the upper, middle and lower. The upper layer differentiates into a medullary plate and an epidermic plate (Remak's *Hornblatt*), and gives origin to the medullary tube with all its evaginations, and to the skin with all its derivatives and pockets. It forms such diverse structures as the brain, the spinal cord, the eye, the ear, the mouth, hairs, feathers, nails, sweat-glands, lacrymal glands, and so forth. All these parts are connected directly or indirectly with sensation, and the upper germ-layer may accordingly be called the *sensory* layer. The lower layer gives rise to the epithelium and the proper tissue of the alimentary canal and its derivatives, as the liver, lungs, pancreas, kidneys, thyroid, thymus, etc. These parts are all concerned in the processes of assimilation and dissimilation, and the lower layer may accordingly be called the *trophic* layer. Now between the upper or sensory layer and the lower or trophic layer there exists, in spite of their very different functions, a close histological likeness, for both are essentially epithelial layers. The resemblance is particularly strong if we compare the lower layer with the *Hornblatt* of the upper layer—both consist of epithelial tissue, and of its derivative, glandular tissue, and form neither vessels nor nerves. The middle layer, on the contrary, forms nerves and muscles, vessels and connective tissue, and little or no epithelium. It does not form all the blood-vessels without exception (and so cannot be called the vessel-layer), for the blood-vessels of the central nervous system are in all probability formed from the upper layer. So, too, it does not form all the nerves and muscles—the optic and auditory nerves and the nerves and muscles of the iris probably arise in the upper layer. But, in spite of these exceptions, its general histological character is so well defined that it may be contrasted with the other two as pre-eminently the layer that forms muscular, nervous, vascular and connective tissue. In view of its functional significance, it may be called the *motory* layer, or better, since it forms also the sexual glands, the *motor-germinative* layer. The middle layer, early in its history, shows a division into dorsal plates (*Urwirbelplatten*) and ventral plates (*Seitenplatten*). The former exhibit almost as soon as they are formed the characteristic proto-vertebral segmentation, the latter split to form the pleuro-peritoneal or body-cavity. Remak describes the latter process as follows:—"In the region of the trunk, where a greater independence of the fate of the alimentary canal and its annexes becomes necessary for the voluntary executive organs, the ventral plates undergo a process of splitting, leading to the formation of the sensitive part of the integument (the *Hautplatten*), the muscular part of the alimentary tube (the *Darmfaserplatten*), and the mother-tissue of the generative organs (the *Mittelplatten*).

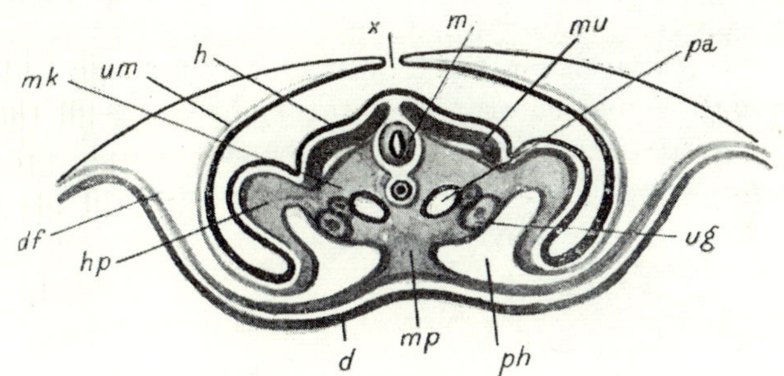

FIG. 12.—Transverse Section of Chick Embryo. (After Remak.)

h. Epidermis.
m. Spinal cord.
mu. Dorsal plate.
ug. Pronephric duct.
pa. Aortic root.
hp. and *um.* "Hautplatte."
mp. "Mittelplatte."
df. "Darmfaserplatte."
x. Edge of amniotic fold.
ph. Pleuro-peritoneal cavity.
d. Epithelium of alimentary canal.

From the *Hautplatten* there develops, without the dorsal plates seeming to take any part in the process, the rudiment of the extremities" (p. 79).

His *Darmfaserplatten* form the nervous and muscular tissue of the alimentary canal and its dependencies, and also the heart; the *Hautplatten* form the general body-wall (exclusive of the skin) and the appendages. In the embryo they line the amniotic cavity. The skeleton and peripheral nerves originate wholly within the middle layer.

Remak's conception of the relations of the three germ-layers to one another and to the body-cavity is well illustrated in Fig. 12.

In his germ-layer theory Remak's standpoint is histological rather than morphological. The distinction which he draws between the sensory and trophic layers on the one hand, and the motor-germinative layer on the other, is entirely a histological one. The greater part of his book, indeed, is devoted to a study of the histogenesis of the different organs of the body; he is bent chiefly upon unravelling the part which each germ-layer takes in the formation of each tissue and organ.

His generalisation that two of the germ-layers give rise exclusively or almost exclusively to one kind of tissue excited great interest at the time, and gave the direction to histogenetic research for quite a number of years, though in the end it turned out to be insufficiently founded.

Though Remak's germ-layer theory had thus principally a histological orientation, it laid down the main lines of the modern morphological treatment of the germ-layers.

CHAPTER XIII

THE RELATION OF LAMARCK AND DARWIN TO MORPHOLOGY.

It is a remarkable fact that morphology took but a very little part in the formation of evolution-theory. When one remembers what powerful arguments for evolution can be drawn from such facts as the unity of plan and composition and the law of parallelism, one is astonished to find that it was not the morphologists at all who founded the theory of evolution.

It is true that the noticeable resemblances of animals to one another, the possibility of arranging them in a system, the vague perception of an all pervading plan of structure, did suggest to many minds the thought that systematic affinities might be due to blood-relationship. Thus Leibniz considered that the cat tribe might possibly be descended from a common ancestor,[1] and another great philosopher, Immanuel Kant, was led by his perception of the unity of type to suggest as possible the derivation of the whole organic realm from one parent form, or even ultimately from inorganic matter. In the course of his masterly discussion of mechanism and teleology,[2] he writes, "The agreement of so many genera of animals in a certain common schema, which appears to be fundamental not only in the structure of their bones, but also in the disposition of their remaining parts—so that with an admirable simplicity of original outline, a great variety of species has been produced by the shortening of one member and the lengthening of another, the involution of this part and the evolution of that—allows a ray of hope, however faint, to penetrate into our minds, that here

[1] Rádl, *loc. cit.*, i., p. 71.

[2] *Kritik der Urtheilskraft*, 1790.

something may be accomplished by the aid of the principle of the mechanism of Nature (without which there can be no natural science in general). This analogy of forms, which with all their differences seem to have been produced according to a common original type, strengthens our suspicions of an actual relationship between them in their production from a common parent, through the gradual approximation of one animal-genus to another—from those in which the principle of purposes seems to be best authenticated, *i.e.*, from man down to the polype, and again from this down to mosses and lichens, and finally to the lowest stage of Nature noticeable by us, viz., to crude matter."[1]

So, too, Buffon's evolutionism was suggested by his study of the structural affinities of animals, and Erasmus Darwin in his *Zoonomia* (1794) brought forward as one of the strongest proofs of evolution, "the essential unity of plan in all warm-blooded animals."[2]

But, as a matter of historical fact, no morphologist, not even Geoffroy, deduced from the facts of his science any comprehensive theory of evolution. The pre-Darwinian morphologists were comparatively little influenced by the evolution-theories current in their day, and it was in the anatomist Cuvier and the embryologist von Baer that the early evolutionists found their most uncompromising opponents.

Speaking generally, and excepting for the moment the theory of Lamarck, we may say that the evolution-theories of the 18th and 19th centuries arose in connection with the transcendental notion of the *Échelle des êtres*, or scale of perfection. This notion, which plays so great a part in the philosophy of Leibniz, was very generally accepted about the middle of the 18th century, and received complete and even exaggerated expression from Bonnet and Robinet. Buffon also was influenced by it. Towards the beginning of the 19th century the idea was taken up eagerly by the transcendental school and by them given, in their theories of the

[1] Eng. Trans. by J. H. Bernard, p. 337, London, 1892.

[2] H. F. Osborn, *From the Greeks to Darwin*, p. 145, New York and London, 1894.

"one animal," a more morphological turn. Their recapitulation theory was part and parcel of the same general idea.

One understands how easily the notion of evolution could arise in minds filled with the thought of the ideal progression of the whole organic kingdom towards its crown and microcosm, man. Their theory of recapitulation led them to conceive evolution as the developmental history of the one great organism.[1] Many of them wavered between the conception of evolution as an ideal process, as a *Vorstellungsart*, and the conception of it as an historical process. Bonnet, Oken, and the majority of the transcendentalists seem to have chosen the former alternative; Robinet, Treviranus, Tiedemann, Meckel, and a few others held evolution to be a real process.

We have already in previous chapters[2] briefly noticed the relation of one or two of the transcendental evolution-theories to morphology, and there is little more to be said about them here. They had as good as no influence upon morphological theory, nor indeed upon biology in general.[3] It is different with the theory of Lamarck, which, although it had little influence upon biological thought during and for long after the lifetime of its author, is still at the present day a living and developing doctrine.

Lamarck's affinity with the transcendentalists was in many ways a close one, but he differed essentially in being before all a systematist. Nor is the direct influence of the German transcendentalists traceable in his work—his spiritual ancestors are the men of his own race, the materialists Condillac and Cabanis, and Buffon, whose friend he was. The idea of a gradation of all animals from the lowest to the highest was always present in Lamarck's mind, and links him up, perhaps through Buffon, with the school of Bonnet. The idea of the *Échelle des êtres* had for him much less a

[1] See Meckel, *supra*, p. 93; *cf.* Tiedemann, *Zoologie*, p. 65, 1808. "Even as each individual organism transforms itself, so the whole animal kingdom is to be thought of as an organism in course of metamorphosis." Also p. 73 of the same book.

[2] Chapters vii. and ix.

[3] On early evolution-theories see, in addition to Osborn and Rádl, J. Arthur Thomson, *The Science of Life*, 1899, and the opening essay in *Darwin and Modern Science*, Cambridge, 1909.

morphological orientation than it had even for the transcendentalists, for he was lacking almost completely in the sense for morphology. Lamarck's scientific, as distinguished from his speculative work, was exclusively systematic, and it was systematics of a very high order. He introduced many reforms into the general classification of animals. He was the first clearly to separate Crustacea (1799), and a little later (1800) Arachnids, from insects. He reduced to a certain orderliness the neglected tribes of the Invertebrates, and wrote what was for long the standard work on their systematics—the *Histoire naturelle des Animaux sans Vertèbres* (1816-22). His speculative work on biology is contained in three publications, the small book entitled *Considérations sur l'organisation des corps vivants* (1802), the larger work of 1809, the *Philosophie zoologique*, and the introductory matter to his *Animaux sans Vertèbres* (vol. i., 1816).

It is no easy matter to give in short compass an account of Lamarck's biological philosophy. He is an obscure writer, and often self-contradictory.

In the first part of the *Philosophie zoologique* Lamarck is largely pre-occupied with the problem of whether species are really distinct, or do not rather grade insensibly into one another. As a systematist of vast experience Lamarck knew how difficult it is in practice to distinguish species from varieties. "The more," he writes, "we collect the productions of Nature, the richer our collections become, the more do we see almost all the gaps filled up and the lines of separation effaced. We find ourselves reduced to an arbitrary determination, which sometimes leads us to seize upon the slightest differences of varieties, and form from them the distinctive character of what we call a species, and at other times leads us to consider as a variety of a certain species individuals a little bit different, which others regard as forming a separate species."[1]

For Lamarck, as for Darwin later, the chief problem was not the evolution and differentiation of types of structure, but the mode of origin of species.

Lamarck is at great pains to show how arbitrary are our

[1] *Phil. zool.*, ed. Ch. Martins, vol. i., p. 75, 1873.

determinations of species, and how artificial the classificatory groups which we distinguish in Nature. Strictly speaking, there are in Nature only individuals, ". . . this is certain, that among her products Nature has in reality formed neither classes, nor orders, nor families, nor genera, nor constant species, but only individuals which succeed one another and resemble those that produced them. Now, these individuals belong to infinitely diversified races, which shade into one another under all the forms and in all the degrees of organisation, and each of which maintains itself without change, so long as no cause of change acts upon it" (p. 41).

But there is a natural order in the animal kingdom, a progression from the simpler to the more complex organisations, a natural *Échelle des êtres.*

This order is shown by the relation to one another of the large classificatory groups, for they can be arranged in series from the simplest to the most complex, somewhat as follows:—

1. Infusoria.	6. Arachnids.	11. Fishes.
2. Polyps.	7. Crustacea.	12. Reptiles.
3. Radiates.	8. Annelids.	13. Birds.
4. Worms.	9. Cirripedes.	14. Mammals.
5. Insects.	10. Molluscs.	

But the order of Nature is essentially continuous, and the limits of even the best defined of these classes are in reality artificial—"if the order of Nature were perfectly known in a kingdom, the classes which we should be forced to establish in it would always constitute entirely artificial sections" (p. 45).

In the same way the lesser classificatory groups represent smaller sections of the one unique order of Nature. Note that Lamarck's *Échelle* is in no way a morphological one, and was not intended to be such. It is a scale of increasing physiological differentiation, and the stages of it are marked by the acquirement of this or that new organ (*cf.* Oken). "Observation of their state convinces one that in order to produce them successively Nature has proceeded gradually from the simpler to the more complex. Now Nature, having had in mind the realisation of a plan of organisation which would permit of the greatest perfecting (that of the Vertebrates), a plan very different from those which she has been obliged to form as a preliminary to reaching it, one understands that, among the multitude of animals, one must necessarily come across not a single system of organisation which has become progressively perfected, but diverse very distinct systems, each of which has come into existence at the moment when each primary organ first put in its appearance" (p. 171).

For Lamarck this order of Nature was not merely ideal—Nature had actually formed the classes successively, proceeding from the simpler to the more complex; she had brought about this evolution by transforming the primitive species of animals, raising them to higher degrees of organisation, and modifying them in relation to the environment in which they found themselves.

Lamarck's theory of evolution is worked out in great detail in his *Philosophie zoologique*, but the exposition is diffuse and disconnected; it is better in giving an account of it to follow the more concise, mature and general exposition which he gives in the Introduction to his *Histoire naturelle des Animaux sans Vertèbres*.[1] Near the beginning of the Introduction Lamarck gives us in a few short "Fundamental Principles" the main lines of his general philosophy. He is a confirmed materialist. Every fact and phenomenon is essentially physical and owes its existence or production entirely to material bodies or to relations between them. All change and all movement is in the last resort due to mechanical causes. Every fact or phenomenon observed in a living body is at once a physical fact or phenomenon and a product of organisation (p. 19). Life, thought and sensation are not properties of matter, but result from particular material combinations.

His thorough-going materialism is most clearly shown in its relation to living things in the first three of the "Zoological Principles and Axioms," which are developed further on in the book.

These are as follows:—"1. No kind or particle of matter

[1] Quotations in the text are from the 2nd Edit. (Deshayes and Milne-Edwards), i., Paris, 1835.

can have in itself the power of moving, living, feeling, thinking, nor of having ideas; and if, outside of man, we observe bodies endowed with all or one of these faculties, we ought to consider these faculties as physical phenomena which Nature has been able to produce, not by employing some particular kind of matter which itself possesses one or other of these faculties, but by the order and state of things which she has constituted in each organisation and in each particular system of organs.

"2. Every animal faculty, of whatever nature it may be, is an organic phenomenon, and results from a system of organs or an organ-apparatus which gives rise to it and upon which it is necessarily dependent.

"3. The more highly a faculty is developed the more complex is the system of organs which produces it, and the higher the general organisation; the more difficult also does it become to grasp its mechanism. But the faculty is none the less a phenomenon of organisation, and for that reason purely physical" (p. 104).

According to these "axioms" function is a direct and mechanical effect of structure.

The curious thing is that in spite of his avowed materialism, Lamarck's conception of life and evolution is profoundly psychological, and from the conflict of his materialism and his vitalism (of which he was himself hardly conscious), arise most of the obscurities and the irreductible self-contradiction of his theory.

Lamarck divided animals (psychologically!) into three great groups—apathetic or insensitive animals, animals endowed with sensation, and intelligent animals. The first group, which comprise all the lower Invertebrates, are distinguished from other animals by the fact that their actions are directly and mechanically due to the excitations of the environment; they have no principle of reaction to external influences, but passively prolong into action the excitations they receive from without. They are *irritable* merely. The second group are distinguished from the first by their possessing, in addition to irritability, a power which Lamarck calls the *sentiment intérieur*. He has some difficulty in defining exactly what he means by it:—"I have no term to express this internal power possessed not only by intelligent animals but also by those that are endowed merely with the faculty of sensation; it is a power which, when set in action by the feeling of a need, causes the individual to act at once, *i.e.*, in the very moment of the sensation it experiences; and if the individual is of those that are endowed with intelligence it nevertheless acts in such a case entirely without premeditation and before any mental operation has brought its *will* into play" (p. 24).

It is the power we call instinct in animals (p. 25), and it implies neither consciousness nor will. It acts by transforming external into internal excitations.

To this second group of animals, possessing the *sentiment intérieur*, belong the higher Invertebrates, notably insects and molluscs. Only animals possessed of a more or less centralised nervous system can manifest this *sentiment*, or principle of (unconscious) reaction to external stimuli.

The higher animals, or the four Vertebrate classes, form the group of "intelligent animals." In virtue of their more complex organisation they possess in addition to the *sentiment intérieur* the faculties of intelligence and will.

Now, broadly put, Lamarck's theory of evolution is that new organs are formed in direct reaction to needs (*besoins*) experienced by the *sentiment intérieur*. The *sentiment intérieur* is therefore the cause not only of instinctive action but also of all morphogenetic processes. Will and intelligence (which are confined to a relatively small number of animals) have little or nothing to do directly with evolution.

To understand the working-out of Lamarck's evolution-theory we must revert to his conception of the *Échelle des êtres*. What he wrote in the *Philosophie zoologique* is here repeated in the work of 1816 with little modification.

There is a real progression from the simpler to the more complex organisations; Nature has gradually complicated her creatures by giving them new organs and therefore new faculties.

It is interesting to note that Lamarck expressly refers to Bonnet (p. 110), but refuses to accept his view of an *Échelle* extending down into the inorganic. Like Bonnet, however, and like the German transcendentalists, Lamarck makes man the goal of evolution (p. 116). He makes it quite clear that his *Échelle* is a functional one, for he links Vertebrates to molluscs even while expressly admitting that they are not connected by any structural intermediates (p. 123). He does not fall into the error of the transcendentalists and assume that Vertebrates and Invertebrates alike are formed upon one common plan of structure.

The progression of organisation shown by the animal kingdom has not been altogether regular and uninterrupted :—"The progression in complexity of organisation shows here and there, in the general animal series, anomalies induced by the influence of environment and by the influence of the habits contracted" (*Phil. zool.*, i., p. 145).

There are thus really two causes at work to produce the variety of organisation as it appears to us, one which tends to produce a regular increase in complexity, and one which disturbs and diversifies this regular advance.

The first cause Lamarck calls the vital power (*pouvoir de la vie*); the other may be called the influence of circumstance (*Anim. s. Vert.*, p. 134). To the latter cause are due the lacunæ, the blind alleys, and the complications which the otherwise simple scale of perfection shows.

To explain both these aspects of evolution Lamarck propounded in his volume of 1816 four laws, which read as follows :—

"*First Law.*—Life, by its own forces, tends continually to increase the volume of every body possessing it, and to extend the dimensions of its parts, up to a limit which it brings about itself.

"*Second Law.*—The production of a new organ in an animal body results from the arisal and continuance of a new need, and from the new movement which this need brings into being and sustains.

"*Third Law.*—The degree of development of organs and their force of action are always proportionate to the use made of these organs.

"*Fourth Law.*—All that has been acquired, imprinted or changed in the organisation of the individual during the course of its life is preserved by generation and transmitted to the new individuals that descend from the individual so modified" (pp. 151-2).

It is mainly but not entirely by reason of the first of these laws that organisation tends to progress, and mainly by reason of the second and third that difference of environment brings about diversity of organisation. In virtue of the fourth law the acquirements of the individual become the property of the race.

Lamarck's exposition of his first law, that life tends by its own powers to enlarge and extend its bodily instrument, is vague and difficult to understand. He has already explained some pages back how the first organisms arose by spontaneous generation in the form of minute gelatinous utricles (*cf.* Oken). He conceives that it is in the movements of the fluids proper to the organism that the power resides to enlarge and extend the body. Nutrition alone is not sufficient to bring about extension; a special force is required, acting from within outwards (p. 153). In the most primitive organisms the movements of the vital fluids are weak and slow, but in the course of evolution they gradually accelerate, and, becoming more rapid, trace out canals in the delicate tissue which contains them, and finally form organs.

Subtle fluids play a great part in Lamarck's biology: they take the place of the soul or entelechy which the vitalists would postulate to explain organic happenings. Lamarck seems in this to follow certain of the old materialists, who conceived the soul to be formed of a matter more subtle than the ordinary.[1]

In his second law Lamarck's essentially vitalistic attitude comes out very clearly, for it states that a psychological moment enters into all new production of form, that the ultimate cause of the development of new form is the need felt by the organism. This need is of course not a conscious one, it is a need perceived by the *sentiment intérieur*.

[1] For instance, Lucretius:—

> "Is tibi nunc animus quali sit corpore et unde
> constiterit pergam rationem reddere dictis.
> Principio esse aio persubtilem atque minutis
> perquam corporibus factum constare."
>
> —*De Rerum Natura*, iii., vv. 177-80.

In the large group of apathetic or insensitive animals, which do not possess this faculty, needs cannot be experienced; accordingly new organs are here formed directly and mechanically, by the movements of the vital fluids set in action by excitations from without—the evolution, like the behaviour, of these animals is due to the direct and physical action of the environment. "But this is not the case with the more highly organised animals which possess *feeling*. They experience needs, and each need felt, acting upon their 'inner feeling,' immediately directs the fluids and the forces to the part of the body where action can satisfy the need. Now, if there exists at this point an organ capable of performing the required action, it is quickly stimulated to act; and if the organ does not exist and the need is pressing and sustained, bit by bit the organ is produced and developed in proportion to the continuity and the energy of its use" (p. 155).

In intelligent animals the *sentiment intérieur* may be moved by thought or will.

As an example of the way in which the law works Lamarck takes the hypothetical case of a gastropod mollusc, which as it creeps along experiences dimly the need to feel the objects in front of it. It makes an effort (unconscious, be it noted) to touch these objects with the anterior portions of its head, and sends forward continually to these parts a great volume of nervous and other fluids. From these efforts and the repeated afflux of fluids there must result a development of the nerves supplying these parts. And as, along with the nervous fluids, nutritive juices constantly flow to the parts, there must result the formation of two or four tentacles in the places to which these fluids are directed. A curious mixture of mechanistic "explanations" and vitalistic hypothesis!

In his third law, that use and disuse are powerful to modify organs, Lamarck is upon more solid ground, and can point to many instances of the visible effect of these factors of change. It is of course rather closely bound up with his second law and may even be regarded as an extension of it.

The law has reference to one of the most powerful means

employed by Nature to diversify species, a means which comes into play whenever the environment changes. The cause of the great diversity shown by animal species is indeed ultimately to be sought in the environment. As the imperfect and earliest forms developed they spread over the earth and invaded the utmost corners of it:—"One can imagine what an enormous variety of habitats, stations, climates, available foods, environing media, etc., animals and plants have had to endure, as the existing species were forced to change their place of abode. And although these changes have taken place with extreme slowness . . . their reality, necessitated by various causes, has none the less induced the species affected by them slowly to change their manner of life and their habitual actions. Through the effects of the second and third of the laws cited above, these induced activity-changes must have brought into being new organs, and must have been able to develop them further if more frequent use was made of them; they must in the same way have been capable of bringing about the degeneration and finally the complete disappearance of existing organs which had become useless" (p. 161).

On the other hand, if the environment does not change, species remain constant.

It is to be noted that change in environment is rather the occasion than the cause of modification; the environment induces the organism to change its habitual way of life; it sets up new needs, to satisfy which the organism must modify its structure. It is the organism that takes the active part in all this, the action of the environment is indirect.

Of Lamarck's fourth law, which asserts the transmission of acquired characters, little need here be said in the way of exposition. Upon the truth of it depends of course Lamarck's whole theory. He himself never dreamed that anyone would ever dispute it.

Lamarck sums up as follows:—"By the four laws which I have just enunciated all the facts of organisation seem to me to be easily explained; the progression in the complexity of organisation of animals, and in their faculties, seems to me easy to conceive; so, too, the means which Nature has employed to diversify animals, and bring them to the state in which we now see them, become easily determinable" (p. 168).

It is never made quite clear, we may note in passing, how far his second and third laws tend to bring about an increase in complexity, in addition to diversifying animals.[1]

"The function creates the organ," this would seem to be the kernel of Lamarck's doctrine. But how does he reconcile this essentially vitalistic conception with his strictly materialistic philosophy?

We have seen that irritability, the *sentiment intérieur*, and intelligence itself, are the effects of organisation. We are told farther on that both the *sentiment* and intelligence are caused by nervous fluids. A great part of both the *Philosophie zoologique* and the introduction to the *Animaux sans Vertèbres* is given up to the exposition of a materialistic psychology of animals and man, based entirely upon this hypothesis of nervous fluids. Thus habits are due to the fluids hollowing out definite paths for themselves.

The *sentiment intérieur* acts by directing the movements of the subtle fluids of the body (which are themselves modifications of the nervous fluids) upon the parts where a new organ is needed. But if it is itself only a result of the movement of nervous fluids? Again, how can a need be "felt" by a nervous fluid? This is an entirely psychological notion and cannot be applied to a purely material system. Whence arises the power of the *sentiment intérieur* to canalise the energies of the organism, so to direct and co-ordinate them that they build up purposive structures, or effect purposive actions (as in all instinctive behaviour)? Either the *sentiment intérieur* is a psychological faculty, or it is nothing.

There is no doubt that, as expressed by Lamarck, the conception conceals a radical confusion of thought. It is not possible to be a thorough-going materialist, and at the

[1] Contrast Treviranus—"In every living being there exists a capability of an endless variety of form-assumption; each possesses the power to adapt its organisation to the changes of the outer world, and it is this power, put into action by the change of the universe, that has raised the simple zoophytes of the primitive world to continually higher stages of organisation, and has introduced a countless variety of species into animate Nature." Quoted by Haeckel in *History of Creation*, i., p. 93, 1876.

same time to believe that new organs are formed in direct response to needs felt by the organism. Lamarck could never resolve this antinomy, and his speculations were thrown into confusion by it. To this cause is due the frequent obscurity of his writings.

Should we be right in laying stress upon the psychological side of Lamarck's theory, and disregarding the materialistic dress in which, perhaps under the influence of the materialism current in his youth, he clothed his essentially vitalistic thought? Everything goes to prove it—his constant preoccupation with psychological questions, his tacit assimilation of organ-formation to instinctive behaviour, his constant insistence on the importance of *besoin* and *habitude*.

Let us not forget the profundity of his main idea, that, exception made for the lower forms, the animal is essentially active, that it always *reacts* to the external world, is never passively acted upon. Let us not forget that he pointed out the essentially psychological moment implied in all processes of individual adaptation. With keen insight he realised that conscious intelligence counts for little in evolution, and focussed attention upon the unconscious but obscurely psychical processes of instinct and morphogenesis.

Not without reason have the later schools of evolutionary thought, who developed the psychological and vitalistic side of his doctrine, called themselves Neo-Lamarckians.

We shall say then that Lamarck, in spite of his materialism, was the founder of the "psychological" theory of evolution.

Lamarck stood curiously aloof and apart from the scientific thought of his day.[1] He took no interest in the morphological problems that filled the minds of Cuvier and Geoffroy; he had indeed no feeling at all for morphology. He did not realise, like Cuvier, the *convenance des parties*, the marvellous co-ordination of parts to form a whole; he had little conception of what is really implied in the word "organism." He was not, like Geoffroy, imbued with a lively sense of the unity of plan and composition, and of the significance of vestigial

[1] There is no evidence that he was influenced by Erasmus Darwin, who forestalled his evolution theory, and was indeed more aware of its vitalistic implications. See S. Butler, *Evolution, Old and New*, London, 1879, for an excellent account of Erasmus Darwin.

organs as witnesses to that unity. He seems not to have known of the recapitulation theory, of which he might have made such good use as powerful evidence for evolution. Even with the German transcendentalists, with whom in the looseness of his generalisations he shows some affinity, he seems not to have been specially acquainted.

He was interested more in the problems suggested to him by his daily work in the museum. He wanted to know why species graded so annoyingly into one another; he wanted to examine critically his haunting suspicion that species were really not distinct, and that classification was purely conventional. The question, too, of the adaptation of species to their environment, the problem of ecological adaptation, in distinction to that of functional adaptation which interested Cuvier so greatly, came vividly before him as he worked through the vast collections of the museum. He was the first systematist to occupy himself in a philosophical manner with the problems of general biology. He introduced new problems and a new way of looking at old. With Lamarck the problem of species and the problem of ecological adaptation enter into general biology.

The one point in which he does definitely carry on the thought of his predecessors is his conception of the animal kingdom as forming a scale of (functional) perfection. He did not go to the same extreme as Bonnet; he did not even consider that the animal series was a continuation of the vegetable series; in his opinion they formed two diverging scales. He recognised, too, that among animals there was no simple and regular gradation from the lowest to the highest, but that the orderly progression was disturbed and diverted by the necessity of adaptation to different environments. It is interesting to note that in developing this idea he arrived at a roughly accurate distinction between homologous and analogous structures. More importance, he thought, was to be attributed in classifying animals to characters which appeared due to the "plan of Nature" than to such as were produced by an external modifying cause (p. 299). But he did not formulate the distinction in any strictly morphological way.

As his ideas developed he laid less stress upon the simplicity and continuity of the scale; in his supplementary remarks to the Introduction of 1816 he admits that the series is really very much branched, and even that there may be two distinct series among animals instead of one. His last schema of the course of evolution shows no little analogy with the genealogical trees of Darwinian speculation. It is headed "The presumed *Order* of the formation of Animals, showing two separate partly-branching series," and it reads as follows:—

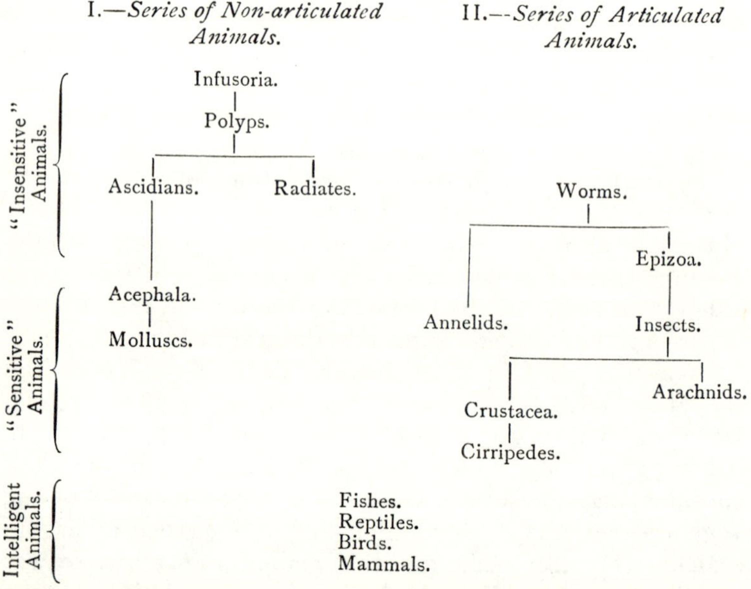

It is interesting to note that Vertebrates are placed between the two series, and are now not linked on directly to any Invertebrate group.

Lamarck's theory had little success. There is evidence, however, that both Meckel and Geoffroy owed a good many of their evolutionary ideas to Lamarck, and Cuvier paid him at least the compliment of criticising his theory,[1] not distinguishing it, however, very clearly from the evolutionary theories of the transcendentalists. But, speaking generally, Lamarck's theory of evolution exercised very little influence upon his

[1] As did also Lyell in his *Principles of Geology*, 1830.

contemporaries. This was probably due partly to the obscurity and confusion of his thought, partly to his lack of sympathy with the biological thought of his day, which was preponderatingly morphological.

It was not that men's minds were not ripe for evolution, for in the early decades of the 19th century evolution was in the air. There were few of von Baer's contemporaries who had not read Lamarck;[1] Erasmus Darwin's *Zoonomia* ran through three editions, and was translated into German, French and Italian;[2] German philosophy was full of the idea of evolution.

There was no unreadiness to accept the derivation of present-day species from a primordial form—if only some solid evidence for such derivation were forthcoming. Cuvier and von Baer as we have seen, combated the current evolution theories on the ground that the evidence was insufficient, but von Baer at least had no rooted objection to evolution. In an essay of 1834, entitled *The Most General Law of Nature in all Development*,[3] von Baer expressed belief in a limited amount of evolution. In this paper he did not admit that all animals have developed from one parent form, and he refused to believe that man has descended from an ape; but, basing his supposition upon the facts of variability and upon the evidence of palæontology, he went so far as to maintain that many species have evolved from parent stocks. In the absence of conclusive proofs he did not commit himself to a belief in any extended or comprehensive process of evolution.

Imbued as he was with the idea of development von Baer saw in evolution a process essentially of the same nature as the development of the individual. Evolution, like development, was due to a *Bildungskraft* or formative force. The ultimate law of all becoming was that "the history of Nature is nothing but the history of the ever-advancing victory of spirit over matter" (p. 71). In a later essay (1835) in the same volume he says that all natural science is nothing but a long commentary on the single phrase *Es werde!* (p. 86).

As we shall see, von Baer adopted in later years the same

[1] K. E. von Baer, *Reden*, i., p. 37, Petrograd, 1864.

[2] Rádl, *loc. cit.*, i., p. 296.

[3] Reprinted in his *Reden*, i., 1864.

attitude to Darwinism as he did to the evolution theories in vogue in his youth.

Although in the twenty or thirty years before the publication of the *Origin of Species* (1859) no evolution-theory of any importance was published, and although the great majority of biologists believed in the constancy of species, there were not wanting some who, like von Baer, had an open mind on the subject, or even believed in the occurrence of evolutionary processes of small scope. Isidore Geoffroy St Hilaire, the son of the great Etienne Geoffroy St Hilaire, seems to have held that species might be formed from varieties. The law which L. Agassiz thought he could establish,[1] of the parallelism between palæontological succession, systematic rank, and embryological development, tended to help the progress of evolutionary ideas. J. V. Carus, who afterwards became a supporter of Darwin, seems already, in 1853, to have inferred from Agassiz's law the probability of evolution.[2]

But no evolution theory was taken very seriously before 1859, when the *Origin of Species* was published.

Like Lamarck, Charles Darwin was, neither by inclination nor by training, a morphologist. In his youth he was a collector, a sportsman and a field geologist. His voyage round the world on the *Beagle* aroused in him keen interest in the problem of species—their variety, their variation according to place and time, their adaptedness to environment. The conviction gradually took possession of his mind that the puzzling facts of geographical range and geological succession which he observed wherever he went were explicable only on the hypothesis that species change. He was not satisfied with the theories of evolution that had been proposed by his grandfather, by Lamarck, and by E. Geoffroy St Hilaire—he did not indeed understand these theories any too well. He resolved to work out the problem in his own way, for his own satisfaction. He tells us all this very clearly in his autobiography. "During the voyage

[1] See Huxley's criticism of it in a Royal Institution lecture of 1851, republished in *Sci. Mem.*, i., pp. 300-4. On its relation to Haeckel's biogenetic law, see below, p. 255.

[2] *System der thierischen Morphologie*, p. 5, 1853.

of the *Beagle* I had been deeply impressed by discovering in the Pampean formation great fossil animals covered with armour like that on the existing armadillos; secondly, by the manner in which closely allied animals replace one another in proceeding southwards over the continent; and thirdly, by the South American character of most of the productions of the Galapagos archipelago, and more especially by the manner in which they differ slightly on each island of the group; some of the islands appearing to be very ancient in a geological sense.

"It was evident that such facts as these, as well as many others, could only be explained on the supposition that species gradually become modified; and the subject haunted me. But it was equally evident that neither the action of the surrounding conditions, nor the will of the organisms (especially in the case of plants) could account for the innumerable cases in which organisms of every kind are beautifully adapted to their habits of life—for instance, a woodpecker or a tree-frog to climb trees, or a seed for dispersal by hooks or plumes. I had always been much struck by such adaptations, and until these could be explained it seemed to me almost useless to endeavour to prove by indirect evidence that species have been modified."[1]

All Darwin's varied subsequent work revolved round these, for him, essential problems—How do species change, and how do they become adapted to their environment? He never ceased to be essentially a field naturalist, and his theory of natural selection would have been an empty and abstract thing if his vast knowledge and understanding of the "web of life" had not given it colour and form. He never lost touch with the living thing in its living, breathing reality—even plants he rightly regarded as active things, full of tricks and contrivances for making their way in the world. No one ever realised more vividly than he the delicacy and complexity of the adaptations to environment which are the necessary condition of success in the struggle for existence. Almost his greatest service to biology was that he made

[1] *Life and Letters of Charles Darwin*, ed. F. Darwin, i., p. 82, 3rd ed., 1887.

biologists realise as they never did before the vast importance of environment. He took biology into the open air, away from the museum and the dissecting-room.

Naturally this attitude was not without its drawbacks. It led him to take only a lukewarm interest in the problems of morphology. It is true he used the facts of morphology with great effect as powerful arguments for evolution, but it was not from such facts that he deduced his theory to account for evolution. It is questionable indeed whether the theory of natural selection is properly applicable to the problems of form. It was invented to account for the evolution of specific differences and of ecological adaptations ; it was not primarily intended as an explanation of the more wonderful and more mysterious facts of the *convenance des parties* and the interaction of structure and function. Perhaps Darwin did not realise this inner aspect of adaptation quite so vividly as he did the more superficial adaptation of organisms to their environment. It was, perhaps, his lack of morphological training and experience that led him to disregard the problems of form, or at least to realise very insufficiently their difficulty.

It is in any case very significant that only a small part of his *Origin of Species* is devoted to the discussion of morphological questions—only one chapter out of the fourteen contained in the first edition.

Though the theory of natural selection took little account of the problems of form, Darwin's masterly vindication of the theory of evolution was of immense service to morphology, and Darwin himself was the first to point out what a great light evolution threw upon all morphological problems. In a few pages of the *Origin* he laid the foundations of evolutionary morphology.

We have here to consider his interpretation of morphological facts and its relation to the current morphology of his time.

The sketch of his theory, written in 1842,[1] shows a very significant division into two parts—the first dealing with the positive facts of variability and the theory of natural selection,

[1] *The Foundations of the Origin of Species, a Sketch written in* 1842. Ed. F. Darwin, Cambridge, 1909.

the second with the general evidence for evolution. It is in the second part that the paragraphs on morphological matters occur. In paragraph 7, on affinities and classification, Darwin points out that on the theory of evolution homological relationship would be real relationship, and the natural system would really be genealogical. In the next paragraph he notes that evolution would account for the unity of type in the great classes, for the metamorphosis of organs, and for the close resemblance which early embryos show to one another. It is of special interest to note that he definitely rejects the Meckel-Serres theory of recapitulation. "It is not true," he writes, "that one passes through the form of a lower group, though no doubt fish more nearly related to fœtal state" (p. 42). The greater divergence which adults show seems to him to be due to the fact that selection acts more on the later than on the embryonic stages. He realises very clearly how illuminative the theory of evolution is when applied to the puzzling facts of embryonic development. "The less differences of fœtus—this has obvious meaning on this view: otherwise how strange that a horse, a man, a bat should at one time of life have arteries, running in a manner which is only intelligibly useful in a fish! The natural system being on theory genealogical, we can at once see why fœtus, retaining traces of the ancestral form, is of the highest value in classification" (p. 45).

Abortive organs, too, gain significance on the evolutionary hypothesis. "The affinity of different groups, the unity of types of structure, the representative forms through which fœtus passes, the metamorphosis of organs, the abortion of others, cease to be metaphorical expressions and become intelligible facts" (p. 50).

In general, organisms can be understood only if we take into account the cardinal fact that they are historical beings. "We must look at every complicated mechanism and instinct as the summary of a long history of useful contrivances much like a work of art" (p. 51).[1]

Already in 1842 Darwin had seized upon the main principles of evolutionary morphology: the indications then given are elaborated in the thirteenth chapter of the *Origin*

[1] *Cf.* a parallel passage in the *Origin*, 1st ed., pp. 485-6.

of Species (1st ed., 1859). A good part of this chapter is given up to a discussion of the principles of classification, only a few pages dealing with morphology proper. But, as Darwin rightly saw, the two things are inseparable.

We note first that there is no hint of the "scale of beings"—Darwin conceives the genealogical tree as many branched. Animals can be classed in "groups under groups," and cannot be arranged in one single series.

He discusses first what kind of characters have the greatest classificatory value. Certain empirical rules have been recognised, more or less consciously, by systematists—that analogical characters are less valuable than homological, that characters of great physiological importance are not always valuable for classificatory purposes, that rudimentary organs are often very useful, and so on. He finds that as a general rule "the less any part of the organisation is concerned with special habits, the more important it becomes for classification" (p. 414), and adduces in support Owen's remark that the generative organs afford very clear indications of affinities, since they are unlikely to be modified by special habits. These rules of classification can be explained "on the view that the natural system is founded on descent with modification; that the characters which naturalists consider as showing true affinity . . . are those which have been inherited from a common parent, and, in so far, all true classification is genealogical; that community of descent is the hidden bond which naturalists have been unconsciously seeking, and not some unknown plan of creation, or the enunciation of general propositions, and the mere putting together and separating objects more or less alike" (p. 420).

In general, then, homological characters are more valuable for classificatory purposes because they have a longer pedigree than analogical characters, which represent recent acquirements of the race.

Coming to morphology proper, Darwin takes up the question of the unity of type, and the homology of parts, for which the unity of type is but a general expression.

He treats this on the same lines as E. Geoffroy St Hilaire, and Owen, referring indeed specifically to Geoffroy's law of connections. "What can be more curious," he asks, "than that the hand of a man, formed for grasping, that of a mole for digging, the leg of a horse, the paddle of the porpoise, and the wing of the bat, should all be constructed on the same pattern, and should include similar bones, in the same relative positions? Geoffroy St Hilaire has strongly insisted on the high importance of relative position or connection in homologous parts; they may differ to almost any extent in form and size, and yet remain connected together in the same invariable order" (p. 434).

The unity of plan cannot be explained on teleological grounds, as Owen has admitted in his *Nature of Limbs*, nor is it explicable on the hypothesis of special creation (p. 435). It can be understood only on the theory that animals are descended from one another and retain for innumerable generations the essential organisation of their ancestors. "The explanation is to a large extent simple on the theory of the selection of successive slight modifications—each modification being profitable in some way to the modified form, but often affecting by correlation other parts of the organisation. In changes of this nature, there will be little or no tendency to alter the original pattern or to transpose the parts. . . . If we suppose that the ancient progenitor, the archetype as it may be called, of all animals, had its limbs constructed on the existing general pattern, for whatever purpose they served, we can at once perceive the plain significance of the homologous construction of the limbs throughout the whole class" (p. 435).

We may note three important points in this passage—first, the identification of the archetype with the common progenitor; second, the view that progressive evolution is essentially adaptive, and dominated by natural selection; and third, the *petitio principii* involved in the assumption that adaptive modification brings inevitably in its train the necessary correlative changes.

In his section on morphology Darwin shows clearly the influence of Owen, and through him of the transcendental anatomists. He refers to the transcendental idea of "metamorphosis," as exemplified in the vertebral theory of the skull and the theory of the plant appendage, and shows how, on the hypothesis of descent with modification, "metamorphosis" may now be interpreted literally, and no longer figuratively merely (p. 439).

Very great interest attaches to Darwin's treatment of development, for post-Darwinian morphology was based to a very large extent on the presumed relation between the development of the individual and the evolution of the race. Just as he kept clear of the notion of the scale of beings, so he avoided the snare of the Meckel-Serres theory of recapitulation, according to which the embryo of the highest animal, man, during its development climbs the ladder upon the rungs of which the whole animal series is distributed, in its gradual progression from simplicity to complexity. The law of development which he adopts is that of von Baer, which states that development is essentially differentiation, and that as a result embryos belonging to the same group resemble one another the more the less advanced they are in development. There can be little doubt that he was indebted to von Baer for the idea, and in the later editions of the *Origin* he acknowledges this by quoting the well-known passage in which von Baer tells how he had two embryos in spirit which he was unable to refer definitely to their proper class among Vertebrates.[1]

Not only are embryos more alike than adults, because less differentiated, but it is in points not directly connected with the conditions of existence, not strictly adaptive, that their resemblance is strongest (p. 440)—think, for instance, of the arrangement of aortic arches common to all vertebrate embryos. Larval forms are to some extent exceptions to this rule, for they are often specially adapted to their particular mode of life, and convergence of structure may accordingly result. All these facts require an explanation. "How, then, can we explain these several facts in embryology —namely, the very general, but not universal, difference in structure between the embryo and the adult—of parts in the same individual embryo, which ultimately become very unlike and serve for different purposes, being at this early period of growth alike—of embryos of different species within the same class, generally but not universally,

[1] In the 1st ed. (p. 439), Darwin makes the curious mistake of attributing this story to Agassiz.

resembling each other—of the structure of the embryo not being closely related to its conditions of existence, except when the embryo becomes at any period of life active and has to provide for itself—of the embryo apparently having sometimes a higher organisation than the mature animal, into which it is developed" (pp. 442-3). Obviously all these facts are formally explained by the doctrine of descent. But Darwin goes further, he tries to show exactly how it is that the embryos resemble one another more than the adults. He thinks that the phenomenon results from two principles—first, that modifications usually supervene late in the life of the individual; and second, that such modifications tend to be inherited by the offspring at a corresponding, not early, age (p. 444).

Thus, applying these principles to a hypothetical case of the origin of new species of birds from a common stock, he writes:—" . . . from the many slight successive steps of variation having supervened at a rather late age and having been inherited at a corresponding age, the young of the new species of our supposed genus will manifestly tend to resemble each other much more closely than do the adults, just as we have seen in the case of pigeons"[1] (pp. 446-7).

Since the embryo shows the generalised type, the structure of the embryo is useful for classificatory purposes. "For the embryo is the animal in its less modified state; and in so far it reveals the structure of its progenitor" (p. 449)—the embryological archetype reveals the ancestral form. "Embryology rises greatly in interest, when we thus look at the embryo as a picture, more or less complete, of the parent form of each great class of animals" (p. 450)—a prophetic remark, in view of the enormous subsequent development of phylogenetic speculation.

We may sum up by saying that Darwin interpreted von Baer's law phylogenetically.

The rest of the chapter is devoted to a discussion of abortive and vestigial organs, whose existence Darwin

[1] In which nestlings of the different varieties are much more alike than adults. Darwin attached much importance to this idea, see *Life and Letters*, i., p. 88, and ii., p. 338.

naturally turns to great advantage in his argument for evolution. Throughout the whole chapter Darwin's preoccupation with the problems of classification is clearly manifest.

On the question as to whether descent was monophyletic or polyphyletic Darwin expressed no dogmatic opinion. "I believe that animals have descended from at most only four or five progenitors, and plants from an equal or lesser number. . . . I should infer from analogy that probably all the organic beings which have ever lived on this earth have descended from one primordial form, into which life was first breathed" (p. 484).

Darwin rightly laid much stress upon the morphological evidence for evolution,[1] which he considered to be weighty. It probably contributed greatly to the success of his theory. Though he himself did little or no work in pure morphology, he was alive to the importance of such work,[2] and followed with interest the progress of evolutionary morphology, incorporating some of its results in later editions of the *Origin*, and in his *Descent of Man* (1871).

In his morphology Darwin was hardly up to date. He does not seem to have known at first hand the splendid work of the German morphologists, such as Rathke and Reichert; he pays no attention to the cell-theory, nor to the germ-layer theory. His sources are, in the main, Geoffroy St Hilaire, Owen, von Baer, Agassiz, Milne-Edwards, and Huxley.

Perhaps his greatest omission was that he did not give any adequate treatment of the problem of functional adaptation and the correlation of parts. It is not too much to say that Darwin not only disregarded these problems almost entirely, but by his insistence upon ecological adaptation and upon certain superficial aspects of correlation, succeeded in giving to the words "adaptation" and "corre-

[1] See his *Letters, passim.*

[2] Writing to Huxley on the subject of the latter's work on the morphology of the Mollusca (1853), he says :—"The discovery of the type or 'idea' (in your sense, for I detest the word as used by Owen, Agassiz & Co.) of each great class, I cannot doubt, is one of the very highest ends of Natural History."—*More Letters*, ed. F. Darwin and A. C. Seward, 1903, i., p. 73.

lation" a new signification, whereby they lost to a large extent their true and original functional meaning.

It is true that Darwin himself, as well as his successors, believed that natural selection was all-powerful to account for the evolution of the most complicated organs, but it may be questioned whether he realised all the conditions of the problem of which he thus easily disposed. He says, rightly, in an important passage, that "It is generally acknowledged that all organic beings have been formed on two great laws —Unity of Type, and the Conditions of Existence. By unity of type is meant that fundamental agreement in structure which we see in organic beings of the same class, and which is quite independent of their habits of life. On my theory, unity of type is explained by unity of descent. The expression of conditions of existence, so often insisted upon by the illustrious Cuvier, is fully embraced by the principle of natural selection. For natural selection acts by either now adapting *the varying parts of each being to its organic and inorganic conditions of life:*[1] or by having adapted them during past periods of time: the adaptations being aided in many cases by the increased use or disuse of parts, being affected by the direct action of the external conditions of life, and subjected in all cases to the several laws of growth and variation. Hence, in fact, the law of the Conditions of Existence is the higher law; as it includes, through the inheritance of former variations and adaptations, that of Unity of Type" (*Origin*, 6th ed., Pop. Impression, pp. 260-1). It is clear that Darwin took the phrase "Conditions of Existence" to mean the environmental conditions, and the law of the Conditions of Existence to mean the law of adaptation to environment. But that is not what Cuvier meant by the phrase: he understood by it the principle of the co-ordination of the parts to form the whole, the essential condition for the existence of any organism whatsoever (see above, Chap. III., p. 34).

Of this thought there is in Darwin little trace, and that is why he did not sufficiently appreciate the weight of the argument brought against his theory that it did not account for the correlation of variations.

[1] Italics mine.

Darwin's conception of correlation was singularly incomplete. As examples of correlation he advanced such trivial cases as the relation between albinism, deafness and blue eyes in cats, or between the tortoise-shell colour and the female sex. He used the word only in connection with what he called "correlated variation," meaning by this expression "that the whole organisation is so tied together during its growth and development, that when slight variations in any one part occur, and are accumulated through natural selection, other parts become modified" (6th ed., p. 177). He took it for granted that the "correlated variations" would be adapted to the original variation which was acted upon by natural selection, and he saw no difficulty in the gradual evolution of a complicated organ like the eye if only the steps were small enough. "It has been objected," he writes, "that in order to modify the eye and still preserve it as a perfect instrument, many changes would have to be effected simultaneously, which, it is assumed, could not be done through natural selection; but as I have attempted to show in my work on the variation of domestic animals, it is not necessary to suppose that the modifications were all simultaneous, if they were extremely slight and gradual" (6th ed., p. 226).

In post-Darwinian speculation the difficulty of explaining correlated variation by natural selection alone became more acutely realised, and it was chiefly this difficulty that led Weismann to formulate his hypothesis of germinal selection as a necessary supplement to the general selection theory.

The change in the conception of correlation which Darwin's influence brought about has been very clearly stated by E. von Hartmann,[1] from whom the following is taken:—"While the correlation of parts in the organism was before Darwin regarded exclusively from the standpoint of morphological systematics, Darwin tried to look at it from the standpoint of physiological and genealogical development, and in so doing he put the standpoint of morphological systematics in the shade. But the more we are now beginning to realise that systematic relationship does not necessarily

[1] *Das Problem des Lebens. Biologische Studien.* Bad Sacha, 1906. See also E. Rádl, *Biol. Centralblatt*, xxi., 1901.

imply genetic affinity the more must the correlation of parts come back into favour as a systematic principle. While Darwin only, as it were, against his will, relied on the law of correlation as a last resort when all other help failed, this law must be regarded, from the standpoint of the orderly inner determination of all organic form-change, as having the rank of the highest principle of all, a principle which rules parallel, divergent and convergent evolution" (pp. 47-8).

Further on, following Rádl, he characterises Darwin's attitude to the law of correlation in these terms:—"Darwin's interest is entirely focussed on the variation, the function, the causes of form-production, in short, upon evolution. Accordingly he regards correlation essentially as correlative variation in the sense of a *departure* from the given type. With morphological correlation in *different* types Darwin troubles himself not at all, nor with correlation in the normal development of a type" (p. 49).

Cuvier's conception of the *convenance des parties*, essential to all biology, remained on the whole foreign to Darwin's thought, and to the thought of his successors.

It was indeed one of their boasts that they had finally eliminated all teleology from Nature. The great and immediate success which Darwinism had among the younger generation of biologists and among scientific men in general was due in large part to the fact that it fitted in well with the prevailing materialism of the day, and gave solid ground for the hope that in time a complete mechanistic explanation of life would be forthcoming. "Darwinismus" became the battle-cry of the militant spirits of that time.

It was precisely this element in Darwinism that was repugnant to most of Darwin's opponents, in whose ranks were found the majority of the morphologists of the old school. They found it impossible to believe that evolution could have come about by fortuitous variation and fortuitous selection; they objected to Darwin that he had enunciated no real *Entwickelungsgesetz*, or law governing evolution. They were not unwilling to believe that evolution was a real process, though many drew the line at the derivation of man from apes, but they felt that if evolution had really taken place, it must have been under the guidance of some principle of development, that there must have been manifested in evolution some definite and orderly tendency towards perfection.[1]

No one expressed this objection with greater force than did von Baer, in a series of masterly essays[2] which the Darwinians, through sheer inability to grasp his point of view, dismissed as the maunderings of old age. In these essays von Baer pointed out the necessity for the teleological point of view, at least as complementary to the mechanistic. His general position is that of the "statical" teleology—to use Driesch's term—of Kant and Cuvier. His attitude to Darwinism is determined by his teleology. He admits, just as in 1834, a limited amount of evolution; he criticises the evolution theory of Darwin on the same lines exactly as forty or fifty years previously he had criticised the recapitulation and evolution-theories of the transcendentalists —principally on the ground that their deductions far outrun the positive facts at their disposal. He rejects the theory of natural selection entirely, on the ground that evolution, like development, must have an end or purpose (*Ziel*)—"A becoming without a purpose is in general unthinkable" (p. 231); he points out, too, the difficulty of explaining the correlation of parts upon the Darwinian hypothesis. His own conception of the evolutionary process is that it is essentially *zielstrebig* or guided by final causes, that it is a true *evolutio* or differentiation, just as individual development is an orderly progress from the general to the special. He believed in saltatory evolution, in polyphyletic descent, and in the greater plasticity of the organism in earlier times.

The idea of saltatory evolution he took from Kölliker, who shortly after the publication of the *Origin* pro-

[1] See the excellent treatment of the difference between the "realism" of Darwin and the "rationalism" of his critics, in Rádl, ii., particularly pp. 109, 135. The most elaborate criticism of Darwinism from the older standpoint was that given by A. Wigand in *Der Darwinismus und die Naturforschung Newtons und Cuviers*, 3 vols., Braunschweig, 1872.

[2] In vol. ii. of his *Reden*, St Petersburg (Petrograd), 1876—*Ueber den Zweck in den Vorgängen der Natur; Ueber Zielstrebigkeit in den organischen Körpern insbesondere;* and *Ueber Darwin's Lehre.*

mulgated in a critical note on Darwinism a sketch of his theory of "heterogeneous generation."[1]

Kölliker's attitude is typical of that taken up by many of the morphologists of the day.[2] He accepts evolution completely, but rejects Darwinism because it recognises no *Entwickelungsgesetz*, or principle of evolution. For the Darwinian theory of evolution through the selection of small fortuitous variations he would substitute the theory of evolution through sudden, large variations, brought about by the influence of a general law of evolution. This is his theory of heterogeneous generation. "The fundamental idea of this hypothesis is that under the influence of a general law of evolution creatures produce from their germs others which differ from them" (p. 181). It is to be noticed that Kölliker laid more stress upon the *Entwickelungsgesetz* than upon the saltatory nature of variation, for he says a few pages further on—"the notion at the base of my theory is that a great evolutionary plan underlies the development of the whole organised world, and urges on the simpler forms towards ever higher stages of complexity" (p. 184). Saltatory evolution was not the essential point of the theory:—"Another difference between the Darwinian hypothesis and mine is that I postulate many saltatory changes, but I will not and indeed cannot lay the chief stress upon this point, for I have not intended to maintain that the general law of evolution which I hold to be the cause of the creation of organisms, and which alone manifests itself in the activity of generation, cannot also so act that from one form others quite gradually arise" (p. 185). He put forward the hypothesis of saltatory variation because it seemed to him to lighten many of the difficulties of Darwinism—the lack of transition forms, the enormous time required for evolution, and so on. It should be noted that Kölliker regarded his principle of evolution as mechanical.

[1] "Ueber die Darwinische Schöpfungstheorie," *Zeits. f. wiss. Zool.*, xiv., pp. 74-86, 1864. Elaborated in *Anat. u. syst. Beschreibung d. Alcyonarien*, 1872.

[2] *Cf.* for instance Nägeli's theory of a perfecting principle, first developed in his *Entstehung u. Begriff der naturhistorischer Art*, München, 1865.

It would take too long to show in detail how a belief in innate laws of evolution was held by the majority of Darwin's critics. A few further examples must suffice.

Richard Owen, who in 1868 [1] admitted the possibility of evolution, held that "a purposive route of development and change, of correlation and interdependence, manifesting intelligent Will, is as determinable in the succession of races as in the development and organisation of the individual. Generations do not vary accidentally, in any and every direction; but in pre-ordained, definite, and correlated courses" (p. 808).

He conceived change to have taken place by abrupt variation, independent of environment and habit, by "departures from parental type, probably sudden and seemingly monstrous, but adapting the progeny inheriting such modifications to higher purposes" (p. 797). He believed spontaneous generation to be a phenomenon constantly taking place, and constantly giving the possibility of new lines of evolution.

E. von Hartmann in his *Philosophie des Unbewussten* (1868) and in his valuable essay on *Wahrheit und Irrtum im Darwinismus* (1874) criticised Darwinism in a most suggestive manner from the vitalistic standpoint. He drew attention to the importance of active adaptation, the necessity for assuming definite and correlated variability, and to the evidence for the existence of an immanent, purposive, but unconscious principle of evolution, active as well in phylogenetic as in individual development.

In France H. Milne-Edwards [2] stated the problem thus:—"In the present state of science, ought we to attribute to modifications dependent on the action of known external agents the differences in the organic types manifested by the animals distributed over the surface of the globe either at the present day, or in past geological ages? Or must the origin of types transmissible by heredity be attributed to causes of another order, to forces whose effects are not apparent in the present state of things, to a creative power

[1] *Anatomy of Vertebrates*, iii., 1868.

[2] *Rapport sur les Progrès récents des Sciences zoologiques en France.* Paris, 1867.

independent of the general properties of organisable matter such as we know them to-day?" (p. 426).

He concluded that the action of environment, direct or indirect, was insufficient to account for the diversity of organic forms, and rejected Darwin's theory completely. He thought it likely that the successive faunas which palæontology discloses have originated from one another by descent. But he thought that the process by which they evolved should rightly be called "creation." The word was of course not to be taken in a crude sense. When the zoologist speaks of the "creation" of a new species, "he in no way means that the latter has arisen from the dust, rather than from a pre-existing animal whose mode of organisation was different; he merely means that the known properties of matter, whether inert or organic, are insufficient to bring about such a result, and that the intervention of a hidden cause, of a power of some higher order, seems to him necessary" (p. 429).

The criticism of Darwinism exercised by the older currents of thought remained on the whole without influence. It was under the direct inspiration of the Darwinian theory that morphology developed during the next quarter of a century.

CHAPTER XIV

ERNST HAECKEL AND CARL GEGENBAUR

AT the time when Darwin's work appeared there already existed, as we have seen, a fully formed morphology with set and definite principles. The aim of this pre-evolutionary morphology had been to discover and work out in detail the unity of plan underlying the diversity of forms, to disentangle the constant in animal form and distinguish from it the accessory and adaptive. The main principle upon which this work was based was the principle of connections, so clearly stated by Geoffroy. The principle of connections served as a guide in the search for the archetype, and this search was prosecuted in two directions—first, by the comparison of adult structure; and second, by the comparative study of developing embryos. It was found that the archetype was shown most clearly by the early embryo, and this embryological archetype came to be preferred before the archetype of comparative anatomy. It became apparent also that the parts first formed (germ-layers) were of primary importance for the establishing of homologies.

While practically all morphologists were agreed as to the main principles of their science, they yet showed, as regards their general attitude to the problems of form, a fairly definite division into two groups, of which one laid stress upon the intimate relation existing between form and function, while the other disregarded function completely, and sought to build up a "pure" or abstract morphology. In opposition to both groups, in opposition really to morphology altogether, a movement had gained strength which tended towards the analysis and disintegration of the organism. This movement took its origin in the current materialism of the day, and found expression particularly in the cell-theory and in materialistic physiology.

The separation between morphology as the science of form and physiology as the science of the physics and chemistry of the living body had by Darwin's day become well-nigh absolute.

The morphology of the 'fifties lent itself readily to evolutionary interpretation. Darwin found it easy to give a formal solution of all the main problems which pre-evolutionary morphology had set—he was able to interpret the natural system of classification as being in reality genealogical, systematic relationship as being really blood-relationship; he was able to interpret homology and analogy in terms of heredity and adaptation; he was able to explain the unity of plan by descent from a common ancestor, and for the concept of "archetype" to substitute that of "ancestral form."

The current morphology, Darwin found, could be taken over, lock, stock and barrel, to the evolutionary camp.

In what follows we shall see that the coming of evolution made surprisingly little difference to morphology, that the same methods were consciously or unconsciously followed, the same mental attitudes taken up, after as before the publication of the *Origin of Species.*

Darwin himself was not a professional morphologist; the conversion of morphology to evolutionary ideas was carried out principally by his followers, Ernst Haeckel and Carl Gegenbaur in Germany, Huxley, Lankester, and F. M. Balfour in England.

It was in 1866 that Haeckel's chief work appeared, a *General Morphology of Organisms,*[1] which was intended by its author to bring all morphology under the sway and domination of evolution.

It was a curious production, this first book of Haeckel's, and representative not so much of Darwinian as of pre-

[1] *Generelle Morphologie der Organismen. Allgemeine Grundzüge der organischen Formenwissenschaft, mechanisch begründet durch die von Ch. Darwin reformierte Descendenztheorie.* Berlin, 1866. Reprinted in part as *Prinzipien der generellen Morphologie der Organismen.* Berlin, 1906.

Darwinian thought. It was a medley of dogmatic materialism, idealistic morphology, and evolution theory; its sources were, approximately, Büchner, Theodor Schwann, Virchow, H. G. Bronn, and, of course, Charles Darwin.

It was scarcely modern even on its first appearance, and many regarded it, not without reason, as a belated offshoot of *Naturphilosophie.*

Its materialism is of the most intransigent character. The form and activities of living things are held to be merely the mechanical result of the physical and chemical composition of their bodies. The simplest living things, the Monera, are nothing more than homogeneous masses of protein substance. "They live, but without organs of life; all the phenomena of their life, nutrition and reproduction, movement and irritability, appear here as merely the immediate outcome of formless organic matter, itself an albumen compound" (p. 63, 1906).

Teleology, the Achilles' heel of Kant's (otherwise sound!) philosophy, is to be regarded as a totally refuted and antiquated doctrine, definitely put out of court by Darwinism.

Haeckel works out his materialistic philosophy of living things very much after the fashion of Schwann. There is the same talk of cells as organic crystals, of crystal trees, of the analogy between assimilation by the cell and the growth of crystals in a mother liquid. Heredity and adaptation are shown equally as well by crystals as by organisms; for heredity, or the internal *Bildungstrieb* (!), is the mechanical effect of the material structure of the crystal or the germ, and adaptation, or the external *Bildungstrieb*, is a name for the modifications induced by the environment. Adaptation so defined comes to be synonymous with the fortuitous variation which plays so great a part in Darwin's theory of natural selection.

It goes without saying that Haeckel allowed to the organism no other nor higher individuality than belongs to the crystal, and took no account at all of that harmonious interaction of the organs which Cuvier called the principle of the "conditions of existence." The concept of correlation had simply no meaning for Haeckel. The analysis and disintegration of the organism was pushed by him to its logical extreme, and in this also he was a child of his time.

A no less important influence clearly visible in the *General Morphology* is the idealistic morphology of men like K. G. Carus and H. G. Bronn. In previous chapters we have seen how K. G. Carus attempted to work out a geometry of the organism, and how Bronn tried in a modest way to found a stereometrical morphology, but had the grace not to push his stereometry *à l'outrance*, recognising very wisely that the greater part of organic form is functionally determined. Haeckel took over this idea [1] and pushed it to wild extremes, founding a new science of "Promorphology" of which he was the greatest—and only—exponent.[2]

This "science" dealt with axes and planes, poles and angles, in a veritable orgy of barbarous technical terms. It was intended to be a "crystallography of the organic," and to lay the foundations of a mechanistic morphology, or morphography at least.

How it was to be linked up with the physics and chemistry of living matter on the one hand and with the ordinary morphology of real animals on the other, was never made quite clear.

The science of Promorphology has no historical significance; it is interesting only because it illustrates Haeckel's close affinity with the idealistic morphologists.

Another abortive science of Haeckel's, the science of Tectology, was equally a heritage from idealistic morphology. Tectology is the science of the composition of organisms from individuals of different orders. There were six orders of individuals:—(1) Plastids (Cytodes and cells); (2) Organs (including cell-fusions, tissues, organs, organ-systems); (3) Antimeres (homotypic parts, *i.e.*, halves or rays); (4) Metameres (homodynamic parts, *i.e.*, segments); (5) Persons (individuals in the ordinary sense); (6) Corms (colonial animals).

The thought is essentially transcendental, and recalls

[1] He mentions as his predecessors in this field, Bronn, J. Müller, Burmeister, and G. Jäger.

[2] In *Grundriss einer Allgemeinen Naturgeschichte der Radiolarien*, Berlin, 1887, and *Kunstformen der Natur*, Suppl. Heft, Leipzig.

the "theory of the repetition of parts," of which so much use was made by the German transcendentalists, such as Goethe,[1] Oken, Meckel and K. G. Carus, as well as by Dugès.

The third, and naturally the most important, ingredient in the *General Morphology* was the doctrine of evolution, in the form given to it by Darwin. We have here no concern with Haeckel's evolutionary philosophy, with the way in which he combined his evolutionism and his materialism to form a queer Monism of his own. We are interested only in the way he applied evolution to morphology, what modifications he introduced into the principles of the science, and in general in what way he interpreted the facts and theories of morphology in the light of the new knowledge.

We find that he repeats very much what Darwin said, giving, of course, more detail to the exposition, and elaborating, particularly in his recapitulation theory or "biogenetic law," certain doctrines not explicitly stated by Darwin.

Like Darwin he held that the natural system is in reality genealogical. "There exists," he writes, "one single connected natural system of organisms, and this single natural system is the expression of real relations which actually exist between all organisms, alike those now in being on the earth and those that have existed there in some past time. The real relations which unite all living and extinct organisms in one or other of the principal groups of the natural system, are genealogical: their relationship in form is blood-relationship; the natural system is accordingly the genealogical tree of organisms, or their genealogema. . . . All organisms are in the last resort descendants of autogenous Monera, evolved as a consequence of the divergence of characters through natural selection. The different subordinate groups of the natural system, the categories of the class, order, family, genus, etc., are larger or smaller branches of the genealogical tree, and the degree of their divergence indicates the degree of genealogical affinity of the

[1] Haeckel had an intense admiration for Goethe's morphological work. It is a curious coincidence that the work of Goethe, Oken and Haeckel was closely associated with the town of Jena.

related organisms with one another and with the common ancestral form" (ii., p. 420).

The degree of systematic relationship is thus the degree of genealogical affinity. It follows that the natural system of classification may be converted straightway into a genealogical tree, and this is actually what Haeckel does in the *General Morphology*. The genealogical trees depicted in the second volume (plates i.-viii.) are nothing more than graphic representations of the ordinary systematic relationships of organisms, with a few hypothetical ancestral groups or forms thrown in to give the whole a genealogical turn.

If the genealogical tree is truly represented by the natural system, it would seem that for each genus a single ancestral form must be postulated, for each group of genera a single more primitive form, and so in general for each of the higher classificatory categories, right up to the phylum. Species of one genus must be descended from a generic ancestral form, genera of one family from a single family *Urform*, and so on for the higher categories.

This consequence was explicitly recognised by Haeckel. "Genera and families," he writes, "as the next highest systematic grades, are extinct species which have resolved themselves into a divergent bunch of forms (*Formenbüschel*)" (ii., p. 420).

The archetype of the genus, family, order, class and phylum was thus conceived to have had at some past time a real existence.

The natural system of classification is based upon a proper appreciation of the distinction between homological and analogical characters. Haeckel, following Darwin, naturally interprets the former as due to inheritance, the latter as due to adaptation, using these words, we may note, in their accepted meaning and not in the abstract empty sense he had previously attributed to them.[1] Similarly the "type of organisation," in von Baer's sense, was due to heredity, the "grade of differentiation" to adaptation.

So far Haeckel merely emphasised what Darwin had already said in the *Origin of Species.* But by his statement of the "biogenetic law," and particularly by the clever use

[1] But he himself would not admit this! See *Gen. Morph.*, ii., p. 11.

he made of it, Haeckel went a step beyond Darwin, and exercised perhaps a more direct influence upon evolutionary morphology than Darwin himself.

Haeckel was not the original discoverer of the law of recapitulation. It happened that a few years before the publication of Haeckel's *General Morphology*, a German doctor, Fritz Müller by name, stationed in Brazil, had been working on the development of Crustacea under the direct inspiration of Darwin's theory, and had published in 1864 a book[1] in which he showed that individual development gave a clue to ancestral history.

He conceived that progressive evolution might take place in two different ways. "Descendants . . . reach a new goal, either by deviating sooner or later whilst still on the way towards the form of their parents, or by passing along this course without deviation, but then instead of standing still advancing still farther" (Eng. trans., p. 111). In the former case the developmental history of descendants agrees with that of the ancestors only up to a certain point and then diverges. "In the second case the entire development of the progenitors is also passed through by the descendants, and, therefore, so far as the production of a species depends upon this second mode of progress, the historical development of the species will be mirrored in its developmental history" (p. 112).

Of course the recapitulation of ancestral history will be neither literal nor extended. "The historical record preserved in developmental history is gradually *effaced* as the development strikes into a constantly straighter course from the egg to the perfect animal, and it is frequently *sophisticated* by the struggle for existence which the free-living larvæ have to undergo" (p. 114).

It follows that "the primitive history of a species will be preserved in its developmental history the more perfectly the longer the series of young stages through which it passes by uniform steps; and the more truly, the less the mode of life of the young departs from that of the adults, and the less the peculiarities of the individual young states can be con-

[1] *Für Darwin*, 1864. Eng. trans. by Dallas as *Facts and Arguments for Darwin*, London, 1869.

ceived as transferred back from later ones in previous periods of life, or as independently acquired" (p. 121).

Applying these principles to Crustacea, he concluded that the shrimp *Peneus* with its long direct development gave the best and truest picture of the ancestral history of the Malacostraca, and that accordingly the nauplius and the zoaea larvæ represented important ancestral stages. He conceived it possible so to link up the various larval forms of Crustacea as to weave a picture of the primeval history of the class, and he made a plucky attempt to work out the phylogeny of the various groups.

The thought that development repeats evolution was already implicit in the first edition of the *Origin*, but the credit for the first clear and detailed exposition of it belongs to F. Müller.

In much the same form as it was propounded by Müller it was adopted by Haeckel, and made the corner-stone of his evolutionary embryology. Haeckel gave it more precise and more technical formulation, but added nothing essentially new to the idea.

It is convenient to use his term for it—the biogenetic law (*Biogenetische Grundgesetz*)—to distinguish it from the laws of Meckel-Serres and von Baer, with which it is so often confused.

Haeckel's statement of it may best be summarised in his own words, "Ontogeny, or the development of the organic individual, being the series of form-changes which each individual organism traverses during the whole time of its individual existence, is immediately conditioned by phylogeny, or the development of the organic stock (phylon) to which it belongs.

"Ontogeny is the short and rapid recapitulation of phylogeny, conditioned by the physiological functions of heredity (reproduction) and adaptation (nutrition). The organic individual (as a morphological individual of the first to the sixth order) repeats during the rapid and short course of its individual development the most important of the form-changes which its ancestors traversed during the long and slow course of their palæontological evolution according to the laws of heredity and adaptation.

"The complete and accurate repetition of phyletic by biontic development is obliterated and abbreviated by secondary contraction, as ontogeny strikes out for itself an ever straighter course ; accordingly, the repetition is the more complete the longer the series of young stages successively passed through.

"The complete and accurate repetition of phyletic by biontic development is falsified and altered by secondary adaptation, in that the bion[1] during its individual development adapts itself to new conditions: accordingly the repetition is the more accurate the greater the resemblance between the conditions of existence under which respectively the bion and its ancestors developed" (ii., p. 300).

The last two propositions, it will be observed, are taken over almost verbally from F. Müller.

Now we have seen that the natural system of classification gives a true picture of the genealogical relationships of organisms, that the smaller and larger classificatory groups correspond to greater or lesser branches of the genealogical tree. If ontogeny is a recapitulation of phylogeny, we must expect to find the embryo repeating the organisation first of the ancestor of the phylum, then of the ancestor of the class, the order, the family and the genus to which it belongs. There must be a threefold parallelism between the natural system, ontogeny and phylogeny (ii., pp. 421-2).

It will be observed that there is here implied an analogy between the biogenetic law and the law of von Baer, for both assert that development proceeds from the general to the special, that the farther back in development you go the more generalised do you find the structure of the embryo; both assert, too, that differentiation of structure takes place not in one progressive or regressive line, but in several diverging directions.

But the analogy between the biogenetic law and the Meckel-Serres law is even more obvious, and the resemblance between the two is much more fundamental. It is a significant fact that in his theory of the threefold

[1] The bion is the physiological, as the morphon is the morphological, individual.

parallelism Haeckel merely resuscitated in an evolutionary form a doctrine widely discussed in the 'forties and 'fifties,[1] and championed particularly by L. Agassiz,[2] a doctrine which must be regarded as a development or expansion of the Meckel-Serres law.[3] It is the view that a parallelism exists between the natural system, embryonic development, and palæontological succession. Actually, as Agassiz stated it, the doctrine applied neither to types, nor as a general rule to classes, but merely to orders. It was well exemplified, he thought, in Crinoids:—"The successive stages of the embryonic growth of Crinoids typify, as it were, the principal forms of Crinoids which characterise the successive geological formations. First, it recalls the Cistoids of the palæozoic rocks, which are represented in its simple spheroidal head; next the few-plated Platycrinoids of the Carboniferous period; next the Pentacrinoids of the Lias and Oolite with their whorls of cirrhi; and finally, when freed from its stem, it stands as the highest Crinoid, as the prominent type of the family in the present period" (p. 171).

The Meckel-Serres law, it will be remembered, expressed the idea that the higher animals repeat in their ontogeny the adult organisation of animals lower in the scale. Since Haeckel recognised clearly that a linear arrangement of the animal kingdom was a mere perversion of reality, and that a branching arrangement of groups more truly represented the real relations of animals to one another, he could not of course entertain the Meckel-Serres theory in its original form. But he accepted the main tenet of it when he asserted that each stage of ontogeny had its counterpart in an adult ancestral form. Such ancestral forms might or might not be in existence as

[1] See Vogt, *Embryologie des Salmones*, p. 259, 1842, and *supra*, p. 230.

[2] *An Essay on Classification*, London, 1859.

[3] It was hinted at by Tiedemann. "It is clear that, proceeding from the earlier to the more recent strata, a gradation in fossil forms can be established from the simplest organised animals, the polyps, up to the most complex, the mammals, and that accordingly the animal kingdom as a whole has its developmental periods just like the single individual organism. The species and genera which have become extinct during the evolutionary process may be compared with the organs which disappear during the development of the individual animal" (p. 73, 1808).

real species at the present day; they might or might not be discoverable as fossils. That they had real existence either now or at some past epoch Haeckel never doubted. In his construction of phylogenetic trees he was so confident in the truth of his biogenetic law that he largely disregarded and consistently minimised the importance of the evidence from palæontology.

The biogenetic law differed from the Meckel-Serres law chiefly in the circumstance that many of the adult lower forms whose organisation was supposed to be repeated in the development of the higher animals were purely hypothetical, being deduced directly from a study of ontogeny and systematic relationships. The hypothetical ancestral forms which the theory thus postulated naturally took their place in the natural system, for they were merely the concrete projections or archetypes of the classificatory groups.

The transcendentalists, of course, conceived evolution, whether real or ideal, as a uniserial process, whereas Haeckel conceived it as multiserial and divergent. It is here that the superficial agreement of the biogenetic law with the law of von Baer comes in.

We might almost sum up the relation of the biogenetic law to the laws of von Baer and Meckel-Serres by saying that it was the Meckel-Serres law applied to the divergent differentiation upheld by von Baer instead of to the uniserial progression believed in by the transcendentalists.

How near in practice Haeckel's law came to the recapitulation theory of the transcendentalists may be seen in passages like the following, with its partial recognition of the *Echelle* idea:[1]—"As so high and complicated an organism as that of man . . . rises upwards from a simple cellular state, and as it progresses in its differentiating and perfecting, it passes through the same series of transformations which its animal progenitors have passed through, during immense spaces of time, inconceivable ages ago. . . . Certain very early and low stages in the development of man, and other vertebrate animals in general, correspond completely in many points of structure with conditions which

[1] *The History of Creation*, vol. i., p. 310, 1876. Translation of the *Natürliche Schöpfungsgeschichte*, 1868.

last for life in the lower fishes. The next phase which follows on this presents us with a change of the fish-like being into a kind of amphibious animal. At a later period the mammal, with its special characteristics, develops out of the amphibian, and we can clearly see, in the successive stages of its later development, a series of steps of progressive transformation which evidently correspond with the differences of different mammalian orders and families.'[1]

The biogenetic law went beyond both the Meckel-Serres law and the law of von Baer in that it recognised that the ancestral history of the species accounts in part for the course which the development of the individual takes, that in a certain sense, though not in the crude way supposed by Haeckel, phylogeny is the cause of ontogeny. This thought, that the organism is before all an historical being, is of course implied in the evolution idea, is indeed the essential core of it. Take away this element from the biogenetic law—not a difficult matter—and it becomes merely a law of idealistic morphology, applicable to evolution considered as an ideal process, as the progressive development in the Divine thought of archetypal models.

As a book, the *General Morphology* suffers a good deal from the arid, schematic, almost scholastic manner of exposition adopted. Haeckel's Prussian mania for organisation, for absolute distinctions, for iron-bound formalism, is here given full scope. A treatment less adequate to the variety, fluidity and changeableness of living things could hardly be imagined.

His doctrine, though it remains essentially unchanged, receives in his later works a less formal and more concrete expression, and, in particular, his views on the biogenetic law undergo some small modification.

Even in the *General Morphology* Haeckel had recognised that ontogeny is neither a complete nor an entirely accurate recapitulation of phylogeny; he had admitted, following F. Müller, that the true course of recapitulation was frequently modified by larval and fœtal adaptations. As time went on, he was forced to hedge more and more on this point, and finally in his *Anthropogenie* (1874) and his second

[1] *Cf.* a parallel passage from Serres, *supra*, p. 82.

paper on the Gastræa theory (1875),[1] he had to work out a distinction between palingenetic and cenogenetic characters, of which much use was made by subsequent writers.

The distinction may be given in Haeckel's own words:—"Those ontogenetic processes," he writes, "which are to be referred immediately, in accordance with the biogenetic law, to an earlier completely developed *independent ancestral form*, and are transmitted from this by *heredity*, obviously possess *primary* importance for the understanding of the casual-physiological relations; on the other hand, those developmental processes which appear subsequently through *adaptation* to the needs of embryonic or larval life, and accordingly can *not* be regarded as repeating the organisation of an earlier independent ancestral form, can clearly have for the understanding of the ancestral history only a quite subordinate and *secondary* importance.

"The first I have named *palingenetic*, the second *cenogenetic*. Considered from this critical standpoint, the whole of ontogeny falls into two main parts:—First, *palingenesis*, or 'epitomised history' (*Auszugsgeschichte*), and second, *cenogenesis*, or 'counterfeit history' (*Fälschungsgeschichte*). The first is the true ontogenetic epitome or short recapitulation of past evolutionary history; the second is the exact contrary, a new foreign ingredient, a falsifying or concealing of the epitome of phylogeny."[2]

As examples of palingenetic processes in the development of Amniotes, for instance, may be quoted the separation of two primary germ-layers, the formation of a simple notochord between medullary tube and alimentary canal, the appearance of a simple cartilaginous cranium, of the gill-arches and their vessels, of the primitive kidneys, the primitive tubular heart, the paired aortæ and the cardinal veins, the hermaphroditic rudiment of the gonads, and so on. Cenogenetic processes, on the other hand, include such phenomena as the formation of yolk and the embryonic membranes, the temporary allantoic circulation, the navel, the curved and contracted shape of the embryo, and the like.

The most important phenomena to be included under

[1] *Jenaische Zeitschrift*, ix., pp. 402-508, 1875.

[2] *Loc. cit.*. ix., p. 409.

the general heading of cenogenesis are, first, the occurrence of food-yolk, and second, those anomalies of development which are classed by Haeckel as heterochronies and heterotopies.

It is to the influence of the different amounts of yolk present in the egg that are due the great differences in the segmentation and gastrulation processes, which almost mask their true significance.

Heterochronic processes are such as arise through the dislocation of the proper phylogenetic order of succession: heterotopic processes in the same way are caused by a wandering of cells from one germ-layer to another. The two classes of phenomena are disturbances either of the proper spatial or of the proper temporal relation of the parts during development.

Heterochrony shows itself, as a rule, either as an acceleration or as a retardation of developmental events, as compared with their relative time of occurrence during phylogeny. Thus the notochord, the brain, the eyes, the heart, appear earlier in the ontogenetic than in the phylogenetic series, while, on the other hand, the septum of the auricles appears in the development of the higher Vertebrates before the ventricular septum, which is undoubtedly a reversal of the phylogenetic order.

Cases of heterotopy, or of organs being developed in a position or a germ-layer other than that in which they originally arose in phylogeny, are not so easy to find. According to Haeckel, the origin of the generative products in the mesoderm is a heterotopic phenomenon, for he considers that they must have originated phylogenetically in one of the two primary layers, ectoderm or endoderm.

It is worthy of note that the help of comparative anatomy is admittedly required in deciding what processes are palingenetic and what cenogenetic (p. 412).

Haeckel's morphological notions, and particularly his biogenetic law, excited a good deal of adverse criticism from men like His, Claus, Salensky, Semper and Goette. Nor was his principal work, the *General Morphology*, received with much favour. Nevertheless, since he did express, though in a crude, dogmatic and extreme manner, the main hypotheses upon which evolutionary morphology is founded, his historical importance is considerable. He cannot perhaps be regarded as typical of the morphologists of his time—he was too trenchantly materialistic, too much the populariser of a crude and commonplace philosophy of Nature. In point of concrete achievement in the field of pure research he fell notably behind many of his contemporaries.

His friend, Carl Gegenbaur, who gained a great and well-deserved reputation by his masterly studies on vertebrate morphology,[1] was a sounder man, and probably exercised a wider and certainly a more wholesome influence upon the younger generation of professional morphologists than the more brilliant Haeckel. It is true that in his famous *Grundzüge der vergleichenden Anatomie*, the second edition of which, published in 1870, soon came to be regarded as the classical text-book of evolutionary morphology, Gegenbaur enunciated very much the same general principles as Haeckel, and referred to the *Generelle Morphologie* as the chief and fundamental work on animal morphology. But in Gegenbaur's pages the Haeckelian doctrines are modified and subdued by the strong commonsense and thorough appreciation of the older classical or Cuvierian morphology that characterise Gegenbaur's work. According to Haeckel,[2] Gegenbaur was greatly influenced by J. Müller, who, as we know, laid as much stress on function as on form.

The "General Part" of Gegenbaur's text-book is in many ways a significant document and deserves close attention.

We note first of all that physiology and morphology are considered by Gegenbaur to be entirely distinct sciences, with different subject-matter and different methods. "The task of physiology is the investigation of the functions of the animal body or of its parts, the referring back of these functions to elementary processes and their explanation by

[1] *Untersuchungen zur vergl. Anatomie d. Wirbelthiere*, Leipzig, i., 1864; ii., 1865; and iii., 1872.

[2] "U. d. Biologie in Jena während des 19 Jahrhunderts," *Jenaische Zeitschrift*, xxxix., pp. 713-26, 1905.

general laws. The investigation of the material substratum of these functions, of the form of the body and its parts, and the explanation of this form, constitute the task of Morphology (2nd ed., p. 3).

Morphology falls naturally into two divisions—comparative anatomy and embryology. The method of comparative anatomy is *comparison* (p. 6), and in employing this method account is to be taken of "the spatial relations of the parts to one another, their number, extent, structure, and texture." Through comparison one is enabled to arrange organs in continuous series, and it comes out very clearly during this proceeding "that the physiological value of an organ is by no means constant throughout the different form-states of the organ, that an organ, through the mere modification of its anatomical relations, can subserve very different functions. Exclusive regard for their physiological functions would place morphologically related organs in different categories. From this it follows that in comparative anatomy we should never in the first place consider the function of an organ. The physiological value comes only in the second place into consideration, when we have to reconstruct the relations to the organism as a whole of the modification which an organ has undergone as compared with another state of it. In this way comparative anatomy shows us how to arrange organs in series; within these series we meet with variations which sometimes are insignificant and sometimes greater in extent; they affect the extent, number, shape, and texture of the parts of an organ, and can even, though only in a slight degree, lead to alterations of position" (p. 6).

Geoffroy St Hilaire would have subscribed to every word of this vindication of his "principle of connections."

Between comparative anatomy and embryology there exists a close connection, for the one throws light on the other. "While in some cases the same organ shows only slight modifications in its development from its early beginnings to its perfect state, in other cases the organ is subjected to manifold modifications before it reaches its definitive form; we see parts appear in it which later disappear, we observe alterations in it in all its anatomical relations, alterations which may even affect its texture. This fact is of great importance, for those changes which an organ undergoes during its individual development lead through states which the organ in other cases permanently shows, or at the least the first appearance of the organ is the equivalent of a permanent state in another organism. If then the fully developed organ is in any special case so greatly modified that its proper relation to some organ-series is obscured, this relation may be cleared up by a knowledge of the organ's development. The earlier state indicated in this way enables one to find with ease the proper place for the organ and so insert it into an already known series. The relations which we observe in an organ-seriation are then the equivalent of processes which in certain cases take place in a similar manner during the individual development of an organ. Embryology enters therefore into the closest connection with comparative anatomy. . . . It teaches us to know organs in their earliest states, and connects them up with the permanent states of others, whereby they fill up the gaps which we meet with in the various series formed by the fully developed organs of the body" (pp. 6-7).

This recognition of the parallelism between comparative anatomy and embryology is, of course, the kernel of the Meckel-Serres law. For Gegenbaur it had a very definite evolutionary meaning—he subscribed to the evolutionary form of it, the biogenetic law. How near his conception of the relation between ontogeny and phylogeny came to the old Meckel-Serres law may be gauged from the following passage, taken from a later work:—"Ontogeny thus represents, to a certain degree, palæontological development abbreviated or epitomised. The stages which are passed through by higher organisms in their ontogeny correspond to stages which are maintained in others as the definitive organisation. These embryonic stages may accordingly be explained by comparing them with the mature stages of lower organisms, since we regard them as forms inherited from ancestors belonging to such lower stages"[1] (p. 6).

It is worth noting that in Gegenbaur's opinion compara-

[1] *Grundriss der vergl. Anatomie*, 1874, 2nd ed., 1878. Trans. by F. Jeffrey Bell, revised by E. Ray Lankester, as *Elements of Comparative Anatomy*, London, 1878.

tive anatomy was prior in importance to embryology, that embryology could hardly exist as an independent science, since it must seek the interpretation of its facts always in the facts of comparative anatomy (*Grundzüge*, pp. 7-8).

While Gegenbaur was at one with all "pure" morphologists, whether evolutionary or pre-evolutionary, in minimising as far as possible the importance of function in the study of form, he was too cautious and sober a thinker not to recognise the immense part which function really plays. Thus he classified organs, according to their function, into those that established relations with the external world and those that had to do with nutrition and reproduction, very much as Bichat had done before him.

Like Darwin, Haeckel and most evolutionists, he interpreted the homological resemblances of animals as being due to heredity, their differences as due to adaptation,[1] but he did not adopt Haeckel's crude and shallow definition of these terms. For Gegenbaur heredity was a convenient expression for the fact of transmission, and was not explained offhand as the mere mechanical result of a certain material structure handed down from germ to germ. Adaptation he defined in a way which took the fullest account of function, and was as far as possible removed from Haeckel's definition of it as the direct mechanical effect of the environment upon the organism. "The organism is altered," writes Gegenbaur, "according to the conditions which influence it. The consequent *Adaptations* are to be regarded as gradual, but steadily progressive, changes in the organisation, which are striven after during the individual life of the organism, preserved by transmission in a series of generations, and further developed by means of natural selection. What has been gained by the ancestor becomes the heritage of the descendant. Adaptation and Transmission are thus alternately effective, the former representing the modifying, the latter the con-

[1] "This theory (evolution) shows that what was formerly called 'structural plan' or 'type' is the sum of the dispositions (*Einrichtungen*) of the animal organisation which are perpetuated by heredity, while it explains the modifications of these dispositions as adaptive states. Heredity and adaptation are thus the two important factors through which both the unity and the variety of organisation can be understood" (*Grundzüge*, p. 19).

servative principle. . . . Adaptation is commenced by a change in the function of organs, so that the *physiological relations* of organs play the most important part in it. Since adaptation is merely the material expression of this change of function, the modification of the function as much as its expression is to be regarded as a gradual process. In Adaptation, the closest connection between the function and the structure of an organ is thus indicated. Physiological functions govern, in a certain sense, structure; and so far what is morphological is subordinated to what is physiological" (*Elements*, pp. 8-9). Gegenbaur recognised also that morphological differentiation depended largely on the physiological division of labour (*Grundzüge*, p. 49).

It is clear that Gegenbaur realised vividly the importance of function, and in this respect, as in others, he is far beyond Haeckel. The same thing comes out markedly in his treatment of correlation. Haeckel had no slightest feeling for the true meaning of correlation. For him, as for Darwin, it reduced itself to a law of correlative variation, according to which "actual adaptation not only changes those parts of the organism which are directly affected by its influence, but other parts also, not directly affected by it."[1] Such "correlative adaptation" was due to nutrition being a "connected, centralised activity."

Gegenbaur, on the contrary, had a firm grasp of the Cuvierian conception, and expressed it in unmistakable terms. "As indeed follows from the conception of life as the harmonious expression of a sum of phenomena rigorously determining one another, no activity of an organ can in reality be thought of as existing for itself. Each kind of function (*Verrichtung*) presupposes a series of other functions, and accordingly every organ must possess close relations with, and be dependent on, all the others" (*Grundzüge*, p. 71). The organism must be regarded as an individual whole which is as much conditioned by its parts as one part is conditioned by the others. For an understanding of correlation a knowledge of functions, and of the functional relations of the organism to its environment, is clearly indispensable.

[1] *History of Creation*, i., pp. 241-2.

Gegenbaur's morphological system was out-and-out evolutionary. "The most important part of the business of comparative anatomy," in Gegenbaur's eyes, "is to find indications of genetic connection in the organisation of the animal body" (*Elements*, p. 67).

The most important clue to discovering this genetic connection is of course that given by homology; it is indeed the main principle of evolutionary morphology that what is common in organisation is due to common descent, what is divergent is due to adaptation. "Homology . . . corresponds to the hypothetical genetic relationship. In the more or the less clear homology, we have the expression of the more or less intimate degree of relationship. Blood-relationship becomes dubious exactly in proportion as the proof of homologies is uncertain" (*Elements*, p. 63).

It is worth noting that while Gegenbaur agrees with Haeckel generally that morphological relationships are really genealogical, that, for instance, each phylum has its ancestral form, he enters a caution against too hastily assuming the existence of a genetic relation between two forms on the basis of the comparison of one or two organs. "In treating comparative anatomy from the genealogical standpoint required by the evolution-theory," he writes, "we have to take into consideration the fact that the connections can almost never be discovered in the real genealogically related objects, for we have almost always to do with the divergent members of an evolutionary series. We derive, for instance, the circulatory system of insects from that of Crustacea . . . but there exists neither a form that leads directly from Crustacea to insects nor any organisatory state (*Organisationszustand*), which as such shows the transition. Even when one point of organisation can be denoted as transitional, numerous other points prevent us from regarding the whole organism strictly in the same light" (*Grundzüge*, p. 75). The real ancestral forms cannot, as a rule, be discovered among living species, nor often as extinct. "When we arrange allied forms in series by means of comparison, and seek to derive the more complex from the simpler, we recognise in the lower and simpler forms only similarities with the ancestral form, which remains essentially hypothetical" (p. 75).

The facts of development, Gegenbaur goes on to say, help us out greatly in our search for ancestral forms, for the early stages in the ontogeny of a highly organised animal give us some idea of the organisation of its original ancestor. Characters common to the early ontogeny of all the members of a large group are particularly important in this respect (*cf.* von Baer's law).

Gegenbaur distinguishes homologous or morphologically equivalent structures from such as are analogous or physiologically eqivalent, just as did Owen and the older anatomists. Like von Baer he recognises homologies, as a rule, only within the type.

He contributed, however, to the common stock a useful analysis of the concept of homology, and established certain classes and degrees of it. He distinguished first between general and special homology, in quite a different sense from Owen.

General homology, in Gegenbaur's sense, relates to resemblances of organs within the organism, and includes four kinds of resemblance, homotypy, homodynamy, homonomy and homonymy. Right and left organs are homotypic, metameric organs are homodynamic; homonomy is the relation exemplified by fin-rays or fingers, which are arranged with reference to a transverse axis of the body; homonymy is a sort of metamerism in secondary parts (not the main axis) of the body, and is shown by the various divisions of the appendages (*Grundzüge*, p. 80).

Special homology, on the other hand, relates to resemblances between organs in different animals. The interesting thing is that Gegenbaur defines it genetically. Special homology is the name we give "to the relations which obtain between two organs which have had a common origin, and which have also a common embryonic history" (*Elements*, p. 64). This is his definition; but, in practice, Gegenbaur establishes homologies by comparison just as the older anatomists did, and infers common descent from homology, not homology from common descent.

"Special homology," he continues, "must be again separated into sub-divisions, according as the organs dealt with are essentially unchanged in their morphological characters, or are altered by the addition or removal of parts" (p. 65). In the former case the homology is said to be "complete," in the latter "incomplete." Thus the bones of the upper arm are completely homologous throughout all vertebrate classes from Amphibia upwards, while the heart of a fish is incompletely homologous with the heart of a mammal.

Independently of Gegenbaur, Sir E. Ray Lankester proposed in 1870 a genetic definition of homology.[1] He proposed, indeed, to do away with the term homology altogether, on the ground that it included many resemblances which were obviously not due to common descent—as, for instance, the resemblance of metameres. So, too, organs which were homologous in the ordinary sense, as the heart of birds and mammals, might have arisen separately in evolution. He proposed, therefore, that "structures which are genetically related, in so far as they have a single representative in a common ancestor," should be called *homogenous* (p. 36). All other resemblances were to be called *homoplastic*. "Homoplasy includes all cases of close resemblance of form which are not traceable to homogeny, all details of agreement not homogenous, in structures which are broadly homogenous, as well as in structures having no genetic affinity" (p. 41). Serial homology, for instance, was a case of homoplasy.

The term "analogy" was to be retained for cases of functional resemblance, whether homogenetic or not.

The attempt was an interesting one, but most morphologists wisely adhered to the old concept of homology, in spite of Lankester's declaration that this belonged to an older "Platonic" philosophy, and ought to be superseded by a term more consonant with the new philosophy of evolution.

[1] "On the use of the term Homology in Modern Zoology, and the distinction between Homogenetic and Homoplastic agreements," *Ann. Mag. Nat. Hist.* (4), vi., pp. 35-43, 1870.

CHAPTER XV

EARLY THEORIES ON THE ORIGIN OF VERTEBRATES

HAECKEL and Gegenbaur set the fashion for phylogenetic speculation, and up to the middle 'eighties, when the voice of the sceptics began to make itself heard, the chief concern of the younger morphologists was the construction of genealogical trees. The period from about 1865 to 1885 might well be called the second speculative or transcendental period of morphology, differing only from the first period of transcendentalism by the greater bulk of its positive achievement. It must be remembered that the later workers (at least towards the end of this period) had immense advantages over their predecessors in the matter of equipment and technique; they possessed well-fitted laboratories in the university towns and by the sea; they had at their command perfected microscopes and microtomes; while the whole new technique of microscopical anatomy with its endless variety of stains and reagents made it possible for the tyro to confirm in a day what von Baer and Müller had taken weeks of painful endeavour to discover.[1] But the democratisation of morphology which followed upon the facilitation of its means of research left an evil heritage of detailed and unintelligent work to counterbalance the very great and real advances which technical improvements alone rendered possible.

This period of rapid development, which set in soon after the coming of evolution and multiplied the concrete

[1] The stages in the development of microscopical technique are well summarised by R. Burckhardt, *Geschichte der Zoologie*, p. 121, Leipzig 1907.

facts of morphology an hundredfold, may for our present purpose be conveniently divided into two somewhat overlapping periods, of which the second may be said to begin with the enunciation by Haeckel of his Gastræa theory. Within the first period fall the evolutionary speculations associated with the names of Kowalevsky, Dohrn, Semper, and others; the characteristic of the second period is the preponderating influence exercised upon phylogenetic speculations by the germ-layer doctrine in its two main evolutionary developments, the Gastræa and Cœlom theories.

In the first period we might again distinguish two main tendencies, according as speculations were based mainly upon anatomical or mainly upon embryological considerations, and it so happens that these two tendencies are very well illustrated by the various theories as to the origin of Vertebrates which began to appear towards the 'seventies. We shall accordingly, in this chapter, consider very briefly the history of the earlier views on the phylogeny of the vertebrate stock.

In the early days, before the other claimants to the dignity of ancestral form to the Vertebrates—*Balanoglossus*, Nemertines and the rest—had put in an appearance, there were two main views on the subject, one upheld by Haeckel, Kowalevsky and others, to the effect that the proximate ancestor of Vertebrates was a form somewhat resembling the ascidian tadpole, the other supported principally by Dohrn and Semper that Vertebrates and Arthropods traced their descent to a common segmented annelid or pro-annelid ancestor. The former view is historically prior, and arose directly out of the brilliant embryological investigations of A. Kowalevsky, who proved himself to be a worthy successor of the great comparative embryologist Rathke. His work was indeed a true continuation of Rathke's. It was not directly inspired by evolution, though it supplied much useful confirmation of the theory—you may read Kowalevsky's earlier memoirs and not realise that they were written several years after the publication of the *Origin of Species.*

His first paper of evolutionary importance was a note in

Russian on the development of Amphioxus, published in 1865. This subject was followed up in two papers which appeared in 1867[1] and 1877.[2] In his papers on Amphioxus Kowalevsky made out the main features in the development of this primitive form, and showed that the chief organs were formed in essentially the same way as in Vertebrates; he described the formation of the archenteron by invagination, the appearance of the medullary folds, which coalesced to form the neural canal, the formation of the notochord and of the gill-slits. At first he made the mistake of supposing that the body-cavity arose from the segmentation-cavity, but in his later paper he rightly surmised that it was formed from the cavities of the "primitive vertebræ," or mesodermal segments. The origin of the notochord from the endoderm was also not made out by Kowalevsky in his paper of 1867.

Although many important details remained to be discovered by later investigators,[3] Kowalevsky's work at once made the development of Amphioxus the key to vertebrate embryology, the typical ontogeny with which all others could be compared.

Meanwhile, in 1866 and 1871, Kowalevsky had communicated memoirs of even greater interest,[4] in which he showed that the simple Ascidians developed in an extraordinarily similar way to Amphioxus and hence to Vertebrates in general. His proof that Ascidians also develop on the vertebrate type aroused great interest at the time, and was naturally acclaimed by the evolutionists as a striking piece of evidence in favour of their doctrine. The systematic position of the Ascidians was at that time quite uncertain; they were grouped, as a rule, with the Mollusca, and certainly no one suspected that their well-

[1] "Entwickelungsgeschichte des Amphioxus lanceolatus," *Mém. Acad. Sci. St Pétersbourg* (Petrograd) (vii.), xi., No. 4, 1867, 17 pp., 3 pls.

[2] "Weitere Studien ü. die Entwickelungsgeschichte des Amphioxus lanceolatus," *Arch. für mikr. Anat.*, xiii., pp. 181-204, 1877.

[3] Particularly by Hatschek (1881) and Boveri (1892).

[4] "Entwickelungsgeschichte der einfachen Ascidien," *Mém. Acad. Sci. St Pétersbourg* (Petrograd), (vii.), x., No. 15, 1866, 19 pp., 3 pls. "Weitere Studien ü. die Entwicklung der einfachen Ascidien," *Arch. f. mikr. Anat.*, vii., pp. 101-130, 1871.

known tailed larvæ, first seen by Savigny, showed any but the most superficial analogy with the tadpoles of Amphibia. Kowalevsky's papers put a different complexion on the matter. In the first of them he showed how the nervous system of the simple Ascidian developed from ectodermal folds just as it did in Amphioxus and Vertebrates, how gill-slits were formed in the walls of the pharynx, and how there existed in the ascidian larva a structure which in position and mode of development was the strict homologue of the vertebrate notochord. In his second paper he entered into much more detail, and published some excellent figures, often reproduced since (see Fig. 13), but the proof of the affinity between Vertebrates and Ascidians was in all essentials complete in his paper of 1866.

Kowalevsky's results were accepted by Haeckel, Gegenbaur, Darwin,[1] and many others as conclusive evidence of the origin of Vertebrates from a form resembling the ascidian tadpole; they were extended and amplified by Kupffer[2] in 1870, later by van Beneden and Julin[3] and numerous other workers; they were adversely criticised by Metschnikoff[4] and von Baer,[5] as well as by H. de Lacaze-Duthiers and A. Giard.[6] Lacaze-Duthiers and von Baer both held fast to the old view that Ascidians were directly comparable with Lamellibranch molluscs; they denied the homology of the ascidian nervous system with that of Vertebrates, von Baer being at great pains to show that the ascidian nerve-centre was really ventral in position. He pointed out also that the "notochord" was confined to the tail of the ascidian larva. Giard's attitude was by no means so uncompromising, and the criticisms he passed on the Kowalevsky theory are both subtle and instructive. He admits that there exists a real homology between, for instance, the notochord of Vertebrates and that of Ascidians. "But," he adds, "it is too often for-

[1] *Descent of Man*, i., p. 205, 1871.

[2] *Arch. f. mikr. Anat.*, vi., 1870, and viii., 1872.

[3] *Archives de Biologie*, 1884, 1885, and 1887.

[4] *Bull. Acad. Sci. St Pétersbourg* (Petrograd) xiii., 1869, and *Zeits. f. wiss. Zool.*, xxii., 1872.

[5] *Mém. Acad. Sci. St Pétersbourg* (Petrograd) (7), xix., 1873.

[6] Giard, *Arch. zool. expér. gén.*, i., 1872, and Lacaze-Duthiers, *ibid.*, iii., 1874.

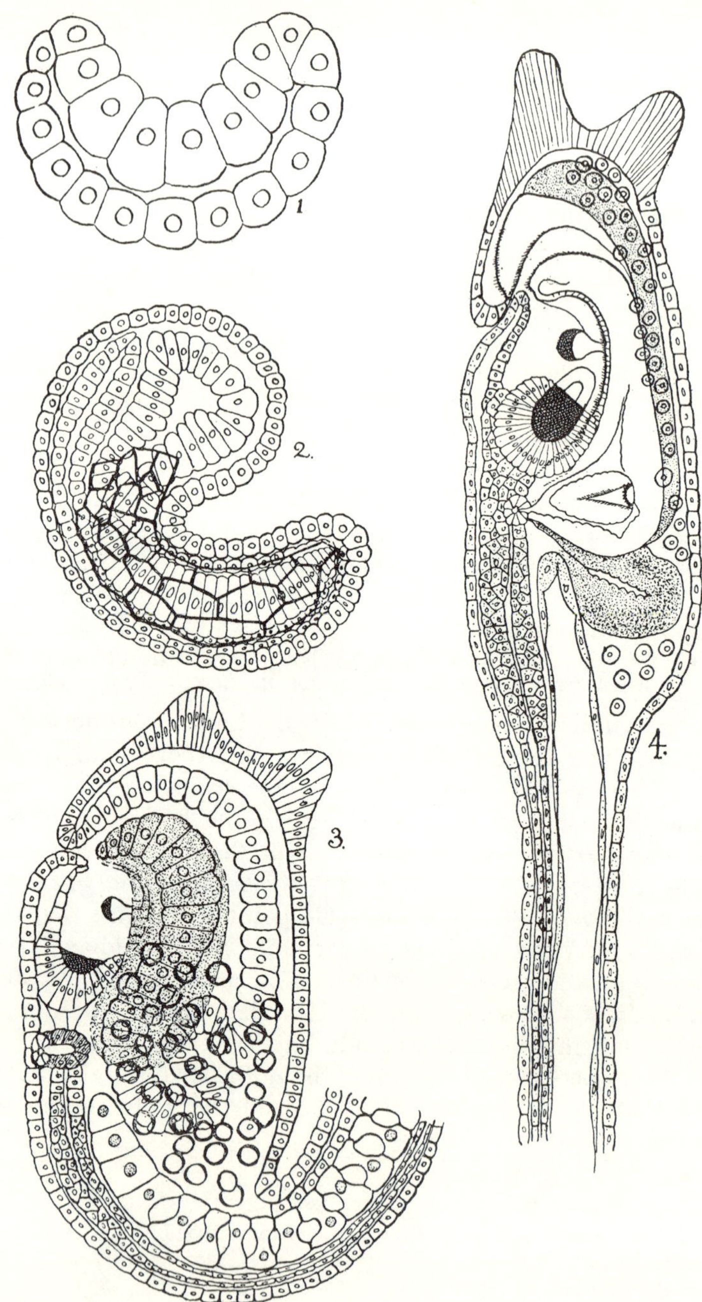

FIG. 13.—Development of the Ascidian Larva. (After Kowalevsky.)

gotten that homology does not necessarily mean an immediate common origin or close relationship. There exist, doubtless, homologies of great atavistic importance—I consider as such, for example, the formation of the cavity of Rusconi [the archenteron] in Ascidians and lower Vertebrates. But there are also adaptive and purely analogical homologies, such as the interdigital palmation of aquatic birds, amphibians and mammals. These are not purely analogous organs, for they can be superposed one on another, which is not the case with simply analogous structures (the bat's wing, for example, cannot be superposed on the bird's wing); they are homologous formations, resulting from the adaptation of the same fundamental organs to identical functions. Such is, in my opinion, the nature of the homology existing between the tail of the ascidian tadpole and that of Amphioxus or of young amphibians. The ascidian larva, having no cilia and being necessarily motile, requires for the insertion of its muscles or contractile organs . . . a central flexible axis, a true chorda dorsalis analogous to that of Vertebrates" (pp. 278-9). This point of view is strengthened by the fact that in *Molgula*, studied by Lacaze-Duthiers, the embryo is practically stationary, and forms no notochord, nor ever develops sense-organs in the cerebral vesicle.

Giard's general conclusion is that "the true homology with Vertebrates ceases after the formation of the cavity of Rusconi and the medullary groove: the homologies established by Kowalevsky for the notochord and the relations of the digestive tube and nervous systems are not atavistic, but adaptive, homologies" (p. 282). There is accordingly no close genetic relationship between Ascidians and Vertebrates.

Giard's criticisms did not avail to check the vogue of the new theory, which soon became an accepted article of faith in most morphological circles.[1] The fall of the Ascidians from their larval high estate provided the text for many a Darwinian sermon.

[1] For the later history of the Amphioxus-Ascidian theory the reader may be referred to A. Willey's well-known work, *Amphioxus and the Ancestry of the Vertebrates*, New York and London, 1894, and to Delage et Hérouard, *Traité de Zoologie concrète*, Tome viii., Paris, 1898.

Some years after the genetic relationship of Ascidians and Vertebrates had been established, a rival theory of the origin of Vertebrates made its appearance—a theory which was practically a rehabilitation in a somewhat altered form of the old Geoffroyan conception that Vertebrates are Arthropods walking on their backs. This was the so-called Annelid theory of Dohrn and Semper. Both Dohrn and Semper started out from the fact that Annelids and Vertebrates are alike segmented animals, and it was an essential part of their theory that this resemblance was due to descent from a common segmented ancestor. Both laid great stress on the fact that the main organs in Vertebrates are arranged in the same way as in an Annelid lying on its back, the nervous system being uppermost, the alimentary system coming next, and below this the vascular.

Dohrn's earlier views are contained in the fascinating little book published in 1875, which bears the title *Der Ursprung der Wirbelthiere und das Princip des Functionswechsel* (Leipzig). He followed this up by a long series of studies on vertebrate anatomy and embryology,[1] in which he modified his views in certain details. We shall confine our attention to the first sketch of his theory.

If the Vertebrate is conceived to have evolved from a primitive Annelid which took to creeping or swimming ventral surface uppermost, a difficulty at once arises with regard to the relative positions of the "brain" and the mouth. In Vertebrates the brain, like the rest of the nervous system, is dorsal to the mouth and the alimentary canal; in an inverted Annelid, however, the brain is ventral to the mouth and is connected with the dorsal nerve cord by commissures passing round the œsophagus. It would seem, therefore, that the primitive Vertebrate must have acquired either a new brain or a new mouth. Dohrn took the latter view. He supposed that the original mouth of the primitive ancestor lay between the *crura cerebelli* in the *fossa rhomboidea*, and that in Vertebrates this mouth has been replaced functionally by a new ventrally placed mouth, formed by the

[1] "Studien zur Urgeschichte des Wirbelthierkörpers," *Mittheil. Zool. Stat. Neapel*, 1882-1907.

medial coalescence of a pair of gill-slits.[1] Probably the two mouths at one period co-existed, and the older one was ousted by the growing functional importance of the newer mouth.

The gill-slits were considered by Dohrn to be derived from the segmental organs of Annelids, which were present originally in every segment of the primitive ancestor. The gills were at first external, like the gills of many Chætopods at the present day. For their support cartilaginous gill-arches naturally arose in the body-wall, and the superficial musculature became attached to these bars. "There existed in all the segments of the Annelid-ancestors of Vertebrates gills with cartilaginous skeleton and gill-arches in the body wall. Each gill had its veins and arteries, each had its branch of the ventral nerve-cord, and between each successive pair of gills a segmental organ opened to the exterior" (p. 14, 1875). The paired fins and limbs of the Vertebrate arose by the functional transformation of two pairs of these gills. The anterior gills became the definitive internal gills of the Vertebrate, for they gradually shifted into the mouths of the anterior segmental organs, which had already acquired an opening into the pharynx and had been transformed into true gill-slits. The posterior gills degenerated and disappeared, but their arches remained as ribs. Gill-arches and ribs were accordingly homologous structures and formed a *parietal* skeleton. The vertebrate anus, like the mouth, was probably secondary and formed from a pair of gill-slits, the post-anal gut of vertebrate embryos hinting that the original anus was terminal as in Annelids. The unpaired fins of fish were originally paired and possibly arose from the coalescence of rows of parapodia. Dohrn assumed also that the primitive Annelid ancestor must have possessed a notochord to give support in swimming.

If Vertebrates arose from primitive Annelid ancestors, how account for Amphioxus and the Ascidians, which seem to

[1] Leydig (*Vom Baue des thierischen Körpers*, Tübingen, 1864), who, in a measure, forestalled Dohrn and Semper by comparing Vertebrates with reversed Arthropods, specially insects, supposed the old mouth to pass between the *crura cerebri*.

be the most primitive living Vertebrates and yet show no particular annelidan affinities? Dohrn tries to answer this awkward question by showing that these forms are not primitive but degenerate. He points out first that Cyclostomes are degenerate fish, half specialised and half degraded in adaptation to a parasitic mode of life. He thinks that if an *Ammocoetes* were to become sexually mature and degenerate still further, forms would result which would resemble Amphioxus, and ultimately, if the process of degeneration went far enough, larval Ascidians. Amphioxus therefore might well be considered an extremely simplified and degenerate Cyclostome, and the ascidian larva the last term of this degeneration-series. Both Amphioxus and the Ascidians would accordingly be descended from fish, instead of fish being evolved from them.

Dohrn conceived that the transformation of the Annelid into the Vertebrate took place mainly by reason of an important transforming principle, which he calls the principle of function-change. Each organ, Dohrn thinks, has besides its principal function a number of subsidiary functions which only await an opportunity to become active. "The transformation of an organ takes place by reason of the succession of the functions which one and the same organ possesses. Each function is a resultant of several components, of which one is the principal or primary function, while the others are the subsidiary or secondary functions. The weakening of the principal function and the strengthening of a subsidiary function alters the total function; the subsidiary function gradually becomes the chief function, the total function becomes quite different, and the consequence of the whole process is the transformation of the organ" (p. 60). Examples of function-change are not difficult to find. Thus the stomach in most Vertebrates performs both a chemical and a mechanical function, but in some forms a part of it specialises in the mechanical side of the work and becomes a gizzard, while the remaining part confines its energies to the secretion of the gastric juice. So, too, it is through function-change that certain of the ambulatory appendages of Arthropods have become transformed into jaws—their function as graspers of food has gradually prevailed over their main function as walking limbs. In the evolution of Vertebrates from Annelids the principle came into action in many connections—in the formation of a new mouth from gill-slits, in the transformation of gills into fins and limbs, of segmental organs into gill-slits, and so on. Dohrn tells us that the principle of function-change was suggested to him by Mivart's *Genesis of Species* (1870), and he points out how it enables a partial reply to be made to the dangerous objection raised against the theory of natural selection that the first beginnings of new organs are necessarily useless in the struggle for existence.

We may note in passing that a somewhat similar idea was later applied by Kleinenberg to the explanation of some of the ancestral features of development. He pointed out in his classical memoir on the embryology of the Annelid *Lopadorhynchus*[1] that many embryonic organs seem to be formed for the sole purpose of providing the necessary stimulus for the development of the definitive organs. Thus the notochord is the necessary forerunner of the vertebral column, cartilage the precursor of bone. "From this point of view," he writes, "many rudimentary organs appear in a different light. Their obstinate reappearance throughout long phylogenetic series would be hard to understand were they really no more than reminiscences of bygone and forgotten stages. Their significance in the processes of individual development may in truth be far greater than is generally recognised. When in the course of the phylogeny they have played their part as intermediary organs (*Vermittelungsorgane*) they assume the same function in the ontogeny. Through the stimulus or by the aid of these organs, now become rudimentary, the permanent parts of the embryo appear and are guided in their development; when these have attained a certain degree of independence, the intermediary organ, having played its part, may be placed upon the retired list."[2]

Dohrn was well aware of the functional, or as he calls

[1] *Zeits. f. wiss. Zool.*, xliv., 1886.

[2] Quoted by E. B. Wilson, *Wood's Holl Biological Lectures for* 1894, p. 121.

it, the physiological, orientation of his principle, and he rightly regarded this as one of its chief merits. He held that morphology became too abstract and one-sided if it disregarded physiology completely; he saw clearly that the evolution of function was quite as important a problem as the evolution of form, and that neither could be solved in isolation from the other. "The concept of function-change is purely physiological;" he writes, "it contains the elements out of which perhaps a history of the evolution of function may gradually arise, and for this very reason it will be of great utility in morphology, for the evolutionary history of structure is only the concrete projection of the content and course of the evolution of function, and cannot be comprehended apart from it" (p. 70).[1]

It is very instructive in this connection to note that Dohrn was not, like so many of his contemporaries, a dogmatic materialist, but upheld the commonsense view that vital phenomena must, in the first instance at least, be accepted as they are. "It is for the time being irrelevant," he writes, "to squabble over the question as to whether life is a result of physico-chemical processes or an original property (*Urqualität*) of all being. . . . Let us take it as given" (p. 75).

Semper's speculations on the genetic affinity of Articulates and Vertebrates are contained in two papers[2] which appeared about the same time as Dohrn's. He openly acknowledges that his work is essentially a continuation of Geoffroy's transcendental speculations, and gives in his second paper a good historical account of the views of his great predecessor. It is a significant fact that evolutionary morphologists very generally held that Geoffroy was right in maintaining against Cuvier[3] the unity of plan of the whole

[1] *Cf.* Metschnikoff, *Quart. Journ. Microsc. Sci.*, xxiv., pp. 89-111, 1884.

[2] "Die Stammesverwandschaft der Wirbelthiere und Wirbellosen," *Arb. zool.-zoot. Instit. Würzburg*, ii., pp. 25-76, 1875; "Die Verwandschaftsbeziehungen der gegliederten Thiere," *Ibid.*, iii., pp. 115-404, 1876-7.

[3] Abuse of Cuvier also dates from the early days of evolution, see Rádl, ii., pp. 12-17.

animal kingdom, for they saw in this a strong argument for the monophyletic descent of all animals from one common ancestral form.

In his first paper Semper does little more than break ground; he insists on the fact that both Annelids and Vertebrates are segmented animals, and he points out how close is the analogy between the nephridia or "segmental organs" of the former and the excretory (mesonephric) tubules of the latter, upon which he published in the same volume an extensive memoir. At this time he considered *Balanoglossus*—by reason of its gill-slits (its notochord he did not know)—to be the nearest living representative of the ancestral form of Vertebrates and Annelida.

His second paper is a more exhaustive piece of work and deals with every aspect of the problem, both from an anatomical and from an embryological standpoint. It is consciously and admittedly an attempt to apply Geoffroy's principle of the unity of plan and composition to the three great metameric groups, the Annelida, Arthropoda, and Vertebrata. Semper follows Geoffroy's lead very closely in maintaining that it is not the position of the organs relative to the ground that must be taken into account in establishing their homologies, but solely their spatial relations one to another. He holds that dorsum and venter are terms of purely physiological import, and he proposes to substitute for them the terms neural and cardial (better, hæmal) surfaces, either of which may be either dorsal or ventral in position.

Having established this primary principle, Semper has little difficulty in showing that the main organs of the body lie to one another in the same relative positions in Annelida, Arthropoda, and Vertebrata; and this, together with the metameric segmentation common to them all, constitutes his first great argument in favour of their genetic relationship. But he has still to show that Annelids possess at least the rudiments of certain organs which seem to be peculiar to Vertebrates, as the gill-slits, the notochord, and a nervous system developed from the ectoderm of the "dorsal" surface. He takes particular cognisance also of the old distinction drawn by von Baer, that Vertebrates show a "double-symmetrical" mode of development (*evolutio bigemina*), the dorsal muscle-plates forming a tube above the notochord, the ventral plates a tube below the notochord, whereas Articulates do not possess this axis, and form only one tube, namely, that round the "vegetative" organs (*evolutio gemina*). Semper is at pains to prove that *evolutio bigemina* is characteristic also of Annelidan development.

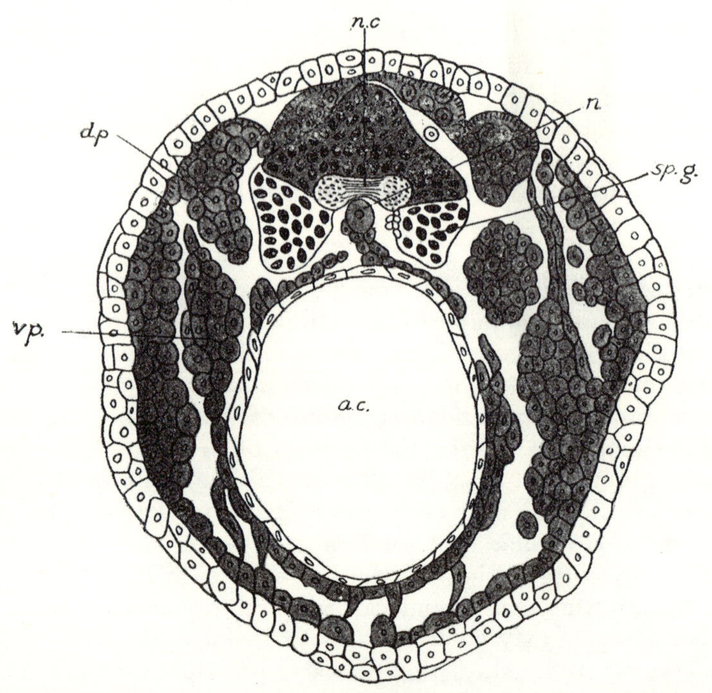

FIG. 14.—Transverse Section (Inverted) of the Worm *Nais*. (After Semper.)

a.c. Alimentary canal. *sp.g.* Spinal ganglion. *d.p.* Neural muscle-plate.
n.c. Nerve cord. *n.* Notochord. *v.p.* Haemal muscle-plate.

He gets his facts from an elaborate study of the process of budding in the *Naidæ*, making the somewhat risky assumption that regeneration takes essentially the same course as embryonic development.

He succeeds in showing—to his own satisfaction at least—that in the formation of new segments in *Nais* and *Chætogaster* a strand of cells appears between the alimentary canal and the nerve-cord, and that from this axial strand the hæmal muscle-plates grow out dorsally round the alimentary canal and the neural muscle-plates ventrally round the nerve-cord (see Fig. 14).

This strand of cells, he concludes, must clearly be the notochord, and the type of development is obviously the double-symmetrical met with in Vertebrates.

The nervous system Semper found to develop in the buds of *Nais* and *Chætogaster* by an ectodermal thickening, just as in some Vertebrates. The cerebral ganglion was formed by the ends of the nerve-cord growing up round the œsophagus and fusing with the paired "sense-plates" which develop from the ectoderm of the head. The cerebral ganglion is accordingly only secondarily hæmal in position, and there is no need therefore to seek in Vertebrates for the homologue of the œsophageal commissures of Annelids, as, for instance, Schneider did.

Since the mouth opens on the neural surface in Annelids and on the hæmal surface in Vertebrates, Semper considers that they cannot be equivalent structures, and he finds the homologue of the Vertebrate mouth in a little pit on the hæmal surface of the head in the leech *Clepsine* (also in the true mouth of Turbellaria and the proboscis-opening in Nemertines). The primitive Annelid mouth, however, does not appear in the embryogeny of Vertebrates, for the great development of the brain crowds it out of existence.

The homologues of the gill-slits Semper finds in two little canals in the head of *Chætogaster*, which open from the pharynx to the exterior. In Sabellids he describes an elaborate system of gill-canals, with a supporting cartilaginous framework which forms a real *Kiemenkorb* or gill-basket, comparable with that of Amphioxus.

Gill-slits, notochord, relation of nervous system, mesonephric tubules, are thus common to Annelids and Vertebrates —what further proof could one desire of the close relationship of these groups? Yet Semper enters into refinements of comparison, seeing, for instance, in the lateral portions of the ventral ganglia (Fig. 14, *sp. g.*) the homologues of the spinal ganglia of Vertebrates, and comparing the lateral line of sense organs in Annelids with the lateral line in Anamnia.

He will not admit that Amphioxus and the Ascidians

show a closer resemblance to Vertebrates than his beloved Annelids. Amphioxus, he thinks, is not a Vertebrate, and Ascidians, though sharing with Annelids the possession of a notochord, gill-slits, and a "dorsal" nervous system, yet are further removed from Vertebrates than the latter by reason of their lacking that essential characteristic of Vertebrates, metameric segmentation.

Not content with establishing the unity of plan of Annelids, Arthropods, and Vertebrates, Semper tries to link on the Annelids, as the most primitive group of the three, to the unsegmented worms, and particularly to the Turbellaria. His speculations on this matter may be summed up somewhat as follows:—The common ancestor of all segmented animals is a segmented worm-like form, not quite like any existing type, resembling the Turbellaria in having two nerve strands on the dorsal side and no œsophageal ring, potentially able to develop either the Vertebrate or the Annelid mouth, and so to give origin both to the Articulate and to the Vertebrate series. The common ancestor alike of unsegmented worms and of all segmented types is probably the trochosphere larva, which in the Vertebrates is represented by the simple *Keimblase* or blastula.

The Annelid theory of Dohrn and Semper was perhaps not so widely accepted as the rival Ascidian theory, but it counted not a few adherents and gave a certain stimulus to comparative morphology. F. M. Balfour, who pointed out about the same time as Semper the analogy between the nephridia of Annelids and the mesonephric tubules of Vertebrates,[1] while not accepting the actual theories of Dohrn and Semper, took up a distinctly favourable attitude to the general idea that Annelids and Vertebrates were descended from a common segmented ancestor. Discussing this question in his classical work on the development of Elasmobranch fishes,[2] Balfour came to the conclusion "that we must look for the ancestors of the Chordata, not in allies

[1] "On the origin and history of the urino-genital organs of Vertebrates," *Journ. Anat. Phys.*, x., 1876. The conclusions of Balfour and Semper were adversely criticised by M. Fürbringer (*Morph. Jahrb.*, iv., 1878), and were negatived by later research.

[2] *A Monograph on the Development of Elasmobranch Fishes*, London, 1878.

of the present Chætopoda, but in a stock of segmented forms descended from the same unsegmented types as the Chætopoda, but in which two lateral nerve-cords, like those of Nemertines, coalesced dorsally instead of ventrally to form a median nervous cord. This group of forms, if my suggestion as to their existence is well founded, appears now to have perished."[1]

He held that while there was much to be said for the interchange of dorsal and ventral surfaces postulated by Dohrn and Semper, the difficulties involved in the supposition were too great; he preferred, therefore, to assume that the present Vertebrate mouth was primitive, and not a secondary formation.

His views as to the phylogeny of the Chordata and the genetic relation of the various classes to one another are exhibited in the following schema,[2] names of hypothetical groups being printed in capitals, names of degenerate groups in italics:—

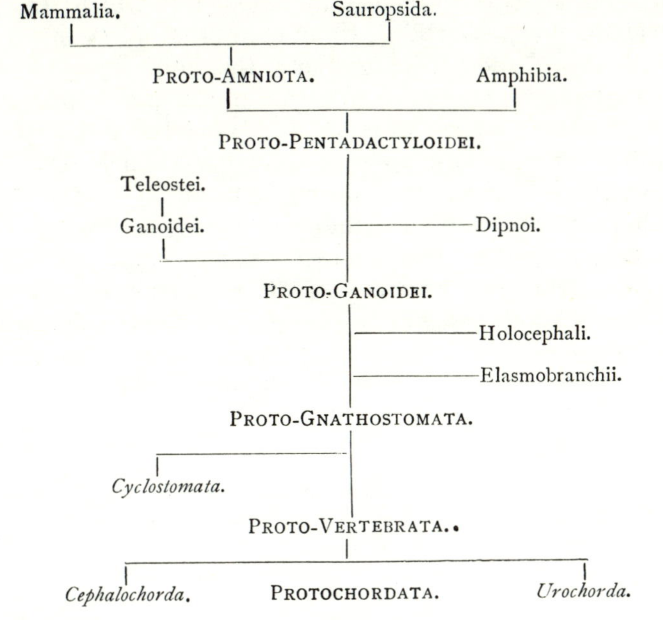

[1] *A Treatise on Comparative Embryology*, vol. ii, p. 311, London, 1881.

[2] *Loc. cit.*, vol. ii., p. 327.

The hypothetical ancestral forms (Protochordata) possessed a notochord, a ventral suctorial mouth and numerous gill-slits, and were presumably descended from the common ancestor of Annelids and Vertebrates. Amphioxus and the Ascidians found their place in this schema as degenerate offshoots of the ancestral Protochordates, while the Cyclostomes were in the same way the degenerate modern representatives of the ancestral Protovertebrates.

Balfour's suggestion, that the nervous system in Annelids and Vertebrates might have arisen by the dorsal or ventral coalescence of the lateral nerve cords found in their common ancestor, bore fruit in the speculations of Hubrecht,[1] on the relation of Nemertines to Vertebrates.

The Annelid theory was firmly supported by Eisig, who in his elaborate monograph on the *Capitellidæ*[2] maintained against Fürbringer the genetic identity of the Annelidan nephridia with the kidney tubules of Vertebrates. The independent discovery by E. Meyer[3] and J. T. Cunningham,[4] of an internal segmental duct in *Lanice*, into which several nephridia opened, seemed to strengthen this view.

Following Ehlers,[5] Eisig found the homologue of the notochord in the accessory intestine of the *Capitellidæ* and *Eunicidæ*, which he supposed might easily be transformed, according to the principle of function-change, from a respiratory to a supporting organ. He finally disposed of the alternative notion that the notochord was represented in Annelids by the "giant-fibres" or neurochordal strands which lie close above the nerve-cord, a view held by Kowalevsky,[6] and for a time by Semper. These strands were

[1] "On the Ancestral Form of the Chordata," *Q.J.M.S.*, xxiii., 1883. "The Relation of the Nemertea to the Vertebrata," *ibid.*, xxvii., 1887. Hubrecht gives the credit for the first indication of the relationship of Nemertines and Vertebrates to Harting (*Leerboek van de Grondbeginselen der Dierkunde*, 1874).

[2] "Monographie der Capitelliden des Golfes von Neapel," *Fauna u. Flora des Golfes von Neapel*, Monog. xvi., Berlin, 1887.

[3] *Mitt. Zool. Stat. Neapel*, vii., 1887. [4] *Nature*, xxxvi., p. 162, 1887.

[5] "Nebendarm und Chorda dorsalis," *Nachr. Ges. Wiss. Göttingen*, p. 390, 1885.

[6] "Embryologische Studien an Würmern u. Arthropoden," *Mém. Acad. Sci. St Pétersbourg* (Petrograd), (7), xvi., 1870. And in *Arch. f. mikr. Anat.*, vii., p. 122, 1871.

shown by Eisig, and by Spengel, to be the neurilemmar sheaths of thick nerve fibres which had in many cases degenerated. The view that the content of the neurochordal tubes was nervous in nature was first promulgated by Leydig in 1864.

Much difference of opinion reigned as to the true homologies of the brain and mouth of Annelids and Vertebrates. Beard[1] and others got over the difficulty of the hæmal position of the cerebral ganglion in Annelids by supposing that it degenerated and disappeared altogether in the Annelidan ancestor of Vertebrates, and that accordingly it had no homologue in the Vertebrate nervous system. Beard put forward also the ingenious theory that the hypophysis represents the old Annelidan mouth.

Van Beneden and Julin[2] assumed that in the ancestors of Vertebrates the œsophagus shifted forward between the still unconnected lobes of the brain to open on the hæmal surface.

The fundamental assumption of the Annelid theory, that dorsal and ventral surfaces are morphologically interchangeable, seemed rather bold to many zoologists, and Gegenbaur[3] voiced a common opinion when he rejected as unscientific the comparison of the ventral nerve cord of Articulates with the dorsal nervous system of Vertebrates.

The *Balanoglossus* theory of Vertebrate descent also belongs, at least in its first form, to the earlier group of evolutionary speculations The gill-slits of *Balanoglossus* were discovered by Kowalevsky as early as 1866.[4] *Tornaria* was discovered by J. Müller in 1850, but by him considered an Asterid larva ; its true nature as the larva of *Balanoglossus* was made out by Metschnikoff in 1870, who also remarked upon its extraordinary likeness to the larvæ of Echinoderms.[5]

[1] "The Old Mouth and the New," *Anat. Anz.*, iii., 1888. *Nature*, xxxix., 1889.

[2] "Recherches sur la Morphologie des Tuniciers," *Arch. de Biol.*, vi., 1887.

[3] "Die Stellung u. Bedeutung der Morphologie," *Morph. Jahrb.*, i., pp. 1-19, 1876.

[4] "Anatomie des Balanoglossus," *Mém. Acad. Sci. St Pétersbourg* (Petrograd), (7), x., 1866.

[5] *Zeit. f. wiss. Zool.*, xx., 1870. For a recent view of the relation of the Enteropneusta to the Echinoderma, see J. F. Gemmill, *Phil. Trans.* B., ccv., pp. 213-94, 1914.

That it had some relationship with Vertebrates was recognised by Semper, Gegenbaur and others, but the full working-out of its Vertebrate affinities is due to Bateson.[1]

Bateson broke completely with the Dohrn-Semper view that the metamerism of Articulates and Vertebrates must be put down to inheritance from a common ancestor. He held that metamerism was merely a special manifestation of the general property of repetition, common to all living things (*cf.* Owen's "vegetative force"), and that accordingly "however far back a segmented ancestor of a segmented descendant may possibly be found, yet ultimately the form has still to be sought for in which these repetitions had their origin" (p. 549). The meaning of the phenomenon was obscure, but he was convinced that the explanation was not to be found in ancestry. "This much alone is clear," he wrote, "that the meaning of cases of complex repetition will not be found in the search for an ancestral form, which, itself presenting this same character, may be twisted into a representation of its supposed descendant. Such forms there may be, but in finding them the real problem is not even resolved a single stage; for from whence was their repetition derived? The answer to this question can only come in a fuller understanding of the laws of growth and of variation, which are as yet merely terms" (pp. 548-9). It was in following up this line of thought that Bateson produced his monumental *Materials for the Study of Variation* (1894).

He found a strong positive argument for his theory that Vertebrates are descended from unsegmented forms in the fact that the notochord arises as an unsegmented structure. With the notochord he homologised the supporting rod in the proboscis of *Balanoglossus*, which like the notochord arises from the dorsal wall of the archenteron, and has a vacuolated structure. The gill-slits of *Balanoglossus*, with their close resemblance in detail to those of Amphioxus, Bateson also used as an argument in favour of the phylogenetic relationship of the Enteropneusta and Vertebrata,

[1] In a series of papers published in 1884-6, the speculative results being discussed in his memoir on "The Ancestry of the Chordata," *Q.J.M.S.* (n. s.), xxvi., pp. 535-71, 1886.

together with the formation from the ectoderm of a dorsal nerve tube.

Bateson's views attracted considerable attention, and were thought by many to lighten appreciably the obscurity in which the origin of Vertebrates was wrapped. Thus Lankester wrote in his article on Vertebrates[1] in the *Encyclopedia Britannica* :—" It seems that in *Balanoglossus* we at last find a form which, though no doubt specialised for its burrowing sand-life, and possibly to some extent degenerate, yet has not to any large extent fallen from an ancestral eminence. The ciliated epidermis, the long worm-like form, and the complete absence of segmentation of the body-muscles lead us to forms like the Nemertines. The great proboscis of *Balanoglossus* may well be compared to the invaginable organ similarly placed in the Nemertines. The collar is the first commencement of a structure destined to assume great importance in *Cephalochorda* and *Craniata*, and perhaps protective of a single gill-slit in *Balanoglossus* before the number of those apertures had been extended. Borrowing, as we may the nephridia from the Nemertines, and the lateral in addition to the dorsal nerve, we find that *Balanoglossus* gives the most hopeful hypothetical solution of the pedigree of Vertebrates."

Much doubt was cast upon the Chordate affinities of the Enteropneusta by Spengel in his monograph of the group,[2] but when the development of the cœlom came to be more thoroughly worked out in *Balanoglossus* and Amphioxus, the striking resemblance in this respect between the two forms gave additional support to the Batesonian view.[3]

[1] Reprinted in *Zoological Articles*, London, 1891.

[2] "Die Enteropneusten des Golfes von Neapel," *Fauna und Flora des Golfes von Neapel*, Monog. xviii., Berlin, 1893.

[3] See Macbride, "A Review of Prof. Spengel's Monograph on Balanoglossus," *Q.J.M.S.*, xxxvi., 1894, and "The Early Development of Amphioxus," *Q.J.M.S.*, xl., 1898.

CHAPTER XVI

THE GERM-LAYERS AND EVOLUTION

In his papers of 1866 and 1867 Kowalevsky had remarked upon the widespread occurrence of a certain type or fundamental plan of early embryonic development, characterised by the formation, through invagination, of a two-layered sac, whose cavity became the alimentary canal. This developmental archetype was manifested in, for instance, *Sagitta*,[1] *Rana*,[2] *Lymnæa*,[3] *Astacus*,[4] *Phoronis*,[5] *Asterias*,[6] *Ascidia*,[5] the *Ctenophora*,[5] and *Amphioxus*.[5] He noticed also that the invagination-opening often became the definitive anus. Further instances of this mode of development were later observed by Metschnikoff[7] and by Kowalevsky[8] himself, but it was left to Haeckel to generalise these observations and build up from them his famous Gastræa theory. This was first enunciated in his monograph of the calcareous sponges,[9] and worked out in detail in a series of papers published in 1874-76.[10]

[1] Gegenbaur, *Zeits. f. wiss. Zool.*, v., 1853.

[2] Remak, *loc. cit.*, p. 183, pl. xii.

[3] Lereboullet, *Ann. Sci. nat.* (4) xviii., pp. 118-9, 1862.

[4] Lereboullet, in Remak, p. 183 f.n.

[5] Kowalevsky, *Mém. Acad. Sci. St Pétersbourg* (Petrograd), (7), x. and xi., 1866 and 1867.

[6] A. Agassiz, *Contrib. Nat. Hist. United States*, v., 1864.

[7] *Mém. Acad. Sci. St Pétersbourg* (Petrograd), (7), xiv., 1869.

[8] "Embryolog. Studien an Würmern u. Arthropoden," *Mém. Acad. Sci. St Pétersbourg* (Petrograd), (7), xvi., 1870.

[9] *Die Kalkschwämme*, 3 vols., Berlin, 1872. General chapters translated in *Ann. Mag. Nat. Hist.* (4), xi., pp. 241-62, 421-30, 1873.

[10] "Die Gastræa-Theorie, die phylogenetische Classification des Thierreichs und die Homologie der Keimblätter." *Jenaische Zeitschrift*, viii., pp. 1-55, 1874. "Die Gastrula und die Eifurchung der Thiere,"

Haeckel maintained that the "gastrula" stage occurred in the development of all Metazoa, and that it was typically formed, by invagination, from a hollow sphere of cells or "blastula." This typical formation might be masked by cenogenetic modifications caused chiefly by the presence of yolk. The gastrula stage was the palingenetic repetition of the ancestral form of all Metazoa, the Gastræa.

From the Gastræa theory there followed at once two consequences, (1) that ectoderm and endoderm, invagination-cavity (*Urdarm*) and gastrula-mouth (*Urmund* or *Protostoma*), were, with all their derivatives, homologous, because homogenous, throughout the Metazoa, and (2) that the descent of the Metazoa had been monophyletic, since all were derived from the ancestral Gastræa. Huxley's suggestion (*supra*, p. 208) that the outer and inner layers in Cœlentera were homologous with the ectoderm and endoderm of the germ was thus fully confirmed and greatly extended.

The great importance of the Gastræa theory lay in the fact that it linked up, by means of the biogenetic law, the germ-layer theory with the doctrine of evolution. It supplied an evolutionary interpretation of the earliest and most important of embryogenetic events, the process of layer-formation. Upon the Gastræa theory or its implications were founded most of the phylogenetic speculations which subsequently appeared.

Upon the Gastræa theory Haeckel based a system of phylogenetic classification which was intended to replace Cuvier's and von Baer's doctrine of Types. This took the form of a monophyletic ancestral tree. Its main outlines are given on p. 290 in graphic form, combined and modified from the table on p. 53 of the 1874 paper and the genealogical tree given in the *Kalkschwämme*.[1]

The scheme is in many respects an interesting and important one. The great contrast between the Protozoa, or animals with neither gut nor germ-layers, and the Metazoa,

ibid., ix., pp. 402-508, 1875. "Die Physemarien, Gastræaden der Gegenwart," and "Nachträge zur Gastræa-Theorie," *ibid.*, x., pp. 55-98, 1876. Republished in *Biologische Studien*, 2nd part, *Studien zur Gastræa-Theorie*, 270 pp., 14 pls., Jena, 1877.

[1] See *Ann. Mag. Nat. Hist.* (4), xi., p. 253.

Monophyletic Genealogical Tree of the Animal Kingdom, based upon the Gastræa Theory and the Homology of the Germ Layers.

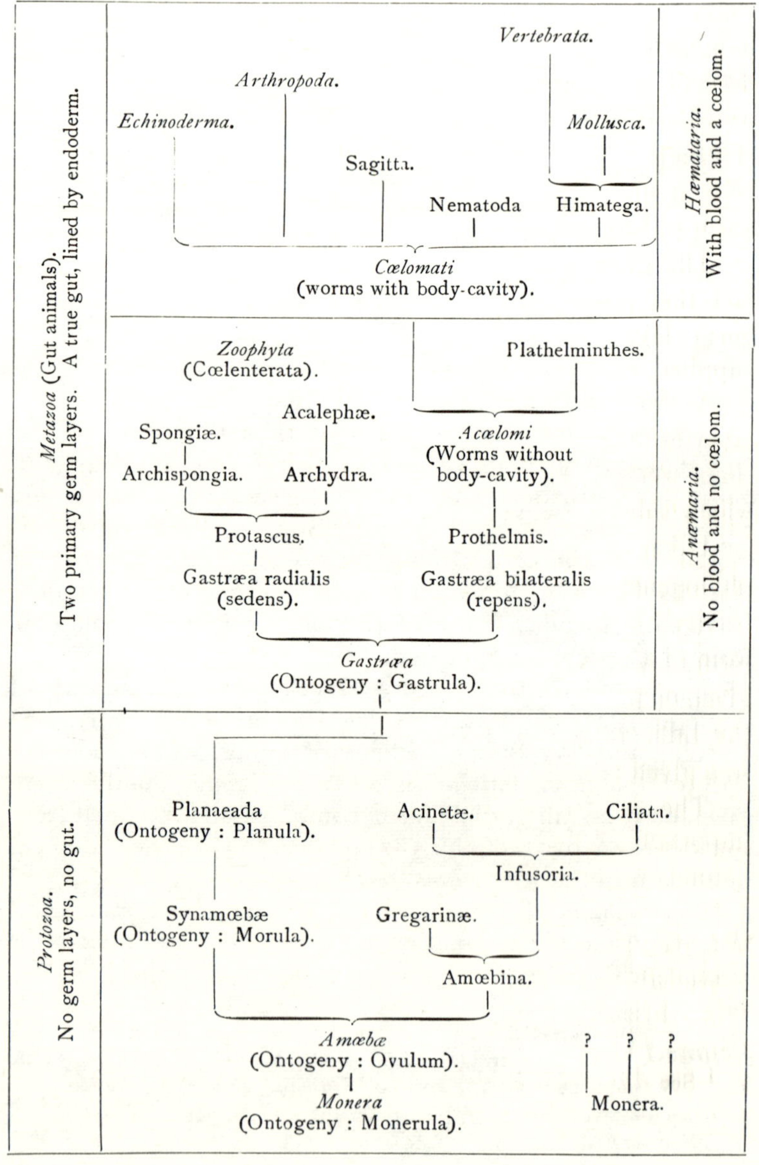

which possess both structures, is for the first time clearly brought out. The derivation of all the Metazoa from a single ancestral form, the Gastræa, leads to the conclusion that the types are not distinct from one another as Cuvier and von Baer supposed, but agree in the one essential point, in the possession of an *archenteron* (Lankester, 1875), and an ectoderm and endoderm which are homologous throughout all the Metazoan phyla. Finally, in the separation of the sponges, Cœlenterata and Acœlomi as animals lacking a body cavity or cœlom[1] from the four higher phyla, which are essentially Cœlomati, there is contained the germ of a conception which later became of importance.

Somewhat similar views as to the importance of the germ-layer theory for the phylogenetic classification of animals were published by Sir E. Ray Lankester in 1873.[2] He distinguished three grades of animals—the Homoblastica, Diploblastica, and Triploblastica The first included the Protozoa, the second the Cœlenterata, the third the other five phyla, distinguished by the possession of a third layer, the mesoderm, and a "blood-lymph" cavity enclosed therein. He used the germ-layer theory to prove the essential unity of type of all the Triploblastica.

The Gastræa theory gave point and substance to the biogenetic law, and enabled Haeckel to state much more concretely the parallelism existing between ontogeny and phylogeny. He was able to assert that five primordial stages, each representing a primitive ancestral form, recurred with regularity in the very earliest development of all Metazoa.[3] These were the monerula, cytula, morula, blastula, and gastrula (see Fig. 15). The monerula was the fertilised ovum after the disappearance of the germinal vesicle;[4] it was the equivalent of the primordial anucleate Monera

[1] Term first introduced in *Die Kalkschwämme*, p. 468, 1872.

[2] "On the Primitive Cell-layers of the Embryo as the Basis of Genealogical Classification of Animals, and on the Origin of Vascular and Lymph Systems," *Ann. Mag. Nat. Hist.* (4), xi., pp. 321-38, 1873.

[3] First distinguished in *Die Kalkschwämme*, i., p. 465.

[4] Even in the 'seventies it was still believed by many that the egg-nucleus disappeared on fertilisation. The true nature of the process was not fully made out till 1875, when O. Hertwig observed the fusion of egg- and sperm-nuclei in *Toxopneustes* (*Morph. Jahrb.*, i., 1876).

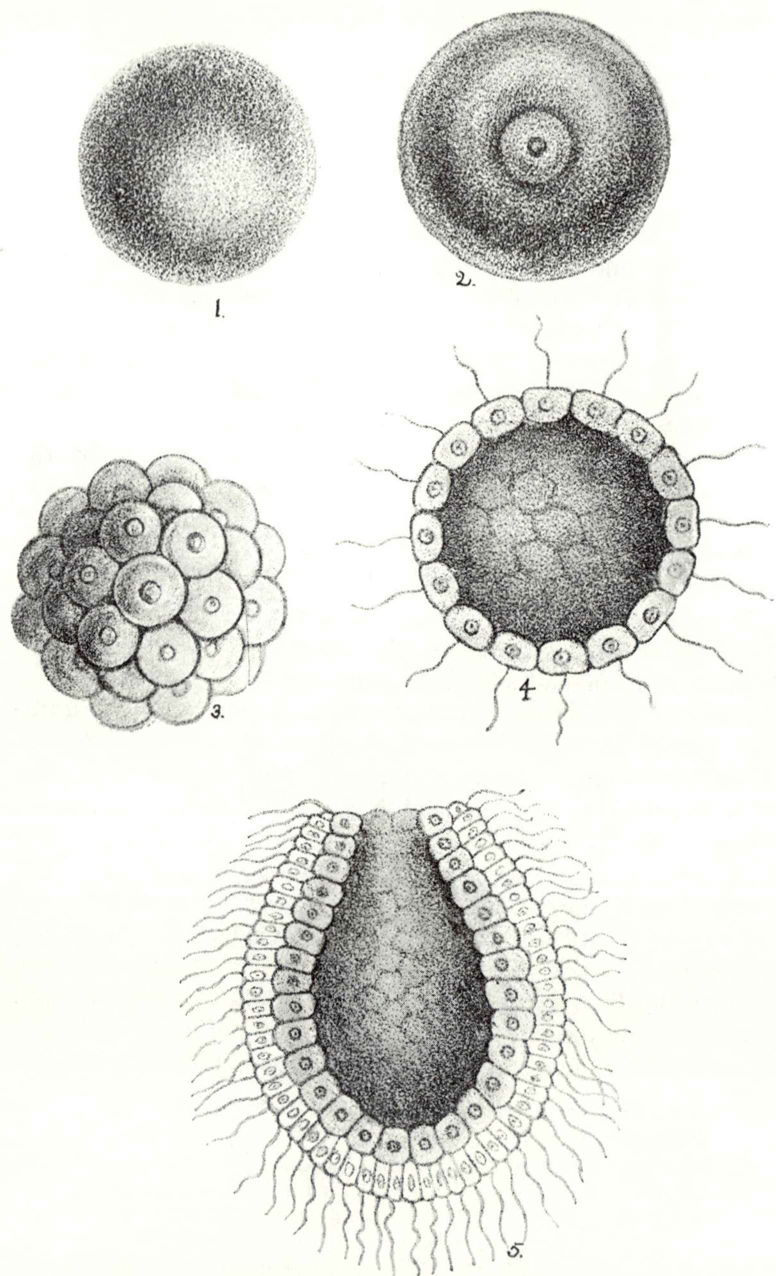

FIG. 15.—The Five Primary Stages of Ontogeny. (After Haeckel.)

1. Monerula. 2. Cytula. 3. Morula. 4. Blastula. 5. Gastrula.

which are the ancestors of all animals. The ovum after the nucleus had been re-formed became the cytula, which was the ontogenetic counterpart of the amœba. The morula, a compact mulberry-like congeries of segmentation-cells, corresponded to the synamœba, or earliest association of undifferentiated amœboid cells to form the first multicellular organism. The blastula, or hollow sphere of segmentation cells, usually ciliated, was reminiscent of the planæa, an ancestral free-swimming form whose nearest living relation is the spherical *Magosphæra.* The gastrula, finally, is the two-layered sac formed from the blastula, typically by invagination of its wall. It repeats the organisation of the gastræa, which is the common ancestor of all Metazoa, and finds its nearest living counterpart in the simple "sponges" *Haliphysema* and *Gastrophysema.*[1] The ancestral line of all the higher animals begins with the five hypothetical forms of the moneron, amœba, synamœba, planæa, and gastræa.

We may take the following account[2] of the phylogeny of the human species, from the gastræa stage onwards, as typical of Haeckel's speculations on the evolution of the higher forms. The progenitors of man are, after the Gastræada :—

1. Turbellaria.
*2. Scolecida. (Worms with a cœlom, probably represented at the present day by *Balanoglossus.*)
*3. Himatega. (Evolved from Scolecida by formation of dorsal nerve-tube and chorda, and resembling tailed larvæ of Ascidians.)
4. Acrania. (With metameric segmentation. Including Amphioxus.)
5. Monorrhina. (Cyclostomes.)
6. Selachia.
7. Dipneusta.
8. Sozobranchia. (Amphibia with permanent gills.)

[1] *Studien z. Gastræa-Theorie*, p. 214, 1877. These forms were known even in 1870 (Carter, *Ann. Mag. Nat. Hist.* (4), vi., pp. 346-7), to be Foraminifera. The figures of supposed collar-cells, etc., do credit to Haeckel's imagination.

[2] *History of Creation*, Eng. Trans., ii., pp. 278 ff.

9. Sozura. (Tailed Amphibia.)
*10. Protamnia.
*11. Promammalia.
12. Marsupialia.
13. Prosimiæ.
14. Menocerca. (Tailed apes.)
15. Anthropoides.
16. Pithecanthropi.
17. Homines.

It will be noticed that except for the hypothetical forms (marked with an asterisk), which are themselves generalised classificatory groups, the ancestral forms belong to long-recognised classes. The whole course of the evolution follows well-worn systematic lines. This is typical of Haeckel's phylogenetic speculations.

A more abstractly morphological scheme of the evolution of Vertebrates is given in the *Systematic Phylogeny* of 1895.[1] The ontogenetic and ancestral stages are arranged in parallel columns thus:—

Cytula.	Cytæa (Protozoa).
Morula.	Moræa (Cœnobium of Protozoa).
Blastula.	Blastæa (*Volvocina*, etc.).
Depula (invaginated blastula).	Depæa.
Gastrula.	Gastræa (cf. *Olynthus*, *Hydra*, and primitive Cœlentera).
Cœlomula (with one pair of cœlom-pockets).	Cœlomæa (cf. *Sagitta*, *Ascidia*, and primitive Helminthes).
Chordula (with medullary tube and chorda).	Chordæa (*cf.* Ascidian larva and larva of Amphioxus).
Spondula (with segmented mesoderm).	Prospondylus (Primitive Vertebrate).

This scheme differs from the earlier one chiefly in taking into account certain advances, notably as regards the cytology of the fertilised ovum and the true nature of the cœlom, which had been made in the interval of some twenty years.

Hæckel's Gastræa theory, though it exercised a great influence upon the subsequent trend of phylogenetic speculation, was by no means universally accepted *telle quelle*. Opinions differed considerably as to the primitive mode of

[1] *Systematische Phylogenie*, iii., p. 41, Berlin, 1895.

origin of the two-layered sac which was very generally admitted to be of constant occurrence in early embryogeny. Ray Lankester, in his paper of 1873, and more fully in 1877,[1] propounded a "Planula" theory, according to which the ancestral form of the Metazoa was a two-layered closed sac formed typically by delamination, less often by invagination. He denied that the invagination opening (which he named the blastopore) represented the primitive mouth,[2] holding that this was typically formed by an "inruptive" process at the anterior end of the planula, which led to the formation of a "stomodæum." A similar process at the posterior end gave rise to the anus and the "proctodæum."

The question as to whether delamination or invagination was to be considered the more primitive process was discussed in detail by Balfour,[3] without, however, any very definite conclusion being reached. He held that both processes could be proved in certain cases to be purely secondary or adaptive, and that accordingly there was nothing to show that either of them reproduced the original mode of transition from the Protozoa to the ancestral two-layered Metazoa (p. 342). He by no means rejected the theory that the Gastræa, "however evolved, was a primitive form of the Metazoa," but, having regard to the great variations shown in the relation of the blastopore to mouth and anus (pp. 340-1), he was inclined to think that if the gastrula had any ancestral characters at all, these could only be of the most general kind. Balfour's attitude perhaps best represents the general consensus of opinion with regard to the Gastræa theory.

From the same origins as the Gastræa theory arose the theory of the cœlom. The term dates back to Haeckel in 1872, and the observations which first led up to the theory were made by the men who supplied the foundations of the Gastræa theory—A. Agassiz, Metschnikoff and Kowalevsky.

[1] "Notes on the Embryology and Classification of the Animal Kingdom," *Q.J.M.S.* (n.s.), xvii., pp. 399-454, 1877

[2] It was "part of the non-historic mechanism of growth" (*loc. cit.*, p. 418).

[3] *Treatise on Comparative Embryology*, ii., chap. xiii., 1881. For a modern discussion of this problem, see Hubrecht, *Q.J.M.S.*, xlix., 1906.

But it was not Haeckel himself who enunciated the cœlom theory.

It will be remembered that Remak introduced in 1855 the conception of the mesoderm as an independent layer derived from the endoderm. The pleuro-peritoneal or body-cavity was formed as a split in the "ventral plates" of the mesoderm. Haeckel's "cœlom" corresponded to the "pleuro-peritoneal cavity" of Remak, but his view of the origin of the mesoderm brought him much closer to von Baer's conception of the origin of *two* secondary layers from ectoderm and endoderm respectively than to Remak's conception of the mesoderm as a single independent layer.

Much uncertainty reigned at the time as to the exact manner of origin of the mesoderm;[1] some held that it developed from the ectoderm, others that it originated in the endoderm, while still others, and among them Haeckel, considered that part of it came from the ectoderm and part from the endoderm (pp. 23-4, 1874).

The solution of the problem came from those observations on the development of the lower forms to which we have just alluded.

The early history of these discoveries and of the theory which grew out of them has been well summarised by Lankester,[2] and may conveniently be given in his own words:—

"As far back as 1864 Alexander Agassiz ("Embryology of the Star-fish," in *Contributions to the Natural History of the United States*, vol. v., 1864) showed in his account of the development of Echinoderma that the great body-cavity of those animals developed as a pouch-like outgrowth of the archenteron of the embryo, whilst a second outgrowth gave rise to their ambulacral system; and in 1869 Metschnikoff (*Mém. de l'Acad. impériale des Sciences de St Pétersbourg*, series vii., vol. xiv., 1869), confirmed the observations of Agassiz, and showed that in Tornaria (the larva of Balanoglossus) a similar formation of body-cavities by pouch-like outgrowths of the archenteron took place.

[1] See Balfour, *loc. cit.*, Chapter xiii.

[2] *A Treatise on Zoology*, Pt. ii., 1900. Introduction by Sir E. Ray Lankester.

Metschnikoff has further the credit of having, in 1874 (*Zeitsch. wiss. Zoologie*, vol. xxiv., p. 15, 1874), revived Leuckart's theory of the relationship of the cœlenteric apparatus of the Enterocœla to the digestive canal and body-cavities of the higher animals. Leuckart had in 1848 maintained that the alimentary canal and the body-cavity of higher animals were united in one system of cavities in the Enterocœla (*Verwandschaftsverhältnisse der wirbellosen Thiere*, Brunswick, 1848). Metschnikoff insisted upon such a correspondence when comparing the Echinoderm larva, with its still continuous enteron and cœlom, to a Ctenophor, with its permanently continuous system of cavities and canals. Kowalevsky, in 1871, showed that the body-cavity of Sagitta was formed by a division of the archenteron into three parallel cavities, and in 1874 demonstrated the same fact for the Brachiopoda. In 1875 (*Quart. Journ. Micr. Sci.*, vol. xv., p. 52) Huxley proposed to distinguish three kinds of body-cavity: the schizocœl, formed by the splitting of the mesoblast, as in the chick's blastoderm; the enterocœl, formed by pouching of the archenteron, as in Echinoderms, Sagitta and Brachiopoda; and the epicœl. . . . Immediately after this I put forward the theory of the uniformity of origin of the cœlom as an enterocœl (*Quart. Journ. Micr. Sci.*, April, 1875). . . . My theory of the cœlom as an enterocœl was accepted by Balfour and was greatly strengthened by his observations on the derivation of both notochord and mesoblastic somites from archenteron in the Elasmobranchs, and by the publication in 1877 by Kowalevsky of his second paper on the development of Amphioxus—in which the actual condition which I had supposed to exist in the Vertebrata was shown to occur, namely, the formation of the mesoblast as paired pouches in which a narrow lumen exists, but is practically obliterated on the nipping-off of the pouch from the archenteron, after which process it opens out again as cœlom" (pp. 16-18).

The enterocœlic theory was taken up by O. and R. Hertwig as an essential part of their *Cœlomtheorie.*[1] In

[1] *Studien zur Blättertheorie*, Jena, 1879-80. "Die Cœlomtheorie, Versuch einer Erklärung des mittleren Keimblattes," *Jenaische Zeitschrift*, xv., pp. 1-150, 1882.

a lengthy series of monographs these workers made a comparative study of the mode of formation of the middle layer, and arrived at a coherent theory of its origin. They distinguished in the middle layer two quite distinct elements, the mesoblast proper, formed by the evagination of the walls of the archenteron, and the mesenchyme, formed by free cells budded off from the germ-layers. The following passage gives a good idea of their views and of the phylogenetic implications involved:—"Ectoblast and entoblast are the two primary germ-layers which arise from the invagination of the blastula; they are always the first to be laid down, and they can be directly referred back to a simple ancestral form, the Gastræa; they form the limits of the organism towards the exterior and towards the archenteron. The parietal and visceral mesoblast, or the two middle layers, are always of later origin, and arise through evagination or plaiting of the entoblast, the remainder of which can now be distinguished as secondary entoblast from the primary. They form the walls of a new cavity, the enterocœl, which is to be regarded as a nipped-off diverticulum of the archenteron. Just as the two-layered animals can be derived from the Gastræa, so can the four-layered animals be derived from a Cœlom form. Embryonic cells, which become singly detached from their epitheliar connections we consider to be something quite different from the germ-layers, and accordingly we call them by the special name of mesenchyme germs or primary cells of the mesenchyme. They may develop both in two-layered and in four-layered animals. Their function is to form between the epithelial limiting layers a secreted tissue (*Secretgewebe*) or connective tissue with scattered cells, which cells can undergo, like the epithelial elements, the most varied modifications. . . . This secreted tissue in its simple or in its differentiated state, with all its derivatives, we call the mesenchyme" (p. 122).

The important point for us is that, just as all Metazoa were considered by Haeckel to be descended from the Gastræa, so all Cœlomati were held by the Hertwigs to be derived from an original cœlomate *Urform.* In both cases an embryological archetype becomes a hypothetical ancestral form.

The Cœlom theory was considerably modified, extended

and developed by later workers, particularly as regards the relations to the cœlom of the genital organs and ducts and the nephridia, but no special methodological interest attaches to these further developments.[1] We shall here focus attention upon one interesting line of speculation followed out in this country particularly by Sedgwick—the theory of the Actinozoan ancestry of segmented animals. Its relation to the Cœlom theory lies in the fact that Sedgwick regarded the segmentation of the body as moulded upon the segmentation of the mesoblast, which in its turn, as Kowalevsky and Hatschek had shown, was a consequence of its mode of origin as a series of pouches of the archenteron. In other respects Sedgwick's speculations link on more closely to the Gastræa theory, for one of his main contentions is that the blastopore or *Urmund* is homologous throughout at least the three metameric phyla. In following up Balfour's observations on the development of *Peripatus*,[2] Sedgwick was struck with the close resemblance existing between the elongated slit-like blastopore of this form (giving rise to both mouth and anus), with its border of nervous tissue, and the slit-like mouth of the Actinozoan (functioning both as mouth and anus), round which, as the Hertwigs had shown, there lies a special concentration of nerve cells and nerve fibres. He found another point of resemblance in the gastric pouches of the Actinozoa, which he homologised directly with the enterocœlic pouches of the Cœlomati. He was led to enunciate the following theses:—[3] (1) that the mouth and anus of Vermes, Mollusca, Arthopoda, and probably Vertebrata, is derived from the elongated mouth of an ancestor resembling the Actinozoa; (2) that somites are derived from a series of archenteric pouches, like those of Actinozoa and Medusæ; (3) that excretory organs (nephridia, segmental organs) are derived from parts of these pouches which in the ancestral form, as in many polyps, were connected by a circular or longitudinal canal, and opened

[1] For an historical account of this work, see Lankester, *loc. cit.*, pp. 21-37.

[2] *Proc. Roy. Soc.*, 1883, and *Q.J.M.S.*, xxiii., 1883.

[3] "Origin of Metameric Segmentation," *Q.J.M.S.*, xxiv., pp. 43-82 1884.

to the exterior by pores. This longitudinal canal was lost in Invertebrates, but persisted in Vertebrates as the pronephric duct, while the pores remained in Invertebrates and disappeared in Vertebrates; (4) that the tracheæ of Arthropods, as well as the canal of the central nervous system in Vertebrates, are to be traced back to certain ectodermal pits in the diploblastic ancestor comparable to the sub-genital pits of the Scyphomedusæ. These ectodermal pits were all originally respiratory organs. "The essence of all these propositions," he writes, "lies in the fact that the segmented animals are traced back not to a triploblastic unsegmented ancestor, but to a two-layered Cœlenterate-like animal with a pouched gut, the pouching having arisen as a result of the necessity for an increase in the extent of the vegetative surfaces in a rapidly enlarging animal (for circulation and respiration)" (p. 47). "I have attempted to show," he writes further on, "that the majority of the Triploblastica . . . are built upon a common plan, and that that plan is revealed by a careful examination of the anatomy of Cœlenterata; that all the most important organ-systems of these Triploblastica are found in a rudimentary condition in the Cœlenterata; and that all the Triploblastica referred to must be traced back to a diploblastic ancestor common to them and the Cœlenterata" (p. 68). The main assumption was that the neural or blastoporal surface must be homologous throughout the Metazoa, though it was dorsal in the Chordata, ventral in the Annelida and Arthropoda. He derived the central nervous system of the Chordata from the circumoral ring of the common ancestor by means of the hypothesis that both the pre-blastoporal and the post-blastoporal parts of it disappeared.[1]

The characteristic relation of the central nervous system to the blastopore in Annelida and Vertebrates had already been pointed out by Kowalevsky,[2] who had also sketched a theory of the common descent of these two phyla from an ancestral form in which the nervous system encircled the blastopore.

[1] See further the same author's article "Embryology" in the *Ency. Brit.*, vol. xi., 11th ed., Cambridge, 1910.

[2] *Arch. f. mikr. Anat.*, xiii., pp. 181-204, 1877.

In 1882, before the publication of Sedgwick's papers, A. Lang[1] had put forward the somewhat similar view that the stomach-diverticula of the Turbellaria, which he had found to be segmentally arranged in certain Triclads, were the morphological equivalents of the enterocœlic pouches of higher animals. This view, however, he soon gave up.[2] Sedgwick's views found a supporter in A. A. W. Hubrecht,[3] who utilised them in connection both with his speculations on the relation of Nemertines to Vertebrates, and with his exhaustive work on the early development of the Mammalia. He postulated as the far-back ancestor of Vertebrates, "an actinia-like, vermiform being, elongated in the direction of the mouth-slit" (p. 410, 1906), and derived the central nervous system from the circum-oral ring of this primitive form, the notochord from its stomodæum, and the cœlom from the peripheral parts of the gastric cavity (p. 169, 1909).

[1] "Der Bau von Gunda segmentata," *Mitth. Zool. Stat. Neap.*, iii., pp. 187-250, 1882.

[2] "Die Polycladen," *Fauna u. Flora des Golfes von Neapel*, Monog. v., Leipzig, 1884, and "Beiträge zu einer Trophocœltheorie," *Jen. Zeits.*, xxxviii., pp. 1-373, 1904 (which see for a modern account of theories of metamerism).

[3] "Die Abstammung der Anneliden u. Chordaten," *Jen. Zeits.*, xxxix., pp. 151-76, 1905. "The Gastrulation of the Vertebrates," *Q.J.M.S.*, xlix., pp. 403-19, 1906. "Early Ontogenetic Phenomena in Mammals," *Q.J.M.S.*, liii., pp. 1-181, 1909.

CHAPTER XVII

THE ORGANISM AS AN HISTORICAL BEING

"Of late the attempt to arrange genealogical trees involving hypothetical groups has come to be the subject of some ridicule, perhaps deserved. But since this is what modern morphological criticism in great measure aims at doing, it cannot be altogether profitless to follow this method to its logical conclusions. That the results of such criticism must be highly speculative, and often liable to grave error, is evident."

The quotation is from Bateson's paper of 1886, and it is symptomatic of the change which was soon to come over morphological thought. New interests, new lines of work, began to usurp the place which pure morphology had held so long.

This is accordingly a convenient stage at which to take stock of what has gone before, to consider the relation of evolutionary morphology to the transcendental and the Cuvierian schools of thought which preceded it, and to make clear what new element evolution-theory added to morphology.

The close analogy between evolutionary and transcendental morphology has already been remarked upon and illustrated in the last three chapters. We have seen that the coming of evolution made comparatively little difference to pure morphology, that no new criteria of homology were introduced, and that so far as pure morphology was concerned, evolution might still have been conceived as an ideal process precisely as it was by the transcendentalists. The principle of connections still remained the guiding thread of morphological work; the search for archetypes, whether anatomical or embryological, still continued in the same way as before, and it was a point of subordinate importance that, under the influence of the evolution-theory, these were considered to represent real ancestral forms rather than purely abstract figments of the intelligence. The law of Meckel-Serres was revived in an altered shape as the law of the recapitulation of phylogeny by ontogeny; the natural system of classification was passively inherited, and, by a *petitio principii*, taken to represent the true course of evolution. It is true that the attempt was made to substitute for the concept of homology the purely genetic concept of homogeny, but no inkling was given of any possible method of recognising homogeny other than the well-worn methods generally employed in the search after homologies.

There was a close spiritual affinity between the speculative evolutionists and the transcendentalists. Both showed the same subconscious craving for simplicist conceptions—the transcendentalists clung fast to the notion of the absolute unity of type, of the ideal existence of the "one animal," and the evolutionists did precisely the same thing when they blindly and instinctively accepted the doctrine of the monophyletic descent of all animals from one primeval form. Geoffroy persisted in regarding Arthropods as being built on the same plan as Vertebrates: Dohrn and Semper did nothing different when they derived both groups from an ancestor combining the main characters of both. The determination to link together all the main phyla of the animal kingdom and to force them all into a single mould was common to evolutionary and pre-evolutionary transcendentalists alike.

From the fact that all Metazoa develop from an ovum which is a simple cell, the evolutionists inferred that all must have arisen from one primordial cell. From the fact that the next step in development is the segmentation of the ovum, they argued that the ancestral Metazoa came into being through the division of the primal Protozoon with aggregation of the division-products. From the fact that a gastrula stage is very commonly formed when segmentation has been completed, they assumed that all germ-layered animals were descended from an ancestral Gastræa.

They quite ignored the possibility that a different explanation of the facts might be given ; they seized upon the simplest and most obvious solution because it satisfied their overwhelming desire for simplification. But is the simplest explanation always the truest—especially when dealing with living things? One may be permitted to doubt it. It is easy to account for the structural resemblance of the members of a classificatory group, by the assumption that they are all descended from a common ancestral form ; it is easy to postulate any number of hypothetical generalised types ; but in the absence of positive evidence, such simplicist explanations must always remain doubtful. The evolutionists, however, had no such scruples.

Phylogenetic method differed in no way from transcendental — except perhaps that it had learnt from von Baer and from Darwin to give more weight to embryology. The criticisms passed by Cuvier and von Baer upon the transcendentalists and their recapitulation theory might with equal justice be applied to the phylogenetic speculations which were based on the biogenetic law. There was the same tendency to fix upon isolated points of resemblance and disregard the rest of the organisation. Thus, on the ground of a presumed analogy of certain structures to the vertebrate notochord, several invertebrate groups, as the Enteropneusta, the Rhabdopleura, the Nemertea, were supposed to be, if not ancestral, at least offshoots from the direct line of vertebrate descent. And if other points of resemblance could in some of these cases be discovered, yet no successful attempt was made to show that the total organisation of any of these forms corresponded with that of the Vertebrate type. With the possible exception of the Ascidian theory, all the numerous theories of vertebrate descent suffered from this irremediable defect, and none carried complete conviction.

In spite of the efforts of the evolutionists, as of those of the transcendentalists, the phyla or "types" remained distinct, or at best connected by the most general of bonds.

The close affinity of transcendentalists and evolutionists is shown very clearly in their common contrast in habits of thought with the Cuvierian school. It is the cardinal principle of pure morphology that function must be excluded from consideration. This is a necessary and unavoidable simplification which must be carried out if there is to be a science of pure form at all. But this limitation of outlook, if carried over from morphology to general biology becomes harmful, since it wilfully ignores one whole side of life—and that the most important. The functional point of view is clearly indispensable for any general understanding of living things, and this is where the Cuvierian school has the advantage over the transcendental—its principles are applicable to biology in general.

Geoffroy and Cuvier in pre-evolutionary times well typified the contrast between the formal and the functional standpoints. For Geoffroy form determined function, while for Cuvier function determined form. Geoffroy held that Nature formed nothing new, but adapted existing "materials of organisation" to meet new needs. Cuvier, on the other hand, was always ready to admit Nature's power to form entirely new organs in response to new functional requirements.

The evolutionists followed Geoffroy rather than Cuvier. They laid great store by homological resemblances, and dismissed analogies of structure as of little interest. They were singularly unwilling to admit the existence of convergence or of parallel evolution, and they held very firmly the distinctively Geoffroyan view that Nature is so limited by the unity of composition that she can and does form no new organs.

By no one has this underlying principle of evolutionary morphology been more explicitly recognised than by Hubrecht, who in his paper of 1887, after summarising the points of resemblance between Nemertines and Vertebrates which led him to assume a genetic connection between them, writes as follows:—"At the base of all the speculations contained in this chapter lies the conviction, so strongly insisted upon by Darwin, that new combinations or organs do not appear by the action of natural selection unless others have preceded, from which they are gradually derived by a slow change and differentiation.

"That a notochord should develop out of the archenteric wall because a supporting axis would be beneficial to the animal may be a teleological assumption, but it is at the same time an evolutional heresy. It would never be fruitful to try to connect the different variations offered, *e.g.*, by the nervous system throughout the animal kingdom, if similar assumptions were admitted, for there would be then quite as much to say for a repeated and independent origin of central nervous systems out of indifferent epiblast just as required in each special case. These would be steps that might bring us back a good way towards the doctrine of independent creations. The remembrance of Darwin's, Huxley's, and Gegenbaur's classical foundations, and of Balfour's and Weismann's brilliant superstructures, ought to warn us away from these dangerous regions" (p. 644).

This same prejudice lies at the root of the idea of *Functionswechsel*, in spite of the general functional orientation of that idea.

Dohrn's constant assumption is that Nature makes shift with old organs wherever possible, instead of forming new ones. He derives gill-slits from segmental organs, fins and limbs from gills, ribs from gill-arches, and so on, instead of admitting that these organs might quite as well have arisen independently. He objects on principle to the origin of organs *de novo*. Thus, rebutting the suggestion that certain organs which are not found in the lower Vertebrates might have arisen as new formations, he writes:—"Against this supposition the whole weight of all those objections can be directed that are to be brought in general against the method of explanation which consists in appealing without imperative necessity to the *Deus ex machina*, 'New formation,' which is neither better nor worse than *Generatio equivoca*" (p. 21).

Of a similar nature was the objection to convergence.[1]

Why, we may ask, were morphologists so unwilling to

[1] The importance of convergence came to be realised after the vogue of phylogenetic speculation had passed—see Friedmann, *Die Konvergenz der Organismen*, Berlin, 1904, and A. Willey, *Convergence in Evolution*, London, 1911. Also L. Vialleton, *Elements de morphologie des Vertébrés*, Paris, 1912.

admit the creative power of life? Dohrn, for instance, was fully aware of the great transforming influence exerted by function upon form—his theory of *Functionswechsel* regards as the most powerful agent of change the activity of the animal, its effort to make the best use of its organs, to apply them at need in new ways to meet new demands. Why then did he not go a step further and admit that the animal could by its own subconscious efforts form entirely new organs? Why did most morphologists join with him in belittling the organism's power of self-transformation?

The reasons seem to have been several. There is first the fundamental reason, that the idea of an active creative organism is repugnant to the intelligence, and that we try by all means in our power to substitute for this some other conception. In so doing we instinctively fasten upon the relatively less living side of organisms—their routine habits and reflexes, their routine structure—and ignore the essential activity which they manifest both in behaviour and in form-change.

We tend also to lay the causes of form-change, of evolution, as far as possible outside the living organism. With Darwin we seek the transforming factors in the environment rather than within the organism itself. We fight shy of the Lamarckian conception that the living thing obscurely works out its own salvation by blind and instinctive effort. We like to think of organisms as machines, as passive inventions[1] gradually perfected from generation to generation by some external agency, by environment or by natural selection, or what you will. All this makes us chary of believing that Nature is prodigal of new organs.

Other causes of the unwillingness of morphologists to admit the new formation of organs are to be sought in the main principle of pure morphology itself, that the unity of plan imposes an iron limit upon adaptation, and in the

[1] From this point of view there is a very profound analogy between artificial and natural selection. Upon the theory of natural selection organisms are lifeless constructs which are mechanically perfected by external agency, just as machines are improved by a process of conscious selection of the most successful among a number of competing models. (*Cf.* passage quoted below, on p. 308.)

powerful influence exercised at the time by materialistic habits of thought. Teleology had become a bugbear to the vast majority of biologists, and all real understanding of the Cuvierian attitude seems, in most cases, to have been lost, although, curiously enough, teleological conceptions were often unconsciously introduced in the course of discussions on the "utility" of organs in the struggle for existence.

Evolutionary morphology, being for the most part a form of pure or non-functional morphology, agreed then in all essential respects with pre-evolutionary or transcendental morphology.

But it contained the germ of a new conception which threw a new light upon the whole science of morphology. This was the conception of the organism as an historical being.

We have seen this thought expressed with the utmost clearness by Darwin himself (*supra*, p. 233). In his eyes the structure and activities of the living thing were a heritage from a remote past, the organism was a living record of the achievements of its whole ancestral line. What a light this conception threw upon all biology! "When we no longer look at an organic being as a savage looks at a ship as something wholly beyond his comprehension; when we regard every production of Nature as one which has had a long history; when we contemplate every complex structure and instinct as the summing-up of many contrivances, each useful to the possessor, in the same way as any great mechanical invention is the summing-up of the labour, the experience, the reason, and even the blunders of numerous workmen; when we thus view each organic being, how far more interesting—I speak from experience—does the study of natural history become!" (*Origin*, 6th ed., pp. 665-6).

Sedgwick expressed the same thing from the morphological point of view when he wrote, with reference to the ancestral significance of the blastopore:—"If there is anything in the theory of evolution, every change in the embryo must have had a counterpart in the history of the race, and it is our business as morphologists to find it out" (p. 49, 1884).

By the evolution-theory the problems of form were linked indissolubly with the problem of heredity. Unity of plan could no longer be explained idealistically as the manifestation of Divine archetypal ideas; it had a real historical basis, and was due to inheritance from a common ancestor. The evolution-theory gave meaning and intelligibility to the transcendental conception of the unity of plan; in particular it supplied a simple and satisfying explanation of those puzzling vestigial organs, whose existence was such a stumbling-block to the teleologists. It enabled the biogenetic law to be substituted for the laws of Meckel-Serres and von Baer, as being in some measure a combination and interpretation of both.

Where the concept of evolution proved itself particularly useful was in the interpretation of structures which were not immediately conditioned by adaptation to present requirements, such as, for instance, the arrangement of gill-slits and aortic arches in the fœtus of land Vertebrates. Such "heritage characters" could only be explained on the hypothesis that they had once had functional or adaptational meaning. Why, for instance, should the blastopore so often appear as a long slit, closing by concrescence, unless this had been the original method of its formation in remote Cœlenterate ancestors?

The point hardly requires elaboration, since it has become an integral part of all our thinking on biological problems. It may be as well, however, for the sake of continuity, to give one or two examples of the historical interpretation of animal structures. The first may conveniently be the phylogenetic interpretation of the contrast between "membrane" and "cartilage" bones.

In his *Grundzüge* of 1870, Gegenbaur made the suggestion that the investing or membrane bones were derived phylogenetically from integumentary ossifications, and this was worked out in detail a few years later by O. Hertwig.[1]

Many years before, several observers—J. Müller, Williamson, and Steenstrup—had been struck with the resemblance existing between the placoid scales and the teeth of Elasmobranch fishes. Hertwig followed up this clue, and came to the conclusion not only that placoid scales and teeth were

[1] *Arch. f. mikr. Anat.*, xi. (suppl.), 1874; *Morph. Jahrb.*, ii., 1876, v. 1879, and vii., 1882.

strictly homologous, but also that all membrane bones were derived phylogenetically from ossifications present in the skin or in the mucous membrane of the mouth, just as cartilage bones were derived from the cartilaginous skeletons of the primitive Vertebrates. In some cases this manner of derivation could even be observed in ontogeny, as Reichert had seen in the Newt, where certain bones in the roof of the mouth are actually formed by the concrescence of little teeth, (*supra*, p. 163). Hertwig considered that the following bones were originally formed by coalescence of teeth—parasphenoid, vomer, palatine, pterygoid, the tooth-bearing part of the pre-maxillary, the maxillary, the dentary and certain bones of the hyo-mandibular skeleton of Teleosts. All the investing bones (*Deckknochen*) of the skull were of common origin, and could be traced back to integumentary skeletal plates, which in the ancestral fish formed a dense carapace.

These conclusions were accepted by Kölliker himself, who wrote in his *Entwickelungsgeschichte* (1879)—"The distinction between the primary or primordial, and the investing or secondary bones is from the morphological standpoint sharp and definite. The former are ossifications of the (cartilaginous) primordial skeleton, the latter are formed outside this skeleton, and are probably all ossifications of the skin or the mucous membrane" (p. 464).

Gegenbaur[1] consistently upheld the phylogenetic derivation of investing bones from dermal ossifications, and even went further and derived substitutionary bones as well from the integument, thus establishing a direct comparison between the skeletal formations of Vertebrates and Invertebrates. Investing bones were actual integumentary ossifications which had gradually sunk beneath the skin to become part of the internal skeleton; substitutionary bones were produced by cells (osteoblasts) which were ultimately derived from the integument.[2]

[1] *Vergleich. Anat. d. Wirbelthiere*, i., pp. 200-1, 1898.

[2] For a full historical account of work on membrane and cartilage bones (as well as on the theory of the skull) see E. Gaupp, "Altere und neuere Arbeiten über den Wirbelthierschädel," *Ergeb. Anat. Entw.*, x., 1901, and "Die Entwickelung des Kopfskelettes," in Hertwig's *Handbuch vergl. exper. Entwickelungslehre d. Wirbelthiere*," iii., 2, pp. 573-874, 1905.

A further instance of the historical interpretation of animal structure, taken from quite a different field, is afforded by the speculations of Dollo[1] on the ancestral history of the Marsupials. In a brilliant paper of 1880[2] Huxley made the suggestion that the ancestors of Marsupials were arboreal forms. "I think it probable," he wrote, "from the character of the pes, that the primitive forms, whence the existing Marsupialia have been derived, were arboreal animals; and it is not difficult, I conceive, to see that, with such habits, it may have been highly advantageous to an animal to get rid of its young from the interior of its body at as early a period of development as possible, and to supply it with nourishment during the later periods through the lacteal glands, rather than through an imperfect form of placenta" (p. 655). Dollo followed up this suggestion, which had in the meantime been strengthened by Hill's discovery of a true allantoic placenta in *Perameles*, by demonstrating in the foot of present-day Marsupials certain features which could only be interpreted as inherited from a time when the ancestors of Marsupials were tree-living animals. These were the occurrence of an opposable big toe (when this was present at all), the great development of the fourth toe, the reduction and partial syndactylism of the second and third toes, and in some cases the regression of the nails. These characters were shown to be typical of arboreal Vertebrates, and their occurrence in forms not arboreal indicated that these were descended from tree-living ancestors. Traces of an arboreal ancestry could be demonstrated even in the marsupial mole *Notoryctes*.

These are only two examples out of hundreds that might be given. Present day structure was interpreted in the light of past history; the common element in organic form was seen to be due to common descent; the existence of vestigial and non-functional organs was no longer a riddle.

There was even a tendency to concentrate attention upon the historical side of structure, upon what the animal passively inherited rather than upon what it personally

[1] "Les Ancêtres des Marsupiaux étaient-ils arboricoles?" *Trav. Stat. zool. Wimereux*, vii., pp. 188-203, pls. xi.-xii., 1899. See also Bensley, *Trans. Linn. Soc.* (2) ix., pp. 83-214, 1903.

[2] *Proc. Zool. Soc.*, pp. 649-62, 1880. *Sci. Mem.*, iv., pp. 457-72.

achieved. Homologies were considered more interesting than analogies, vestigial organs more interesting than fœtal and larval adaptations. Convergence was anathema. The dead-weight of the past was appreciated at its full and more than its full value; and the essential vital activity of the living thing, so clearly shown in development and regeneration, was ignored or forgotten.

But evolutionary morphology for all practical purposes was a development of pure or idealistic morphology, and was powerless to bring to fruit the new conception with which evolution-theory had enriched it. The reason is not far to seek. Pure morphology is essentially a science of comparison which seeks to disentangle the unity hidden beneath the diversity of organic form. It is not immediately concerned with the causes of organic diversity—that is rather the task of the sciences of the individual, heredity and development. To take an example—the recapitulation theory may legitimately be used as a law of pure morphology, as stating the abstract relation of ontogeny to phylogeny, and the probable line of descent of any organism may be deduced from it, as a mere matter of the ideal derivation of one form from another; but an explanation of the reason for the recapitulation of ancestral history during development can clearly not be given by pure morphology unaided. From the fact that the common starfish shows in the course of its development distinct traces of a stalk[1] it is possible to infer, taking other evidence also into consideration, that the ancestors of the starfish were at one stage of their existence stalked and sessile organisms. But this leaves unanswered the question as to how and why the starfish does still repeat after so many millions of years part of the organisation of one of its remote ancestors. Why is this feature retained, and by what means has it been conserved through countless generations? It is clear that the answer can be given only by a science of the causes of the production and retention of form, by a causal morphology, based upon a study of heredity and development.

From the point of view of the pure morphologist the recapitulation theory is an instrument of research enabling

[1] J. F. Gemmill, *Phil. Trans. B*, ccv., p. 255, 1914.

him to reconstruct probable lines of descent; from the standpoint of the student of development and heredity the fact of recapitulation is a difficult problem whose solution would perhaps give the key to a true understanding of the real nature of heredity.

To make full use of the conception of the organism as an historical being it is necessary then to understand the causal nexus between ontogeny and phylogeny.

We shall see in the next chapter that the transformation of morphology from a comparative to a causal science did take place towards the end of the century, and that some progress was made towards an understanding of the relation between individual development and ancestral history, particularly by Roux and Samuel Butler, working with the fruitful Lamarckian conception of the transforming power of function.

CHAPTER XVIII

THE BEGINNINGS OF CAUSAL MORPHOLOGY

UNTIL well into the 'eighties animal morphology remained a purely descriptive science, content to state and summarise the relations between the coexistent and successive form-states of the same and of different animals. No serious attempt had been made to discover the causes which led to the production of form in the individual and in the race.

It is true that evolution-theory had offered a simple solution of the great problem of the unity in diversity of animal forms, but this solution was formal merely, and went little beyond that abstract deduction of more complex from simpler forms, which had been the main operation of pre-evolutionary morphology. Little was known of the actual causes of ontogeny, and nothing at all of the causes of phylogeny; it was, for instance, mere rhetoric on Haeckel's part to proclaim that phylogeny was the mechanical cause of ontogeny.

Animal physiology, on its side, had developed in complete isolation from morphology into a science of the functioning of the adult and finished animal, considered as a more or less stable physico-chemical mechanism. Since the days of Ludwig, Claude Bernard and E. du Bois Reymond, the physiologists' chief care had been to analyse vital activities into their component physical and chemical processes, and to trace out the interchange of matter and energy between the organism and its environment. Physiologists had left untouched, perhaps wisely, the much more difficult problem of the causes of the development of form. For all practical purposes they took the animal-machine as given, and did not trouble about its mode of origin. They held indeed that form-production was due to a complex of physico-chemical causes, which they hoped some day to unravel;[1] but this future physiology of development remained quite embryonic.

Physiology then had not really come into contact with the problems of form, and it could give the morphologist no direct help when he turned to investigate the causes of form-production. It had, however, a determining influence upon the methods of those who first broke ground in this No Man's Land between morphology proper and physiology. But it is significant that it was a morphologist and not a physiologist that did the first spade-work.

The pioneer in this field, both as investigator and as thinker, was W. Roux, who sketched in the 'eighties the main outlines of a new science of causal morphology, to which he gave the name of *Entwicklungsmechanik.* The choice of name was deliberate, and the word implied, first, that the new science was essentially an investigation of the development of form, not of the mode of action of a formed mechanism, and second, that the methods to be adopted were mechanistic.[2]

Though Roux was the only begetter of the science of *Entwicklungsmechanik*, he was, of course, not the first tc investigate experimentally the formative processes of animal life. Study of regeneration dates back to Trembley (1740-44), Reaumur (1742), Bonnet (1745), and Spallanzani (1768-82),[3] and in the years preceding Roux's activity good work was done by Philipeaux. A beginning had been made with experimental teratology by E. Geoffroy St Hilaire and others, and the work of C. Dareste[4] remains classical. Back in the 18th century, some of John Hunter's experiments had a bearing upon the problems of form; his work on transplantation was followed up in the 19th century by Flourens, P. Bert, Ollier and many others. In founding in 1872 the *Archives de Zoologie expérimentale et générale* H. de Lacaze-

[1] See Carus's remark, referred to on p. 194, above.

[2] Roux, *Die Entwicklungsmechanik*, p. 26, Leipzig, 1905.

[3] T. H. Morgan, *Regeneration*, p. 1, New York and London, 1901.

[4] *Recherches sur la production artificielle des Monstruosités*, Paris, 1877, and many later papers.

Duthiers put forward in his introduction a powerful plea for the use of the experimental method in zoology.

In some ways more directly connected with *Entwicklungsmechanik* was His's attempt in 1874[1] to explain on mechanical principles the formation of certain of the embryonic organs by the bendings and foldings of tubes or plates of cells. "His compared the various layers of the chick embryo to elastic plates and tubes; out of these he suggested that some of the principal organs might be moulded by mere local inequalities of growth—the ventricles of the brain, for instance, the alimentary canal, the heart—and he further succeeded in imitating the formation of these organs by folding, pinching, and cutting india-rubber tubes and plates in various ways."[2]

But Roux was undoubtedly the first to make a systematic survey of the problems to be solved and to work out an organised method of attack. His earliest work deals with the important problem of functional adaptation—its importance to the organism, and its possible mechanistic explanation. The first paper[3] was a study of the branching and distribution of the arteries in the human body (1878), and a second paper on the same subject followed in 1879.[4]

In these papers Roux showed how the development of the blood-vascular system was largely determined by direct adaptation to functional requirements, and he inferred the existence in the vascular tissues of certain vital properties, in virtue of which the functional adaptation of the blood-vessels came about. Thus the intima or inner lining must possess the faculty of so reacting to the friction set up by the blood-current as to oppose the least possible resistance to its flow; the muscular coats must react to increased pressure by growing thicker, and so on.

These papers were followed in 1881 by his well-known

[1] *Unsere Körperform und das physiologische Problem ihrer Entstehung*, Leipzig, 1874.

[2] J. W. Jenkinson, *Experimental Embryology*, p. 3, Oxford, 1909.

[3] "Ueber die Verzweigungen der Blutgefässe des Menschen," *Jen. Zeit.*, xii., 1878.

[4] "Ueber die Bedeutung der Ablenkung des Arterienstammes bei der Astabgabe," *Jen. Zeit.*, xiii., 1879.

book, *Der Kampf der Theile im Organismus*, which contained the working-out of his mechanistic explanation of functional adaptation, and most of the elements of his general "causal-analytical" theory of form production. The significance of the book was popularly considered at the time to lie in its supposed application of the selection idea to the explanation of the internal adaptedness of animal structure—in the theory of "cellular selection," and the book owed its success to its fitting in so well with the prevalent Darwinism of the day. But its real importance, as a big step towards causal morphology, was naturally not so fully appreciated.

During the next few years Roux continued his studies on functional adaptation,[1] and at the same time made a new departure by inaugurating, almost contemporaneously with the physiologist Pflüger, the study of experimental embryology. Isolated observations had previously been made upon the development of single blastomeres or parts of blastulæ, by Haeckel and Chun for instance,[2] but Roux[3] and Pflüger[4] were the first to investigate the subject systematically, choosing for their work the egg of the frog.[5] Roux continued for many years to follow up this line of work.[6]

In 1890 he drew up a programme and manifesto[7] of *Entwicklungsmechanik* as "an anatomical science of the

[1] "Beiträge zur Morphologie der funktionellen Anpassung. I. Struktur eines hochdifferenzierten bindgewebigen Organes (der Schwanzflosse des Delphin)," *Arch. Anat. Physiol.* (*Anat. Abt.*) for 1883. II. "Ueber die Selbstregulation der 'morphologischen' Länge der Skeletmuskeln des Menschen," *Jen. Zeit.*, xvi., 1883. III. "Beschreibung . . . einer Kniegelenkeknochenankylose," *Arch. Anat. Physiol.* (*Anat. Abt.*) for 1885.

[2] In 1869 and 1877 respectively (Roux, p. 53, 1905).

[3] *Ueber die Zeit. der Bestimmung der Hauptrichtungen des Froschembryo*, Leipzig, 1883.

[4] "Ueber den Einfluss der Schwerkraft auf die Teilung der Zellen," Pflüger's *Archiv*, xxxi., 1883. Also subsequent papers in same journal.

[5] For an account of the classical experiments on the frog's egg, see T. H. Morgan, *The Development of the Frog's Egg*, New York, 1897.

[6] In a series of "Beiträge zur Entwicklungsmechanik des Embryo," published in various journals from 1884 to 1891, all dealing with the frog's egg. Also in many papers in the *Archiv f. Entw.-mech.*, from 1895 onwards.

[7] *Die Entwicklungsmechanik der Organismen, eine anatomische Wissenschaft der Zukunft*, Wien, 1890.

future," and in 1895 he founded the famous *Archiv für Entwicklungsmechanik*,[1] publishing in the same year the two large volumes of his collected papers,[2] of which the first volume dealt with functional adaptation, the second with experimental embryology.

His subsequent work includes several important general papers;[3] besides a number of special memoirs dealing with the factors of development, and with his original subject, functional adaptation.[4]

In our sketch of his views we shall have occasion to refer particularly to his publications of 1881, 1895 (the *Einleitung*), 1902, 1905, and 1910.

Although Roux's biological philosophy is out-and-out mechanistic, he yet recognises the difficulty, even the impossibility, of straightway reducing development to the physico-chemical level. He tries to steer a course midway between the simplicist conceptions of the materialists and the "metaphysics" of the neo-vitalist school, which the experimental study of development and regeneration soon brought into being. In 1895 he writes:—"The too simple mechanistic conception on the one hand, and the metaphysical conception on the other represent the Scylla and Charybdis, between which to sail is indeed difficult, and so far by few satisfactorily accomplished; it cannot be denied that with the increase of knowledge the seduction of the second has lately notably increased" (p. 23).

The *via media* adopted by Roux is the analysis of development, not directly into simple physico-chemical processes, but into more complex organic processes dependent

[1] The first volume contains the important *Einleitung* or general Introduction.

[2] *Gesammelte Abhandlungen über Entwicklungsmechanik der Organismen*, 2 vols., Leipzig, 1895.

[3] "Für unser Programm und seine Verwirklichung," *A.E.M.*, v., pp. 1-80 and 219-342, 1897. "Ueber die Selbstregulation der Lebewesen," *A.E.M.*, xiii., pp. 610-5, 1902. "Die Entwicklungsmechanik, ein neuer Zweig der biologischen Wissenschaft," Heft I. of the *Vorträge u. Aufsätze über Entwicklungsmechanik der Organismen*, Leipzig, 1905. Oppel and Roux, "Ueber die gestaltliche Anpassung der Blutgefässe," Heft x., of the *Vorträge u. Aufsätze*, Leipzig, 1910.

[4] "Ueber d. funkt. Anpassung des Muskelmagens der Gans," *A.E.M.*, xxi., pp. 461-99, 1906.

upon the fundamental properties of living matter. The aim of *Entwicklungsmechanik* is defined by Roux to be the reduction of developmental events to the fewest and simplest *Wirkungsweisen*, or causal processes.[1] Two classes of causal processes may be distinguished, as "complex components" and "simple components" of development. The latter are directly explicable by the laws of physics and chemistry; the former, while in essence physico-chemical, are yet so very complicated that they cannot at present be reduced to physico-chemical terms. The ultimate aim of *Entwicklungsmechanik* is to reduce development to its "simple components," but its main task at the present day and for many years to come is the analysis of development into its "complex components."

These complex components must be accepted as having much of the validity of physical and chemical laws. They are mysterious in the sense that they cannot yet be explained mechanistically, but they are constant in their action, and under the same conditions produce always the same effect—hence they may be made the subject of strictly scientific study. They represent biological generalisations, in their way of equal validity with the generalisations of physics and chemistry.

The principal "complex components" which Roux recognises are somewhat as follows:—First come the elementary cell-functions of assimilation and dissimilation, growth, reproduction and heredity, movement and self-division (as a special co-ordination of cell-movements). Then at a somewhat higher level, self-differentiation, and the trophic reaction to functional stimuli. Components of even greater complexity may also be distinguished, as, for instance, the biogenetic law. The various tropisms exhibited in development may be regarded as "directive" complex components. There must be added, not as being itself a component, but rather as a mode or peculiar property of all functioning, the omnipresent faculty of self-regulation.

It will be noticed that Roux's "complex components" are

[1] The exact quantitative formulation of a *Wirkungsweise* constitutes a law. The word itself is perhaps most conveniently rendered as "causal process."

simply the general properties or functions of organised matter.

Expressing Roux's thought in another way, we might say that life can only be defined functionally, *i.e.*, by an enumeration of the "complex components" or elementary functions which all living beings manifest, even down to the very simplest. "Living beings," writes Roux, "can at present be defined with any approach to completeness only functionally, that is to say, through characterisation of their activities, for we have an adequate acquaintance with their functions in a general way, though our knowledge of particulars is by no means complete" (p. 105, 1905). Defined in the most general and abstract way, living things are material objects which persist in spite of their metabolism, and, by reason of their power of self-regulation, in spite also of the changes of the environment. This is the "functional minimum-definition of life" (pp. 106-7, 1905).

We may now go on to consider the relation of function to form throughout the course of development. Roux distinguishes in all development two periods, in the first of which the organ is formed prior to and independent of its function, while in the second the differentiation and growth of the organ are dependent on its functioning. Latterly (1906 and 1910) Roux has distinguished three periods, counting as the second the transition period when form is partly self-determined, partly determined by functioning. As this conception of Roux's is of the greatest importance we shall follow it out in some detail.

The idea was first elaborated in the *Kampf der Theile* (1881), where he wrote:—"There must be distinguished in the life of all the parts two periods, an embryonic in the broad sense, during which the parts develop, differentiate and grow of themselves, and a period of completer development, during which growth, and in many cases also the balance of assimilation over dissimilation, can come about only under the influence of stimuli" (p. 180). There is thus a period of self-differentiation in which the organs are roughly formed in anticipation of functioning, and a period of functional development in which the organs are perfected through functioning and only through functioning. The two periods cannot be sharply separated from one another, nor does the transition from the one to the other occur at the same time in the different tissues and organs.

The conception is more fully expressed in 1905 as follows:—"This separation (of development into two periods) is intended only as a first beginning. The first period I called the embryonic period κατ' ἐξοχήν, or the period of organ-rudiments. It includes the 'directly inherited' structures, *i.e.*, the structures which are directly predetermined in the structure of the germ-plasm, as, for instance, the first differentiation of the germ, segmentation, the formation of the germ-layers and the organ-rudiments, as well as the next stage of 'further differentiation,' and of *independent* growth and maintenance, that is, of growth and maintenance which take place without the functioning of the organs.

"This is accordingly the period of direct fashioning through the activity of the formative mechanism implicit in the germ-plasm, also the period of the self-conservation of the formed parts without active functioning.

"The second period is the period of 'functional form-development.' It includes the further differentiation and the maintenance in their typical form of the organs laid down in the first period; and this is brought about by the exercise of the specific functions of the organs. This period adds the finishing touches to the finer functional differentiation of the organs, and so brings to pass the 'finer functional harmony' of all organs with the whole. The formative activity displayed during this period depends upon the circumstance that the functional stimulus, or rather the exercise by the organs of their specific functions, is accompanied by a subsidiary formative activity, which acts partly by producing new form and partly by maintaining that which is already formed. . . . Between the two periods lies presumably a transition period, an intermediary stage of varying duration in the different organs, in which both classes of causes are concerned in the further building-up of the already formed, those of the first period in gradually decreasing measure, those of the second in an increasing degree" (pp. 94-6, 1905).

In the first period the organ forms or determines the

function, in the second period the function forms the organ, or at least completes its differentiation. It is characteristic that in the first period functionally adapted structure appears in the complete absence of the functional stimulus.

The explanation of the difference between the two periods is to be found in the different evolutionary history of the characters formed during each. First-period characters are *inherited* characters, and taken together constitute the historical basis of the organism's form and activity; second-period characters are those of later acquirement which have not yet become incorporated in the racial heritage.

Inherited characters appear in development in the absence of the stimulus that originally called them forth; acquired characters are those that have not yet freed themselves from this dependence upon the functional stimulus. First-period characters were originally, like second-period characters, entirely dependent for their development upon the functional stimuli in response to which they arose, and only gradually in the course of generations did they gain that independence of the functional stimulus which stamps them as true inherited characters. Speaking of the formative stimuli which are active in second-period development, Roux writes:—"These stimuli can also produce new structure, which if it is constantly formed throughout many generations finally becomes hereditary, *i.e.*, develops in the descendants in the absence of the stimuli, becomes in our sense embryonic" (p. 180, 1881). Again, "form-characteristics which were originally acquired in post-embryonic life through functional adaptation may be developed in the embryo without the functional stimulus, and may in later development become more or less completely differentiated, and retain this differentiation without functional activity or with a minimum of it. But in the continued absence of functional activity they become atrophied . . . and in the end disappear" (p. 201, 1881).

This conception of the nature of hereditary transmission is an important one, and constitutes the first big step towards a real understanding of the historical element in organic form and activity. It supplies a practical criterion for the distinguishing of "heritage" characters from acquired characters, of palingenetic from cenogenetic — a criterion which descriptive morphology was unable to find.[1] The introduction of a functional moment into the concept of heredity was a methodological advance of the first importance, for it linked up in an understandable way the problems of embryology, and indirectly of all morphology, with the problem of hereditary transmission, and gave form and substance to the conception of the organism as an historical being.

It is this element in Roux's theories that puts them so far in advance of those of Weismann. Weismann did not really tackle the big problem of the relation of form to function, and he left no place in his mechanical system of preformation for functional or second-period development; he conceived all development to be in Roux's sense embryonic, and due to the automatic unpacking of a complex germinal organisation. Roux himself was to a certain extent a preformationist, for the development of his first-period characters is conditioned by the inherited organisation of the germ-plasm, and is purely automatic. It was indeed his experiments on the frog's egg (1888) that supplied some of the strongest evidence in favour of the mosaic theory of development. The number of *Anlagen* which he postulates in the germ is however small, and the germ-plasm in his conception of it has a relatively simple structure (p. 103, 1905).

The transmission of acquired characters forms, of course, an integral part of Roux's conception of heredity and development, for without this transmission second-stage characters could not be transformed into first-stage characters. He discusses this difficult question at some length in the *Kampf der Theile*, coming to the conclusion that such transmission takes place in small degree and gradually, and that many generations are required before a new character can become hereditary. He thinks that acquired characters are probably transmitted at the chemical level. It is conceivable that acquired form-changes are dependent on

[1] M. Fürbringer, perhaps under the influence of Roux, emphasised the importance, from a morphological point of view, of studying post-embryonic (functional) development, *Unters. z. Morph. u. Syst. der Vögel*, ii., Amsterdam, p. 925, 1888.

chemical changes, or are correlative with such, and that, since the germ-cells stand in close metabolic relations with the soma, these chemical changes may soak through to the germ-cells and so modify them that a predisposition will appear in the descendants towards similar form-changes.[1] From this point of view the problem of transmission might be merged in the broader problem of the production of form through chemical processes—the central problem of all development.

Inherited characters develop by an automatic process of self-differentiation, and the separate parts of the embryo show during this first period a surprising functional independence of one another. But this state of things changes progressively as the second period is reached, until finally all form-production and maintenance and all correlation depend upon functioning. It is in the first period of automatic development through internal "determining" factors that the "developmental" functions in the strict sense, *e.g.* automatic growth, division and self-differentiation, are most clearly shown. In the second or "functional" period the formative influence of function upon structure comes into play, and development becomes largely a matter of "functional adaptation" to functional requirements.

All structure, according to Roux, is either functional or non-functional. The former includes all structure that is adapted to subserve some function. "Such 'functional structures' are, for example, the composition of striated muscle fibres out of fibrillæ and these out of muscle-prisms, or again the length and thickness of the muscles, the static structure of the bones, the composition of the stomach and the blood-vessels out of longitudinal and circular fibres, the external shape of the vertebral centra and of the cuneiform bones of the foot" (p. 73, 1910). Indeed, as Cuvier had already pointed out, practically every organ in the body shows a functional structure which is accurately and minutely adjusted to the function it is intended to perform. Thus, to take some further examples, the arteries are admirably adapted as regards size of lumen, elasticity of wall, direction of branching, to conduct the blood to all parts of the body

[1] See, for the development of this idea, Oppel, in Roux-Oppel, 1910.

with the least possible waste of the propelling power through frictional resistance. So, too, the spongy substance of the long bones is arranged in lamellæ which take the direction of the principal stresses and strains which fall upon the bones in action.

Functional structure may be formed either in the first or in the second period of development, may be either inherited or acquired, but it reaches its full differentiation only in the second period, *i.e.*, under the influence of functioning. Practically speaking, functional structure is directly dependent for its full development and for its continued conservation upon the exercise of the particular function which it serves. In the second period, but not in the first, increased use leads to hypertrophy of the functional structure, disuse to atrophy.

From functional structure is to be distinguished non-functional structure, which has no relation to the bodily functions—is neither adapted to perform any of these, nor has arisen as a by-product of functional activity. "To this category belong, for example, among typical structures, the triangular form of the cross-section of the tibia, the dolicocephalic or brachycephalic shape of the skull, most of the external characters distinguishing genera and species, many of the external features of the embryo which change in the course of development, besides most of the abnormal forms shown by monstrosities, tumours, etc." (p. 74, 1910). Non-functional structure is not affected by functional adaptation, and may accordingly be left out of consideration here.

Now the influence of functioning upon the form and structure of an organ is twofold. There is first the immediate change brought about by the very act of functioning—for example, the shortening and thickening of skeletal muscles when they act. This is a purely temporary change, for the organ at once returns to its normal quiescent state as soon as it ceases to function. Such temporary functional change, brought about in the moment of functioning, is usually dependent for its initiation upon some neuro-muscular mechanism, though it may be elicited also by a chemical stimulus. It is thus always a phenomenon of "behaviour."

"From such temporary changes are sharply to be distinguished all permanent alterations which first appear in perceptible fashion through oft-repeated or long-continued, enhanced functional activity. These produce a new and lasting internal equilibrium of the organ, consisting in an insertion of new molecules or a rearrangement of old. For this reason they outlast the periods of functional form-change, or, if as in the case of the muscles they themselves alter during functional activity, they regain their state when the organ ceases to function" (p. 72, 1910). "Oft-repeated exercise or heightened exercise of the specific functions, or repeated action of the functional stimuli which determine them, produces, as we have said before, true form-changes as a by-product. These are of two kinds. In so far as these form-changes facilitate the repetition of the specific functions, I have called them *functional adaptations*. . . . Such as do not improve the functioning of the organ are indeed by-products of functioning, but without adaptive character; they do not belong to the class of functional adaptations at all" (p. 75, 1910).

We may now enquire in what way functional adaptations can arise as by-products of functioning.

It is clear that natural selection in the sense of individual or "personal" selection cannot adequately explain the origin of functional structure and the functional harmony of structure, for thousands of cells would have to vary together in a purposive way before any real advantage could be gained in the struggle for existence, and it is in the highest degree unlikely that this should come about by chance variation.[1] The development of purposive internal structure is only to be explained by the properties of the tissues concerned.

In illustration and proof of the statement that functional adaptation is due to the properties of the tissues we may adduce the development and regulation of the blood-

[1] *Cf.* the controversy between Herbert Spencer and Weismann on the subject of "coadaptation" in the *Contemporary Review* for 1893 and 1894. See also Weismann's paper in *Darwin and Modern Science*, Cambridge, 1909.

vascular system, which has been thoroughly studied from this point of view by Roux and Oppel (1910).

It appears that only the very first rudiments of the vascular system are laid down in the short first period of automatic non-functional development. All the subsequent growth and differentiation of the blood-vessels falls into the second period, and is due wholly or in great part to direct functional adaptation to the requirements of the tissues. Thus from the rudiments formed in the first period there sprout out the definitive vessels in direct adaptation to the food-consumption of the tissues they are to supply. The size, direction and intimate structure of these vessels are accurately adjusted to the part they play in the economy of the whole, and this adjustment is brought about in virtue of the peculiar properties or reaction-capabilities of the different tissues of which the blood-vessels are composed.

The properties which Roux finds himself compelled to postulate in the vascular tissues, after a thorough-going analysis of the different kinds of functional adaptation shown by the blood-vessels, are summarised by him as follows:—

"(1) The faculty—depending on a direct sensibility possessed by the endothelium and perhaps also by the other layers of the intima—of yielding to the impact of the blood, so far as the external relations of the vessel permit. In this way the wall adapts itself to the hæmodynamically conditioned 'natural' shape of the blood-stream, and reaches this shape as nearly as possible." Through this faculty of the lining tissue of the blood-vessels, the size of the lumen and the direction of branching are so regulated as to oppose the least possible resistance to the flow of the blood.

"(2) The faculty possessed by the endothelium of the capillaries of each organ of adapting itself qualitatively to the particular metabolism of the organ." This adaptedness of the capillaries is, however, more usually an inherited state, *i.e.*, brought about in the first period of development.

"(3) The faculty possessed by the capillary walls of being stimulated to sprout out and branch by increased functioning, *i.e.*, by increased diffusion, and their power to exhibit a chemically conditioned cytotropism, which causes the sprouts to find one another and unite. A similar process can be directly observed in isolated segmentation-cells, which tend to unite in consequence of a power of mutual attraction.

"(4) The faculty of developing normal arterial walls in response to strong intermittent pressure, and normal venous walls in response to continuous lesser pressure." It has been shown, for instance, by Fischer and Schmieden that in dogs a section of vein transplanted into an artery takes on an arterial structure, at least as regards the circular musculature, which doubles in thickness.

"(5) The power to regulate the normal[1] length of the arteries and veins, in adaptation to the growth of the surrounding tissues, in such a way that the stretching action of the blood-stream brings the vessel to its proper functional length.

"(6) The power to form, in response to slight increases in longitudinal tension, new structural parts which take their place alongside the existing longitudinal fibres.

"(7) The power to regulate the width of the circular musculature according to the degree of food-consumption by the tissues, in response to nerve impulses initiated in these tissues.

"(8) The power possessed by the circular musculature of responding to such continuous functional widening, by the formation of new structural parts in the circular musculature, and so of widening the vessel permanently or by this new formation of muscular fibres thickening the circular musculature.

"(9) The faculty of being stimulated by increased blood-pressure to produce the same structural changes as mentioned in par. 8, though here the response is otherwise conditioned" (pp. 126-7, 1910).

It is by virtue of the tissue-properties detailed above that the complex functional adaptations of the blood-vessels come about.

The development of the vascular system is no mere automatic and mechanical production of form, apart from

[1] That is, the length they take up when separated from the body.

and independent of functioning; it implies a living and co-ordinated activity of the tissues and organs concerned, a power of active response to foreseen and unforeseen contingencies. Form is then not something fixed and congealed—it is the ever-changing manifestation of functional activity. "Since most of the structure and form of the blood-vessels arises in direct adaptation to function, the vessels of adult men and animals are no fixed structures, which, once formed, retain their form and structural build unchanged throughout life; on the contrary, they require even for their continued existence the stimulus of functional activity. . . . The fully formed blood-vessels are no static structures, such as they appear to be according to the teaching of normal histology, and such as they have long been taken to be. Observation and description of normal development never shows us anything but the visible side of organic happenings, the *products* of activity, and leaves us ignorant of the real processes of form-development and form-conservation, and of their causes" (p. 125, 1910).

The real thing in organisation is not form but activity. It is in this return to the Cuvierian or functional attitude to the problems of form that we hold Roux's greatest service to biology to consist. The attitude, however, seems to smack of vitalism, and Roux, as we have seen, is no vitalist. He holds that the marvellous and apparently purposive tissue-qualities which underlie all processes of functional adaptation have arisen "naturally," in the course of evolution, by the action of natural selection upon the various properties, useful and useless, which appeared fortuitously in the primary living organisms. He is, moreover, deeply imbued with the materialistic philosophy of his youth, and it is indeed one of the chief characteristics of his system that he states the fundamental properties or qualities of life in terms of metabolism. A vital quality is for Roux a special process or mode of assimilation. The faculty of "morphological assimilation" whereby form is imposed upon formless chemical processes is the ultimate term of Roux's analysis—"the most general, most essential, and most characteristic formative activity of life" (p. 631, 1902).

We have now to consider very briefly the early results

achieved by Roux's fellow-workers in the field of causal morphology. As D. Barfurth points out,[1] the years 1880-90 saw a general awakening of interest in experimental morphology, and it is hard to say whether Roux's work was cause or consequence. "There fall into this period," writes Barfurth, "the experimental investigations by Born and Pflüger on the sexual difference in frogs (1881), by Pflüger on the parthenogenetic segmentation of Amphibian ova, on crossing among the Amphibia, and on other important subjects (1882). In the following year (1883) appeared two papers of fundamental importance, by E. Pflüger and W. Roux: Pflüger publishing his researches on 'the influence of gravity on cell-division,' Roux his experimental investigations on 'the time of the determination of the chief planes in the frog-embryo.' . . . In the same year appeared A. Rauber's experimental studies 'on the influence of temperature, atmospheric pressure, and various substances on the development of animal ova,' which have brought many similar works in their train. The following year (1884) saw a lively controversy on Pflüger's gravity-experiments with animal eggs, in which took part Pflüger, Born, Roux, O. Hertwig and others, and in this year appeared work by Roux dealing with the experimental study of development, and in particular giving the results of the first definitely localised pricking-experiments on the frog's egg (in the *Schles. Gesell. f. vaterl. Kultur*, 15th Feb. 1884), also the important researches of M. Nussbaum and Gruber (followed up later by Verworn, Hofer and Balbiani) on Protozoa, and other experimental work" (pp. xi.-xii.).

In 1888 appeared a famous paper by W. Roux,[2] in which he described how he had succeeded in killing by means of a hot needle one of the two first blastomeres of the frog's egg, and how a half-embryo had developed from the uninjured cell. Some years before[3] he had enunciated, at about the same time as Weismann, the view that development

[1] "Wilhelm Roux zum 60. Geburtstage," *Arch. f. Entw.-Mech.*, xxx. *Festschrift für Prof. Roux*, Pt. i, 1910.

[2] Virchow's *Archiv*, cxiv., 1888. First announced in Sept. 1887.

[3] *Ueber die Bedeutung der Kernteilungsfiguren*, Leipzig, 1883.

was brought about by a qualitative division of the germ-plasm contained in the nucleus, and that the complicated process of karyokinetic or mitotic division of the nucleus was essentially adapted to this end. He conceived that development proceeded by a mosaic-like distribution of potencies to the segmentation-cells, that, for instance, the first segmentation furrow separated off the material and potencies for the right half of the embryo from those for the left half. He had tried to show experimentally that the first furrow in the frog's egg coincided with the sagittal plane of the embryo,[1] and his later success in obtaining a half-embryo from one of the first two blastomeres seemed to establish the "mosaic theory" conclusively.

Roux's needle-experiment aroused much interest, especially as Weismann's theory of heredity was then being keenly discussed. Chabry had published in 1887 some interesting results on the Ascidian egg,[2] which strongly supported the Roux-Weismann theory. Considerable astonishment was therefore caused by Driesch's announcement in 1891 [3] that he had obtained complete larvæ from single blastomeres of the sea-urchin's egg isolated at the two-celled stage. He followed this up in the next year [3] by showing that whole embryos could be produced from one or more blastomeres isolated at the four-cell stage. Similar or even more striking results were obtained by E. B. Wilson on *Amphioxus*,[4] and Zoja on medusæ.[5] Driesch succeeded also in disturbing the normal course and order of segmentation by compressing the eggs of the sea-urchin between glass plates, and yet obtained normal embryos. Similar pressure-experiments were carried out on the frog by O. Hertwig,[6] and on *Nereis* by E. B. Wilson,[7] with analogous results.

In 1895 O. Schultze [8] showed that if the frog's egg is held between two plates and inverted at the two-celled stage

[1] *Bresl. ärtz. Zeitschr.*, 1885.

[2] *Journ. de l'Anat. et de la Physiologie*, xxiii., 1887.

[3] *Zeits. f. wiss. Zool.*, liii., 1891 and 1892.

[4] *Journ. Morph.*, viii., 1893.

[5] *Arch. f. Ent.-Mech.*, i., 1895 ; ii., 1896.

[6] *Arch. f. mikr. Anat.*, xliii., 1893.

[7] *Arch. f. Ent.-Mech.*, iii., 1896.

[8] *Arch. f. Ent.-Mech.*, i., 1895.

there are formed two embryos instead of one. In the same year T. H. Morgan[1] repeated Roux's fundamental experiment of destroying one of the two blastomeres, but inverted the egg immediately after the operation—a whole embryo of half size resulted. A year or two later Herlitzka[2] found that if the first two blastomeres of the newt's egg were separated by constriction, two normal embryos of rather more than half normal size were formed.

The main result of the first few years' work on the development of isolated blastomeres was to show that the mosaic theory was not strictly true, and that the hypothesis of a qualitative division of the nucleus was on the whole negatived by the facts.

Evidence soon accumulated that the cytoplasm of the egg stood for much in the differentiation of the embryo. A number of years previously Chun had made the discovery that single blastomeres of the Ctenophore egg, isolated at the two-celled stage, gave half-embryos. This was in the main confirmed by Driesch and Morgan in 1896,[3] and they made the further interesting discovery that the same defective larvæ could be obtained by removing from the unsegmented egg a large amount of cytoplasm. Conclusive proof of the importance of the cytoplasm was obtained soon after by Crampton,[4] who removed the anucleate "yolk-lobe" from the egg of the mollusc *Ilyanassa* at the two-celled stage, and obtained larvæ which lacked a mesoblast. This result was brilliantly confirmed and extended some years later by E. B. Wilson,[5] working on the egg of *Dentalium*. He found that if the similar anucleate "polar lobe" of this form is removed at the two-celled stage, deficient larvæ are formed, in which the post-trochal region and the apical organ are absent. He further showed that in the unsegmented but mature egg prelocalised cytoplasmic regions can be distinguished, which later become separated from one another through the segmentation of the egg. The segmentation-cells into which these cytoplasmic substances are thus segregated show a marked specificity of develop-

[1] *Anat. Anz.*, x., 1895.
[2] *Arch. f. Ent.-Mech.*, iv. 1897.
[3] *Arch. f. Ent.-Mech.*, ii., 1896.
[4] *Arch. f. Ent.-Mech.*, iii., 1896.
[5] *Journ. exper. Zool.*, i., 1904.

ment, giving rise, even when isolated, to definite organs of the embryo. Wilson concluded that the cytoplasm of the egg contains a number of specific organ-forming stuffs, which have a definite topographical arrangement in the egg. Development is thus due in part to a qualitative division not of the nucleus but of the cytoplasm. Corroborative evidence of the existence of cytoplasmic organ-forming stuffs has been supplied for several other species, *e.g.*, *Patella* (Wilson), *Cynthia* (Conklin), *Cerebratulus* (Zeleny), and *Echinus* (Boveri).

It is interesting to recall that so long ago as 1874 W. His[1] put forward the theory that there exist in the blastoderm and even in the egg prelocalised areas, which contain the formative material for each organ of the embryo, and from which the embryo is developed by a simple process of unequal growth.

The experimental study of form was prosecuted in many other directions besides that of experimental embryology. The study of regeneration and of regulatory processes attracted many workers, among whom may be mentioned T. H. Morgan, C. M. Child, and H. Driesch. In an interesting series of papers C. Herbst applied the principles of the physiology of stimulus to the interpretation of development.[2] The formative power of function was studied in Germany by Roux and his pupils, Fuld, O. Levy, Schepelmann and others, particularly by E. Babák. In France, F. Houssay inaugurated[3] an important series of memoirs by himself and his pupils on "dynamical morphology," the most important memoir being his own valuable discussion of the functional significance of form in fishes.[4] The principles of his dynamical morphology were first laid down in his book *La Forme et la Vie* (1900).

The famous experiments of Loeb, Delage and others on

[1] *Unsere Körperform*, p. 19, Leipzig, 1874.

[2] *Biolog. Centrlbl.*, xiv., 1894, xv., 1895. *Formative Reize in der thierischen Ontogenese*, Leipzig, 1901.

[3] "La Morphologie dynamique," No. i. of the *Collection de Morphologie dynamique*, Paris, 1911.

[4] "Forme, Puissance et Stabilité des Poissons," No. iv. of the *Collection*, Paris, 1912.

artificial parthenogenesis may also be mentioned, though their connection with morphology is somewhat remote.

The period was characterised also by the lively discussion of first principles, in which Driesch took a leading part. Materialistic methods of interpretation were upheld by perhaps the majority of biologists, but vitalism found powerful support.

CHAPTER XIX

SAMUEL BUTLER AND THE MEMORY THEORIES OF HEREDITY

WE have laid stress upon the distinction established by Roux between the two stages of development—the automatic and the functional—because of the light which it seems to throw upon the phylogenetic relation of form to function. We have pointed out, too, the paramount rôle that function plays in Roux's theories of development and heredity, and we have brought out the close kinship existing between his theory and that of Lamarck. For Roux, as for Lamarck, the function creates the organ, and it is only after long generations that the organ appears before the function.

It so happened that just about the time when Roux's papers were beginning to appear a brilliant attempt was made by Samuel Butler to revive and complete the Lamarckian doctrine.

A man of singular freshness and openness of mind, combining in an extraordinary degree extreme intellectual subtlety with a childlike simplicity of outlook, Butler was one of the most fascinating figures of the 19th century. He was not a professional biologist, and much of his biological work is, for that reason, imperfect. But he brought to bear upon the central problems of biology an unbiassed and powerful intelligence, and his attitude to these problems, just because it is that of a cultivated layman, is singularly illuminating.

He was not well acquainted with biological literature; he seems to have hit upon the main ideas of his theory of life and habit in complete independence of Lamarck, and only later to have become aware that Lamarck had in a measure forestalled him. He puts this very beautifully in the following passage from his chief biological work *Life and Habit* (1877[1]):—"I admit that when I began to write upon my subject I did not seriously believe in it. I saw, as it were, a pebble upon the ground, with a sheen that pleased me; taking it up, I turned it over and over for my amusement, and found it always grow brighter and brighter the more I examined it. At length I became fascinated, and gave loose rein to self-illusion. The aspect of the world changed; the trifle which I had picked up idly had proved to be a talisman of inestimable value, and had opened a door through which I caught glimpses of a strange and interesting transformation. Then came one who told me that the stone was not mine, but that it had been dropped by Lamarck, to whom it belonged rightfully, but who had lost it; whereon I said I cared not who was the owner, if only I might use it and enjoy it. Now, therefore, having polished it with what art and care one who is no jeweller could bestow upon it, I return it, as best I may, to its possessor" (p. 306). In one of his later works, however, Butler made up for his first neglect of his predecessors by giving what is undeniably the best account in English literature of the work of Buffon, Lamarck, and Erasmus Darwin—in his *Evolution, Old and New* (1879). Many of his facts he took from Charles Darwin, whose theory of natural selection he bitterly opposed, in the two books just mentioned and in *Unconscious Memory* (1880) and *Luck or Cunning* (1887).

Butler's main thesis is that living things are active, intelligent agents, personally continuous with all their ancestors, possessing an intense but unconscious memory of all that their ancestors did and suffered, and moving through habit from the spontaneity of striving to the automatism of remembrance.

The primary cause of all variation in structure is the active response of the organism to needs experienced by it, and the indispensable link between the outer world and the creature itself is that same "sense of need" upon which

[1] The quotations are taken from the 1910 reprint, London, Fifield.

Lamarck insisted. "According to Lamarck, genera and species have been evolved, in the main, by exactly the same process as that by which human inventions and civilisations are now progressing; and this involves that intelligence, ingenuity, heroism, and all the elements of romance, should have had the main share in the development of every herb and living creature around us" (*Life and Habit*, p. 253). Variations are indubitably the raw material of evolution—"The question is as to the origin and character of these variations. We say they mainly originate in a creature through a sense of its needs, and vary through the varying surroundings which will cause those needs to vary, and through the opening-up of new desires in many creatures, as the consequence of the gratification of old ones; they depend greatly on differences of individual capacity and temperament; they are communicated, and in the course of time transmitted, as what we call hereditary habits or structures, though these are only, in truth, intense and epitomised memories of how certain creatures liked to deal with protoplasm" (p. 267).

Butler's theory then is essentially a bold and enlightened Lamarckism, completed and rounded off by the conception that heredity too is a psychological process, of the same nature as memory.

In seeking to establish a close analogy between memory and heredity Butler starts out from the fact of common experience, that actions which on their first performance require the conscious exercise of will and intelligence, and are then carried out with difficulty and hesitation, gradually through long-continued practice come to be performed easily and automatically, without the conscious exercise of intelligence or will.

He tries to show that this is a general law—that knowledge and will become intense and perfect only when through long-continued exercise they become automatic and unconscious—and he applies this conception to the elucidation of development.

Developmental processes, especially the early ones (of Roux's first stage) are automatic and unconscious, and yet imply the possession by the embryo of a wonderfully perfect knowledge of the processes to be gone through, and an assured power of will and judgment. Is it conceivable, says Butler, that the embryo can do all these things without knowing how to do them, and without having done them before? "Shall we say . . . that a baby of a day old sucks (which involves the whole principle of the pump, and hence a profound practical knowledge of the laws of pneumatics and hydrostatics), digests, oxygenises its blood (millions of years before Sir Humphrey Davy discovered oxygen), sees and hears—all most difficult and complicated operations, involving a knowledge of the facts concerning optics and acoustics, compared with which the discoveries of Newton sink into utter insignificance? Shall we say that a baby can do all these things at once, doing them so well and so regularly, without being even able to direct its attention to them, and without mistake, and at the same time not know how to do them, and never have done them before?" (p. 54). Assuredly not.

The only possible explanation is that the embryo's ancestors have done these things so often, throughout so many millions of generations, that the embryo's knowledge of how to do them has become unconscious and automatic by reason of this age-long practice. This implies that there is in a very real sense actual personal continuity between the embryo and all its ancestors, so that their experiences are his, their memory also his. "We must suppose the continuity of life and sameness between living beings, whether plants or animals, to be far closer than we have hitherto believed; so that the experience of one person is not enjoyed by his successor, so much as that the successor is *bona fide* but a part of the life of his progenitor, imbued with all his memories, profiting by all his experiences—which are, in fact, his own—and only unconscious of the extent of his own memories and experiences owing to their vastness and already infinite repetitions" (p. 50). It is very suggestive in this connection, he continues—"I. That we are *most conscious of, and have most control over*, such habits as speech, the upright position, the arts and sciences, which are acquisitions peculiar to the human race, always acquired after birth, and not common to ourselves and any ancestor who had not become entirely human.

"II. That we are *less conscious of, and have less control over*, eating and drinking, swallowing, breathing, seeing and hearing, which were acquisitions of our prehuman ancestry, and for which we had provided ourselves with all the necessary apparatus before we saw light, but which are, geologically speaking, recent, or comparatively recent.

"III. That we are *most unconscious of, and have least control over*, our digestion and circulation, which belonged even to our invertebrate ancestry, and which are habits, geologically speaking, of extreme antiquity. . . . Does it not seem as though the older and more confirmed the habit, the more unquestioning the act of volition, till, in the case of the oldest habits, the practice of succeeding existences has so formulated the procedure, that, on being once committed to such and such a line beyond a certain point, the subsequent course is so clear as to be open to no further doubt, to admit of no alternative, till the very power of questioning is gone, and even the consciousness of volition" (pp. 51-2).

The hypothesis then, that heredity and development are due to unconscious memory, finds much to support it—"the self-development of each new life in succeeding generations —the various stages through which it passes (as it would appear, at first sight, without rhyme or reason), the manner in which it prepares structures of the most surpassing intricacy and delicacy, for which it has no use at the time when it prepares them, and the many elaborate instincts which it exhibits immediately on, and indeed before, birth—all point in the direction of habit and memory, as the only causes which could produce them" (p. 125). The hypothesis explains, for instance, the fact of recapitulation:—"Why should the embryo of any animal go through so many stages —embryological allusions to forefathers of a widely different type? And why, again, should the germs of the same kind of creature always go through the same stages? If the germ of any animal now living is, in its simplest state, but part of the personal identity of one of the original germs of all life whatsoever, and hence, if any now living organism must be considered without quibble as being itself millions of years old, and as imbued with an intense though unconscious memory of all that it has done sufficiently often to have made a permanent impression; if this be so, we can answer the above questions perfectly well. The creature goes through so many intermediate stages between its earliest state as life at all, and its latest development, for the simplest of all reasons, namely, because this is the road by which it has always hitherto travelled to its present differentiation; this is the road it knows, and into every turn and up or down of which it has been guided by the force of circumstances and the balance of considerations" (pp. 125-6).

The hypothesis explains also the way in which the orderly succession of stages in embryogeny is brought about, for we can readily understand that the embryo will not remember any stage until it has passed through the stage immediately preceding it. "Each step of normal development will lead the impregnated ovum up to, and remind it of, its next ordinary course of action, in the same way as we, when we recite a well-known passage, are led up to each successive sentence by the sentence which has immediately preceded it. . . . Though the ovum immediately after impregnation is instinct with all the memories of both parents, not one of these memories can normally become active till both the ovum itself and its surroundings are sufficiently like what they respectively were, when the occurrence now to be remembered last took place. The memory will then immediately return, and the creature will do as it did on the last occasion that it was in like case as now. This ensures that similarity of order shall be preserved in all the stages of development in successive generations" (pp. 297-8).

Abnormal conditions of development will cause the embryo to pause and hesitate, as if at a loss what to do, having no ancestral experience to guide it. Abnormalities of development represent the embryo's attempt to make the best of an unexpected situation. Or, as Butler puts it, "When . . . events are happening to it which, if it has the kind of memory we are attributing to it, would baffle that memory, or which have rarely or never been included in the category of its recollections, *it acts precisely as a creature acts when its recollection is disturbed, or when it is required to do something which it has never done before*" (p. 132). "It is certainly noteworthy that the embryo is never at a loss, unless something happens to it which has not usually happened to its forefathers, and which in the nature of things it cannot remember" (p. 132).

Butler's teleological conception of organic evolution was of course completely antagonistic to the naturalistic conceptions current in his time. In one of his later books he repeats Paley's arguments in favour of design, and to the question, "Where, then, is your designer of beasts and birds, of fishes, and of plants?" he replies: "Our answer is simple enough; it is that we can and do point to a living tangible person with flesh, blood, eyes, nose, ears, organs, senses, dimensions, who did of his own cunning, after infinite proof of every kind of hazard and experiment, scheme out and fashion each organ of the human body. This is the person whom we claim as the designer and artificer of that body, and he is the one of all others the best fitted for the task by his antecedents, and his practical knowledge of the requirements of the case —for he is man himself. Not man, the individual of any given generation, but man in the entirety of his existence from the dawn of life onwards to the present moment" (*Evolution, Old and New*, p. 30, 1879).

Butler's theory of life and habit remained only a sketch, and he was perhaps not fully aware of its philosophical implications. Since Butler's time, a new complexion has been put upon biological philosophy by the profound speculations of Bergson.

But it is not impossible that the future development of biological thought will follow some such lines as those which he tentatively laid down.

Butler was not the first to suggest that there is a close connection between heredity and memory—it is a thought likely to occur to any unprejudiced thinker. The first enunciation of it which attracted general attention was that contained in Hering's famous lecture "On Memory as a general Function of organised Matter."[1] Butler was not

[1] *Ueber das Gedächtnis als eine allgemeine Funktion der organisierten Materie*, Wien, 1870.

aware of Hering's work when he published his *Life and Habit*, but in *Unconscious Memory* (1880) he gave full credit to Hering as the first discoverer, and supplied an admirable translation of Hering's lecture. As far as the assimilation of heredity to memory is concerned Hering and Butler have much in common, but Hering did not share Butler's Lamarckian and vitalistic views, preferring to hold fast, for the practical purposes of physiology at all events, to the general accepted theory of the parallelism between psychical and physical processes. He was inclined to regard memory in the ordinary sense as a function of the brain, and memory in general as a function of all organised matter. Speaking of the psychical life, he says, "Thus the cause which produces the unity of all single phenomena of consciousness must be looked for in unconscious life. As we know nothing of this except what we learn from our investigations of matter, and since in a purely empirical consideration, matter and the unconscious must be regarded as identical, the physiologist may justly define memory in a wider sense to be a faculty of the brain, the results of which to a great extent belong to both consciousness and unconsciousness."[1] Hering's views were supported by Haeckel.[2]

In 1893 an American, H. F. Orr,[3] tried to work out a theory of development and heredity based upon the fundamental idea "that the property which is the basis of bodily development in organisms is the same property which we recognise as the basis of psychic activity and psychic development." He tried also to explain the recapitulation of phylogeny by ontogeny as due to habit.

The neo-Lamarckian school of American palæontologists were also in sympathy with the memory idea, and this was expressed most clearly perhaps by Cope.[4]

In 1904 appeared the work on this subject which has attracted the most attention—R. Semon's *Die Mneme*.[5]

[1] Eng. trans. in E. Hering, *Memory*, p. 9, Chicago and London, 1913.

[2] *Die Perigenesis der Plastidule*, Jena, 1875.

[3] *A Theory of Development and Heredity*, New York, 1893.

[4] *The Primary Factors of Organic Evolution*, Chicago, 1896.

[5] *Die Mneme als erhaltendes Prinzip im Wechsel des organischen Geschehens*, Leipzig, 1904; 2nd ed., 1908.

This was an elaborate treatment of the question from the materialistic point of view, the main assumption of Semon's theory being that the action of a stimulus upon the organism leaves a more or less permanent material trace or "engramm," of such a nature as to modify the subsequent action of the organism.

Applied to the explanation of heredity and development, Semon's theory comes to very much the same as Weismann's, with engramms substituted for determinants, but it has the great advantage of allowing for the transmission of acquired characters. The application of the concept of stimulus is valuable and suggestive, but it seems to us that the memory theory of heredity can be properly utilised only by adopting a frankly Lamarckian and vitalistic standpoint, and this standpoint Semon expressly combats. As Ward[1] points out in his illuminating lecture on heredity and memory—"Records or memoranda alone are not memory, for they presuppose it. *They* may consist of physical traces, but memory, even when called 'unconscious,' suggests mind; for, as we have seen, the automatic character implied by this term 'unconscious' presupposes foregone experience. . . . The mnemic theory then, if it is to be worth anything, seems to me clearly to require not merely physical records or 'engrams,' but living experience or tradition. The mnemic theory will work for those who can accept a monadistic or pampsychist interpretation of the beings that make up the world, who believe with Spinoza and Leibniz that 'all individual things are animated albeit in divers degree'" (pp. 55-6).

Perhaps the best and most ingenious treatment of memory and heredity from a physical standpoint is that offered by E. Rignano in his book, *Sur la transmissibilité des caractères acquis.*[2] Rignano seeks to construct a physico-chemical "model" which will explain both heredity and memory.

His system, which is based more firmly upon the facts of experimental embryology than Semon's, postulates the existence of "specific nervous accumulators." The essential

[1] *Heredity and Memory*, Cambridge, 1913.

[2] Paris, 1906. Also in Italian and German. Eng. trans. by B. C. H. Harvey, Chicago, 1911.

hypothesis set up is that every functional stimulus is transformed into specific vital energy, and deposits in the nucleus of the cell a specific substance which is capable of discharging, in an inverse direction, the nervous current which has formed it, as soon as the dynamical equilibrium of the organism is restored to the state in which it was when the original stimulus acted upon it. These specific nuclear substances, different for each cell, are accumulated also in the nuclei of the germinal substance, constituting what Rignano calls the central zone of development. That is to say, each functional adaptation changes slightly the dynamical equilibrium of the organism, and this change in the system of distribution of the nervous currents leads to the deposit in the central zone of development of a new specific substance. In the development of the next individual this new specific element enters into activity, and reproduces the nervous current which has formed it, as soon as the organism reaches the same conditions of dynamical equilibrium as those obtaining when the stimulus acted on the parent.

Development can thus be regarded as consisting of a number of stages, at each of which new specific elements enter automatically into play and lead the embryo from that stage to the stage succeeding. The germinal substance on this theory of Rignano's is to be regarded as being composed of a large number of specific elements, originally formed as a result of each new functional adaptation, but now forming part of the hereditary equipment.

The theory represents an advance upon the more static conceptions of Semon. It owes much to Roux's influence.

In this country, the mnemic theories have been championed particularly by M. Hartog[1] and Sir Francis Darwin.[2]

[1] See *Problems of Life and Reproduction*, London, 1913.
[2] *Presidential Address to the British Association*, 1908.

CHAPTER XX

THE CLASSICAL TRADITION IN MODERN MORPHOLOGY

To write a history of contemporary movements from a purely objective standpoint is well recognised to be an impossible task. It is difficult for those in the stream to see where the current is carrying them: the tendencies of the present will only become clear some twenty years in the future.

I propose, therefore, in this concluding chapter to deal only with certain characteristics of modern work on the problems of form which seem to me to be derived directly from the older classical tradition of Cuvier and von Baer.

The present time is essentially one of transition. Complete uncertainty reigns as to the main principles of biology. Many of us think that the materialistic and simplicist method has proved a complete failure, and that the time has come to strike out on entirely different lines. Just in what direction the new biology will grow out is hard to see at present, so many divergent beginnings have been made—the materialistic vitalism of Driesch, the profound intuitionalism of Bergson, the psychological biology of Delpino, Francé, Pauly, A. Wagner and W. Mackenzie. But if any of these are destined to give the future direction to biology, they will in a measure only be bringing biology back to its pre-materialistic tradition, the tradition of Aristotle, Cuvier, von Baer and J. Müller. It may well be that the intransigent materialism of the 19th century is merely an episode, an aberration rather, in the history of biology—an aberration brought about by the over-rapid development of a materialistic and luxurious civilisation, in which man's material means have outrun his mental and moral growth.

Two movements seem significant in the morphology of the last decade or so of the 19th century—first, the experimental study of form, and second, the criticism of the concepts or prejudices of evolutionary morphology.

The period was characterised also by the great interest taken in cytology, following upon the pioneer work of Hertwig, van Beneden and others on the behaviour of the nuclei in fertilisation and maturation.[1] This line of work gained added importance in connection with contemporary research and speculation on the nature of hereditary transmission, and it has in quite recent years received an additional stimulus from the re-discovery of Mendelian inheritance. Its importance, however, seems to lie rather in its possible relation to the problems of heredity than in any meaning it may have for the problems of form. More significant is the revolt against the cell-theory started by Sedgwick[2] and Whitman,[3] on the ground that the organism is something more than an aggregation of discrete, self-centred cells.

The experimental work on the causes of the production and restoration of form infused new life into morphology. It opened men's eyes to the fact that the developing organism is very much a living, active, responsive thing, quite capable of relinquishing at need the beaten track of normal development which its ancestors have followed for countless generations, in order to meet emergencies with an immediate and purposive reaction. It was cases of this kind, cases of active regulation in development and regeneration, that led men like G. Wolff and H. Driesch to cast off the bonds of dogmatic Darwinism and declare boldly for vitalism and teleology.

There was the famous case of the regeneration of the lens in Amphibia from the edge of the iris—an entirely novel mode of origin, not occurring in ontogeny. The fact seems to have been discovered first by Colucci in 1891, and independently by G. Wolff in 1895.[4] The experiment was later repeated and confirmed by Fischel and other workers.

[1] See E. B. Wilson's masterly book, *The Cell in Development and Inheritance*, New York and London, 1900. [2] *Q.J.M.S.*, xxvi., 1886.

[3] *Wood's Holl Biological Lectures* for 1893.

[4] *Arch. f. Ent.-Mech.*, i., pp. 380-90, 1895.

Wolff drew from this and other facts the conclusion that the organism possesses a faculty of "primary purposiveness" which cannot have arisen through natural selection.[1] And, as is well known, Driesch derived one of his most powerful arguments in favour of vitalism from the extraordinary regenerative processes shown by *Tubularia* and *Clavellina*, in the course of which the organism actually demolishes and rebuilds a part or the whole of its structure. But under the influence of physiologists like Loeb many workers held fast to materialistic methods and conceptions.

The great variety of regulative response of which the organism showed itself capable made it very difficult for the morphologist to uphold the generalisations which he had drawn from the facts of normal undisturbed development. The germ-layer theory was found inadequate to the new facts, and many reverted to the older criterion of homology based on destiny rather than origin. The trend of opinion was to reject the ontogenetic criterion of homology, and to refuse any morphological or phylogenetic value to the germ-layers.[2]

The biogenetic law came more and more into disfavour, as the developing organism more and more showed itself to be capable of throwing off the dead-weight of the past, and working out its own salvation upon original and individual lines.[3] A. Giard in particular called attention to a remarkable group of facts which went to show that embryos or larvæ of the same or closely allied species might develop in most dissimilar ways according to the conditions in which they found themselves.[4] His classical case of

[1] *Beiträge zur Kritik der Darwinschen Lehre*, Leipzig, 1898.

[2] See E. B. Wilson, "The Embryological Criterion of Homology," *Wood's Holl Biological Lectures*, Boston, pp. 101-24, 1895; Braem, *Biol. Centrblt.*, xv., 1895; T. H. Morgan, *Arch. f. Ent.-Mech.*, xviii.; J. W. Jenkinson, *Mem. Manchester Lit. Phil. Soc.*, 1906, and *Vertebrate Embryology*, Oxford, 1913; A. Sedgwick, article "Embryology" in *Ency. Brit.*, p. 318, vol. xi., 11th Ed. (1910).

[3] For a detailed treatment of this important point see the remarkable volume of E. Schulz (Petrograd), *Prinzipien der rationellen vergleichenden Embryologie*, Leipzig, 1910.

[4] "La Pœcilogonie," *Bull. Sci. France et Belgique*, xxxix., pp. 153-87, 1905.

"pœcilogeny" was that of the shrimp *Palæmonetes varians*, the fresh-water form of which develops in an entirely different way from the salt-water form.

Experimental workers indeed were inclined to rule the law out of account, to disregard completely the historical element in development, and this was perhaps the chief weakness of the neo-vitalist systems which took their origin in this experimental work.

From the side also of descriptive morphology the biogenetic law underwent a critical revision. It was studied as a fact of embryology and without phylogenetic bias by men like Oppel, Keibel, Mehnert, O. Hertwig and Vialleton,[1] and they arrived at a critical estimate of it very similar to that of von Baer.

Theoretical objections to the biogenetic law had been raised from time to time by many embryologists, but the positive testing of it by the comparison of embryos in respect of the degree of development of their different organs starts with Oppel's work of 1891.[2] He studied a large number of embryos of different species at different stages of their development, and determined the relative time of appearance of the principal organs and their relative size. His results are summarised in tabular form and have reference to all the more important organs. He was led to ascribe a certain validity to the biogenetic law, but he drew particular attention to the very considerable anomalies in the time of appearance which are shown by many organs, anomalies which had been classed by Haeckel under the name of heterochronies.

Oppel's main conclusions were as follows:—"There are found in the developmental stages of different Vertebrates 'similar ontogenetic series,' that is to say, Vertebrates show at definite stages similarities with one another in the degree of development of the different organs. Early stages resemble one another, so also do later stages; equivalent stages of closely allied species resemble one another, and older stages of lower animals resemble younger stages of

[1] *Un problème de l'évolution. La loi biogénétique fondamentale.* Paris and Montpellier, 1908.

[2] *Vergleichung des Entwickelungsgrades der Organe zu verschiedenen Entwickelungszeiten bei Wirbeltieren*, Jena, 1891.

higher animals; young stages are more alike than old stages. . . . The differences which these similar series show (for which reason they cannot be regarded as identical) may be designated as temporal disturbances in the degree of development of the separate organs or organ-systems. Some organs show very considerable temporal dislocations, others a moderate amount, others again an inconsiderable amount. Among the developmental stages of various higher animals can be found some which correspond to the ancestral forms and also to the lower types which resemble these ancestral forms. On the basis of the tabulated data here given there can be distinguished with certainty in the ontogeny of Amniotes a pro-fish stage, a fish-stage, a land-animal stage, a pro-amniote stage, and following on these a fully developed reptile, bird or mammal stage."[1]

Oppel's methods were employed by Keibel[2] in his investigations on the development of the pig, which formed the model for the well-known series of *Normentafeln* of the ontogeny of Vertebrates which were issued in later years under Keibel's editorship. Keibel was more critical of the biogenetic law than Oppel, and he held that the ancestral stages distinguished by Oppel could not be satisfactorily established. He suggested an interesting explanation of heterochrony in development, according to which the premature or retarded appearance of organs in ontogeny stands in close relation with the time of their entering upon functional activity. Thus in many mammals the mesodermal part of the allantois often appears long before the endodermal part, though this is phylogenetically older. This Keibel ascribes to the fact that the endodermal part is almost functionless. "One can directly affirm," he writes, "that the time of appearance of an organ depends in an eminent degree upon the time when it has to enter upon functional

[1] Quoted by Keibel, *Ergebn. Anat. Entwick.*, vii., p. 741.

[2] "Studien zur Entwickelungsgeschichte des Schweines," Schwalbe's *Morphol. Arbeiten*, iii., 1893, and v., 1895.

Normentafeln zur Entwickelungsgeschichte des Schweines, Jena, 1897.

"Das biogenetische Grundgesetz und die Cenogenese," *Ergebn. Anat. Entw.*, vii., pp. 722-92, 1897.

"U. d. Entwickelungsgrad der Organe," *Handb. vergl. exper. Entwick. der Wirbelthiere*, iii., 3, pp. 131-48, 1906.

activity. This moment is naturally dependent upon the external conditions. Among the highest Vertebrates, the mammals, the traces of phylogeny shown in ontogeny are to a great extent obliterated through the adaptation of ontogeny to the external conditions, and through the modifications which the germs of more highly organised animals necessarily exhibit from the very beginning as compared with germs which do not reach such a high level of development" (p. 754, 1897).

Study of individual variation in the time of appearance of the organs in embryos of the same species was prosecuted with interesting results by Bonnet,[1] Mehnert,[2] and Fischel.[3] Fischel found that variability was greatest among the younger embryos, and became progressively less in later stages. Like von Baer (*supra*, p. 114) he inferred that regulatory processes were at work during development which brought divergent organs back to the normal and enabled them to play their part as correlated members of a functional whole.

Important theoretical views were developed by Mehnert[4] in a series of publications appearing from 1891 to 1898. Like Keibel, Mehnert emphasised the importance of function in determining the late or early appearance of organs, but he conceived the influence of function to be exerted not only in ontogeny, but also throughout the whole course of phylogeny, by reason of the transmission to descendants of the effects of functioning in the individual life.

In his paper of 1897 Mehnert details the results of an extensive examination of the development of the extremities throughout the Amniote series. He finds that in all cases a pentadactylate rudiment is formed, even in those forms in

[1] "Beiträge zur Embryologie der Wiederkäuer," *Arch. Anat. Entw.*, 1889.

[2] "Die individ. Variation d. Wirbeltierembryo," *Morph. Arbeit.*, v., 1895.

[3] "U. Variabilität u. Wachstum d. embryonalen Körpers," *Morph. Jahrb.*, xxiv., 1896.

[4] "Gastrulation u. Keimblätterbildung der *Emys lutaria taurica*," *Morph. Arbeit.*, i., 1891. "Kainogenese," *Morph. Arbeit.*, vii., pp. 1-156, 1897, and also separately. *Biomechanik, erschlossen aus dem Prinzipe der Organogenese*, Jena, 1898.

which only a few of the elements of the hand or foot come to full development. But whereas in forms with a normally developed hand, *e.g.* the tortoise and man, all the digits develop and differentiate at about the same rate, in forms which have in the adult reduced digits, *e.g.* the ostrich and the pig, these vestigial digits undergo a very slow and incomplete differentiation, while the others develop rapidly and completely. He draws a general distinction between organs that are phylogenetically progressive and such as are phylogenetically regressive, and seeks to prove that progressive organs show an ontogenetic acceleration and regressive organs a retardation.[1] The acceleration or retardation affects not only the mass-growth of the organs, but also their histological differentiation.

Now between progression and functioning and between regression and functional atrophy there is obviously a close connection. Loss of function is well known to be one of the chief causes of the degeneration of organs in the individual life, and on the other hand, as Roux has pointed out, all post-embryonic development is ruled and guided by functioning. It is thus in the long run functioning that brings about phylogenetic progression, absence of functional activity that causes phylogenetic regression. This comes about through the transmission of acquired functional characters, a transmission which Mehnert conceives to be extraordinarily accurate and complete.

In general Mehnert adopts the functional standpoint of Cuvier, von Baer, and Roux. His considered judgment as to the phylogenetic value of the biogenetic law closely resembles that formed by von Baer, for he admits recapitulation only as regards the single organs, not as regards the organism as a whole. He has, however, much more sympathy with the

[1] This law was foreshadowed by Reichert in 1837, when he wrote :—"We notice in our investigation of embryos of different animal forms that it is those organs, those systems, which in the fully developed individual are peculiarly perfect, that in their earliest rudiments and also throughout the whole course of their development appear with the most striking distinctness" (Müller's *Archiv*, p. 135, 1837). See also his *Entwick. Kopf. nackt. Amphib.*, p. 198, 1838. So, too, Rathke notes how the elongated shape of the snake appears even in very early embryonic stages (*Entwick. Natter.*, p. 111, 1839).

law than either Keibel or Oppel, though he agrees that it cannot be used for the construction of ancestral trees. But he ascribes to it as a fact of development considerable importance. The following passage gives a good summary of his view as to the scope and validity of the law. "The biogenetic law has not been shaken by the attacks of its opponents. The assertion is still true that individual organogenesis is exclusively dependent on phylogeny. But we must not expect to find that all the stages in the development of the separate organs, which coexisted in any member of the phylogenetic series, appear *at the same time* in the individual ontogeny of the descendants, because each organ possesses its own specific rate of development. In this way it comes about naturally that organs which become differentiated rapidly, as, for example, the medullary tube, as a rule dominate earlier periods of ontogeny than do the organs of locomotion. For the same reason the cerebral hemispheres of man are almost as large in youth as in maturity. The picture which an embryo gives is not a repetition in detail of one and the same phylogenetic stage; it consists rather of an assemblage of organs, some of which are at a phyletically early stage of development, while others are at a phyletically older stage."[1]

A different line of attack was that adopted by O. Hertwig in a series of papers, which contain also what is perhaps the best critical estimate of the present position and value of descriptive morphology.[2]

It had not escaped the notice of many previous observers that quite early embryos not infrequently show specific characters even before the characters proper to their class, order and genus are developed—in direct contradiction of the law of von Baer. Thus L. Agassiz[3] had remarked

[1] Quoted by Keibel (p. 790, 1897) from the *Biomechanik.*

[2] *Die Zelle und die Gewebe*, Jena, 1898, and the subsequent editions of this text-book, published under the title of *Allgemeine Biologie. Die Entwickelung der Biologie im neunzehnten Jahrhundert*, Jena, 1900, 2nd ed., 1908. "Ueber die Stellung der vergl. Entwickelungslehre zur vergl. Anatomie, zur Systematik und Descendenztheorie," *Handb. vergl. exper. Entwickelungslehre der Wirbeltiere*, iii., 3, pp. 149-80, Jena, 1906. (1906, b). Also in Pt. I. of Vol. I. (1906, a).

[3] *An Essay on Classification*, London, 1859.

in 1859 that specific characteristics were often developed precociously. "The Snapping Turtle, for instance, exhibits its small crosslike sternum, its long tail, its ferocious habits, even before it leaves the egg, before it breathes through lungs, before its derm is ossified to form a bony shield, etc.; nay, it snaps with its gaping jaws at anything brought near, when it is still surrounded by its amnion and allantois, and its yolk still exceeds in bulk its whole body" (p. 269).

Wilhelm His,[1] in the course of an acute and damaging criticism of the biogenetic law as enunciated by Haeckel, showed clearly that by careful examination the very earliest embryos of a whole series of Vertebrates could be distinguished with certainty from one another. "An identity in external form of different animal embryos, despite the common affirmation to the contrary, does not exist. Even at early stages in their development embryos possess the characters of their class and order, nay, we can hardly doubt, of their species and sex, and even their individual characteristics" (201).

This specificity of embryos was affirmed with even greater confidence by Sedgwick in a paper critical of von Baer's law.[2] He wrote:—"If v. Baer's law has any meaning at all, surely it must imply that animals so closely allied as the fowl and duck would be indistinguishable in the early stages of development; and that in two species so closely similar that I was long in doubt whether they were distinct species, viz., *Peripatus capensis* and *Balfouri*, it would be useless to look for embryonic differences; yet I can distinguish a fowl and a duck embryo on the second day by the inspection of a single transverse section through the trunk, and it was the embryonic differences between the Peripatuses which led me to establish without hesitation the two separate species. . . . I need only say . . . that a species is distinct and distinguishable from its allies from the very earliest stages all through the development, although these embryonic differences do not necessarily implicate the same organs as do the adult differences" (p. 39).

[1] *Unsere Körperform*, Leipzig, 1874.

[2] *Q.J.M.S.*, xxxvi., pp. 35-52, 1894.

Hertwig interprets this fact of the specific distinctness of closely allied embryos in the light of the preformistic conception of heredity. According to this view the whole adult organisation is represented in the structure of the germ-plasm contained in the fertilised ovum, from which it follows that the ova of two different species, and also their embryos at every stage of development, must be as distinct from one another as are the adults themselves, even though the differences may not be so obvious. If this be the case there can be no real recapitulation in ontogeny of the phylogeny of the race. for the egg-cell represents not the first term in phylogeny, but the last. The egg-cell *is* the organism in an undeveloped state; it has a vastly more complicated structure than was possessed by the primordial cell from which its race has sprung, and it can in no way be considered the equivalent of this ancestral cell.

Hertwig puts this vividly when he says that "the hen's egg is no more the equivalent of the first link in the phylogenetic chain than is the hen itself" (p. 160, 1906, b).

If ontogeny is not a recapitulation of phylogeny, how is it that the early embryonic stages are so alike, even in animals of widely different organisation? Hertwig's answer to this is very interesting. He takes the view that many of the processes characterising early embryonic development are the means necessarily adopted for attaining certain ends. Such are the processes of segmentation, the formation of a blastula, of cell-layers, of medullary folds where the nervous system is a closed tube, the formation of the notochord as a necessary condition of the development of the vertebral column, and so on. "Looked at from this standpoint it cannot surprise us that in all animal phyla the earliest embryonic processes take place in similar fashion, so that we observe the occurrence both in Vertebrates and Invertebrates of a segmentation-process, a morula-stage, a blastula and a gastrula. If now these developmental processes do not depend on chance, but, on the contrary, are rooted in the nature of the animal cell itself, we have no reason for inferring from the recurrence of a similar segmentation-process, morula, blastula, and gastrula in all classes of the animal kingdom the common descent of all animals from one blastula-like or gastrula-like ancestral form. We recognise rather in the successive early stages of animal development only the manifestation of special laws, by which the shaping of animal forms (as distinct from plant forms) is brought about" (p. 178, 1906, b).

"The principal reason why certain stages recur in ontogeny with such constancy and always in essentially the same manner is that they provide under all circumstances the necessary pre-conditions through which alone the later and higher stages of ontogeny can be realised. The unicellular organism can by its very nature transform itself into a multicellular organism only by the method of cell-division. Hence, in all Metazoa, ontogeny must start with a segmentation-process, and a similar statement could be made with regard to all the later stages" (p. 57, 1906, a).

Similarities in early development are therefore no evidence of common descent, and in the same way the resemblances of adult animals, subsumed under the concepts of homology and the unity of plan, are not necessarily due to community of descent, but may also be brought about by the similarity or identity of the laws which govern the evolution of these animals. In the absence, therefore, of positive evidence as to the actual lines of descent (to be obtained only from palæontology), homological resemblance cannot be taken as proof of blood relationship, for homology is a wider concept than homogeny. The only valid definition of homology is that adopted in pre-evolutionary days, when those organs were considered homologous "which agree up to a certain point in structure and composition, in position, arrangement, and relation to the neighbouring organs, and accordingly possess identical functions and uses in the organism" (p. 151, 1906, b).

The concept of homology has thus a value quite independent of any evolutionary interpretation which may be superadded to it. "Homology is a mental concept obtained by comparison, which under all circumstances retains its validity, whether the homology finds its explanation in common descent or in the common laws that rule organic development" (p. 151, 1906, b). As A. Braun long ago pointed out, "It is not descent which decides in matters of morphology, but, on the contrary, morphology which has to decide as to the possibility of descent."[1]

Hertwig, in a word, reverts to the pre-evolutionary conception of homology. "We see in homology," he writes, "only the expression of regularities (*Gesetzmässigkeiten*) in the organisation of the animals showing it, and we regard the question, how far this homology can be explained by common descent and how far by other principles, as for the present an open one, requiring for its solution investigations specially directed towards its elucidation" (p. 179, 1906, b).

Holding, as he does, that no definite conclusions can be drawn from the facts of comparative anatomy and embryology as to the probable lines of descent of the animal kingdom, Hertwig accords very little value to phylogenetic speculation. It is, he admits, quite probable that the archetype of a class represents in a general sort of way the ancestral form, but this does not, in his opinion, justify us in assuming that such generalised types ever existed and gave origin to the present-day forms. "It is not legitimate to picture to ourselves the ancestral forms of the more highly organised animals in the guise of the lower animals of the present day—and that is just what we do when we speak of Proselachia, Proamphibia and Proreptilia" (p. 155, 1906, b).

He rejects on the same general grounds the evolutionary dogma of monophyletic or almost monophyletic descent, and admits with Kölliker, von Baer, Wigand, Naegeli and others that evolution may quite well have started many times and from many different primordial cells.

There is indeed a great similarity between the views developed by O. Hertwig and those held by the older critics of Darwinism—von Baer, Kölliker, Wigand, E. von Hartmann and others. It is true the philosophical standpoint is on the whole different, for while many of that older generation were vitalists Hertwig belongs to the mechanistic school.

But both Hertwig and the older school agree in pointing out the *petitio principii* involved in the assumption that the

[1] Quoted by Hertwig. See also K. Goebel, "Die Grundprobleme der heutigen Pflanzenmorphologie," *Biol. Centrbl.*, xxv., pp. 65-83, 1905.

archetype represents the ancestral form; both reject the simplicist conception of a monophyletic evolution (which may be likened to the "one animal" idea of the transcendentalists); both admit the possibility that evolution has taken place along many separate and parallel lines, and explain the correspondences shown by these separate lines by the similarity of the intrinsic laws of evolution; finally, both emphasise the fact that we know nothing of the actual course of evolution save the few indications that are furnished by palæontology, and both insist upon the unique importance of the palæontological evidence.[1]

It was a curious but very typical characteristic of evolutionary morphology that its devotees paid very little attention to the positive evidence accumulated by the palæontologists,[2] but shut themselves up in their tower of ivory and went on with their work of constructing ideal genealogies. It was perhaps fortunate for their peace of mind that they knew little of the advances made by palæontology, for the evidence acquired through the study of fossil remains was distinctly unfavourable to the pretty schemes they evolved.

As Neumayr, Zittel, Depéret, Steinmann and others have pointed out, the palæontological record gives remarkably little support to the ideal genealogies worked out by morphologists. There is, for instance, a striking absence of transition forms between the great classificatory groups. A few types are known which go a little way towards bridging over the gaps—the famous *Archæopteryx*, for example—but these do not always represent the actual phylogenetic links. There is an almost complete absence of the archetypal ancestral forms which are postulated by evolutionary morphology. Amphibia do not demonstrably evolve from an archetypal Proamphibian, nor do mammals derive from a single generalised Promammalian type. Few of the hypothetical ancestral types imagined by Haeckel have ever

[1] This is also emphasised by Fleischmann in his critical study of evolutionary morphology entitled *Die Descendenztheorie*, Leipzig, 1901.

[2] The same remark applies to the bulk of speculation as to the factors of evolution, with the exception of the contributions made to evolution theory by the palæontologists by profession, such as Cope.

been found as fossils. The great classificatory groups are almost as distinct in early fossiliferous strata as they are at the present day. As Depéret says in his admirable book,[1] in the course of a presentation of the matured views of the great Karl von Zittel, "We cannot forget that there exist a vast number of organisms which are not connected by any intermediate links, and that the relations between the great divisions of the animal and vegetable kingdoms are much less close than the theory [of evolution] demands. Even the Archæopteryx, the discovery of which made so much stir and appeared to establish a genetic relation between classes so distinct as Birds and Reptiles, fills up the gap only imperfectly, and does not indicate the point of bifurcation of these two classes. Intermediate links are lacking between Amphibia and Reptiles. Mammals, too, occupy an isolated position, and no zoologist can deny that they are clearly demarcated from other Vertebrates; indeed, no fossil mammal is certainly known which comes nearer to the lower Vertebrates than does Ornithorhynchus at the present day" (p. 115).

To take a parallel from the Invertebrata, B. B. Woodward,[2] after discussing the phylogeny of the Mollusca as worked out by the morphologists and comparing it with the probable actual course of the evolution of the group, as evidenced by fossil shells, sums up as follows:—"The lacunæ in our knowledge of the interrelationships of the members of the various families and orders of Mollusca are slight, however, compared with the blank caused by the total absence from palæontological history of any hint of passage forms between the classes themselves, or between the Mollusca and their nearest allies. Nor is this hiatus confined to the Molluscan phylum; it is the same for all branches of the animal kingdom. There is circumstantial evidence that transitional forms must have existed, but of actual proof none whatever. All the classes of Mollusca appear fully fledged, as it were. No form has as yet been discovered of which it could be said that it in any way approached the

[1] *Les Transformations du Monde animal*, Paris, 1907.

[2] "Malacology *versus* Palæoconchology," *Proc. Malacological Soc.*, viii., pp. 66-83, 1908.

hypothecated prorhipidoglossate mollusc, still less one linking all the classes" (p. 79).

Pointing in the same direction as the absence of transitional forms is the undeniable fact that all the great groups of animals appear with all their typical characters at a very early geological epoch. Thus, in the Silurian age a very rich fauna has already developed, and representatives are found of all the main Invertebrate groups—sponges, corals, hydroid colonies, five types of Echinoderms, Bryozoa, Brachiopods, Worms, many types of Mollusca and Arthropoda. Of Vertebrates, at least two types of fish are present—Ganoids and Elasmobranchs. In the very earliest fossiliferous rocks of all, the Precambrian formation, there are remains of Molluscs, Trilobites and Gigantostraca, similar to those which flourished in Cambrian and Silurian times.

The contributions of palæontology to the solution of the problems of descent posed by morphology are, however, not all of this negative character. The law of recapitulation is in some well-controlled cases triumphantly vindicated by palæontology. Thus Hyatt and others found that in Ammonites the first formed coils of the shell often reproduce the characters belonging to types known to be ancestral, and what is more they have demonstrated the actual occurrence of the phenomenon known as acceleration or tachygenesis, often postulated by speculative morphologists.[1] This is the tendency universally shown by embryos to reproduce the characters of their ancestors at earlier and earlier stages in their development.

The most valuable contribution made by palæontologists to morphology and to the theory of evolution arose out of the careful and methodical study of the actual succession of fossil forms as exemplified in limited but richly represented groups. Classical examples were the researches of Hilgendorf[2] on the evolution of *Planorbis multiformis* in the lacustrine deposits of Steinheim, those of Waagen[3] on the

[1] Particularly by E. Perrier, "La Tachygenèse," *Ann. Sci. nat.* (*Zool.*) (8), xvi., 1903.

[2] *Monatsber. k. Akad. Wiss.*, Berlin, pp. 474-504, 1866.

[3] *Geognost. u. Palæont. Beiträge*, ii., Heft 2, pp. 181-256, 1869.

phylogeny of *Ammonites subradiatus*, and the work of Neumayr and Paul[1] on *Paludina* (*Vivipara*).

These investigations demonstrated that it was possible to follow out step by step in superjacent strata the actual evolution of fossil species and to establish the actual "phyletic series."

To take an example from among the Vertebrates, Depéret has shown (*loc. cit.*, pp. 184-9), that the European Proboscidea, belonging to the three different types of the Elephants, Mastodons and Dinotheria, have evolved since the Oligocene epoch along five distinct but continuous lines. The Dinotherian stock is represented at the beginning of the Miocene by the relatively small form *D. cuvieri;* this changes progressively throughout Miocene times into *D. laevius*, *D. giganteum*, and *D. gigantissimum*. Among the Mastodons two quite distinct phyletic series can be distinguished, the first commencing with *Palæomastodon beadnelli* of the Oligocene, and evolving between the Miocene and Pliocene into *Mastodon arvernensis*, after traversing the forms *M. angustidens* and *M. longirostris*, the second starting with the *M. turicensis* of the Lower Miocene and evolving through *M. borsoni* into the *M. americanus* of the Quaternary. The phyletic series of the true elephants in Europe are relatively short, and go back only to the Quaternary, *Elephas antiquus* giving origin to the Indian elephant, *E. priscus* to the African.

The careful study of phyletic series brought to light the significant fact that these lines of filiation tend to run for long stretches of time parallel to, and distinct from one another, without connecting forms. This is clearly exemplified in the case of the Proboscidea, and many other examples could be quoted. Almost all rich genera are polyphyletic in the sense that their component species evolve along separate and parallel lines of descent.[2] "Such great genera as the genus *Hoplites* among the Ammonites, the genus *Cerithium* among the Gastropoda, the genus *Pecten* or the genus

[1] *Abhand. k.k. Geol. Reichsanstalt*, vii., Wien, 1875.

[2] The case for polyphyletism is very strongly put by G. Steinmann in his book, *Die geologischen Grundlagen der Abstammungslehre*, Leipzig, 1908.

Trigonia among the Lamellibranchs, each comprise perhaps more than twenty independent phyletic series" (Depéret, p. 200).

Variation along the phyletic lines is gradual[1] and determinate, and appears to obey definite laws. The earliest members of a phyletic series are usually small in size and undifferentiated in structure, while the later members show a progressive increase in size and complexity. Rapid extinction often supervenes soon after the line has reached the maximum of its differentiation.

The general picture which palæontology gives us of the evolution of the animal kingdom is accordingly that of an immense number of phyletic lines which evolve parallel to one another, and without coalescing, throughout longer or shorter periods of geological times. "Each of these lines culminates sooner or later in mutations of great size and highly specialised characters, which become extinct and leave no descendants. When one line disappears by extinction it hands the torch, so to speak, to another line which has hitherto evolved more slowly, and this line in its turn traverses the phases of maturity and old age which lead it inevitably to its doom. The species and genera of the present day belong to lines that have not reached the senile phase; but it may be surmised that some of them, *e.g.* elephants, whales, and ostriches, are approaching this final phase of their existence" (Depéret, p. 249).

It is one of the paradoxes of biological history that the palæontologists have always laid more stress upon the functional side of living things than the morphologists, and have, as a consequence, shown much more sympathy for the Lamarckian theory of evolution. The American palæontologists in particular—Cope, Hyatt, Ryder, Dall, Packard, Osborn—have worked out a complete neo-Lamarckian theory based upon the fossil record.

The functional point of view was well to the fore in the works of those great palæontologists, L. Rütimeyer (1825-1895) and V. O. Kowalevsky (1842-83), who seem to have carried on the splendid tradition of Cuvier. Speaking of

[1] The steps in this chronological variation were termed by Waagen "mutations."

Kowalevsky's classical memoir, *Versuch einer natürlichen Classification der fossilen Hufthiere*, Osborn[1] writes :—" This work is a model union of the detailed study of form and function with theory and the working hypothesis. It regards the fossil not as a petrified skeleton, but as having belonged to a moving and feeding animal; every joint and facet has a meaning, each cusp a certain significance. Rising to the philosophy of the matter, it brings the mechanical perfection and adaptiveness of different types into relation with environment, with changes of herbage, with the introduction of grass. In this survey of competition it speculates upon the causes of the rise, spread, and extinction of each animal group. In other words, the fossil quadrupeds are treated *biologically*—so far as is possible in the obscurity of the past" (p. 8). The same high praise might with justice be accorded to the work of Cope on the functional evolution of the various types of limb-skeleton in Vertebrates, and on the evolution of the teeth as well as to the work of other American palæontologists, including Osborn himself.

Osborn's law of "adaptive radiation," which links on to Darwin's law of divergence,[2] constitutes a brilliant vindication of the functional point of view. "According to this law each isolated region, if large and sufficiently varied in its topography, soil, climate, and vegetation, will give rise to a diversified mammalian fauna. From primitive central types branches will spring off in all directions, with teeth and prehensile organs modified to take advantage of every possible opportunity of securing food, and in adaptation of the body, limbs and feet to habitats of every kind, as shown in the diagram [on p. 363]. The larger the region and the more diverse the conditions, the greater the variety of mammals which will result.

"The most primitive mammals were probably small insectivorous or omnivorous forms, therefore with simple, short-crowned teeth, of slow-moving, ambulatory, terrestrial, or arboreal habit, and with short feet provided with claws. In seeking food and avoiding enemies in different habitats

[1] *The Age of Mammals in Europe, Asia, and North America*, New York, 1910.

[2] *Origin of Species*, 6th ed., Chap. IV.

the limbs and feet radiate in four diverse directions; they either become *fossorial* or adapted to digging habits, *natatorial* or adapted to *amphibious* and finally to *aquatic* habits, *cursorial* or adapted to swift-moving, terrestrial progression, *arboreal* or adapted to tree life. Tree life leads, as its final stage, into

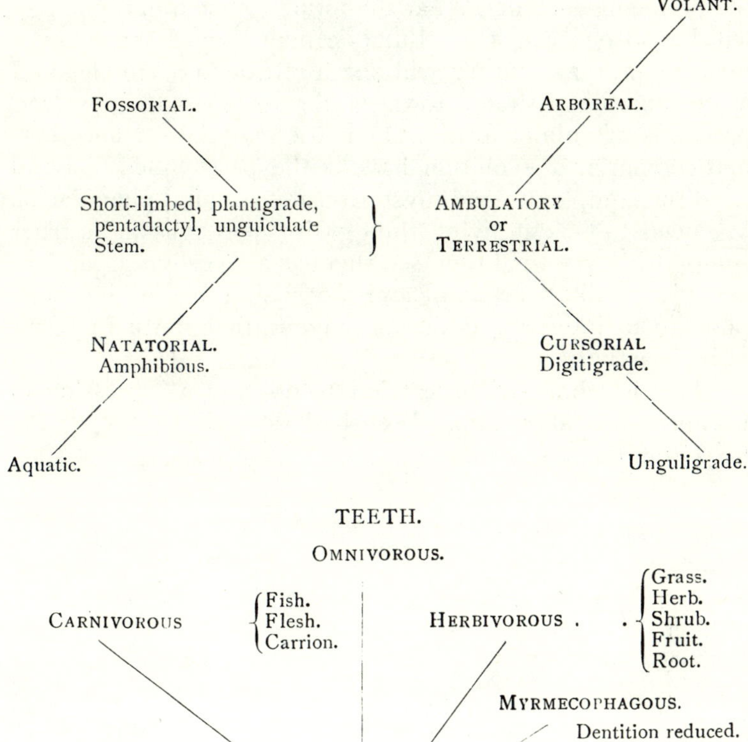

TEETH.

OMNIVOROUS.

CARNIVOROUS { Fish. Flesh. Carrion.

HERBIVOROUS . . { Grass. Herb. Shrub. Fruit. Root.

MYRMECOPHAGOUS. Dentition reduced.

Stem: INSECTIVOROUS.

the parachute types of the flying squirrels and phalangers, or into the true flying types of the bats. . . . Similarly in the case of the teeth, insectivorous and omnivorous types appear to be more central and ancient than either the exclusively carnivorous or herbivorous types. Thus the extremes of carnivorous adaptation, as in the case of the cats, of omnivorous adaptation, as in the case of the bears, of herbivorous adaptation, as in the case of the horses, or myrmecophagous adaptation, as in the case of the ant-eaters, are all secondary" (*loc. cit.*, pp. 23-4).

We have now reached the end of our historical survey of the problems of form. What the future course of morphology will be no one can say. But one may hazard the opinion that the present century will see a return to a simpler and more humble attitude towards the great and unsolved problems of animal form. Dogmatic materialism and dogmatic theories of evolution have in the past tended to blind us to the complexity and mysteriousness of vital phenomena. We need to look at living things with new eyes and a truer sympathy. We shall then see them as active, living, passionate beings like ourselves, and we shall seek in our morphology to interpret as far as may be their form in terms of their activity.

This is what Aristotle tried to do, and a succession of master-minds after him. We shall do well to get all the help from them we can.

INDEX